T0201609

Fundamentals of Nanomedicine

The first introductory book on the subject, this volume will provide a complete grounding to this pioneering field for students and professionals across biomedical engineering, biology, and medicine. It features a comprehensive overview of original work in this revolutionary field. Topics discussed include drug delivery, cell–nanomaterial interaction, and gene therapy, accompanied by real-world examples and 149 illustrations. The book teaches readers how to design and test their own nanomedical systems for real-world applications in biomedical engineering, medicine, and pharmacy.

Presenting a thorough discussion of the science and engineering of nanomedicine, it discusses vital environmental, social, and ethical impacts of this revolutionary technology. With 250 thought-provoking study questions, allowing the reader to self-assess their understanding, this book is a rich source of information that will be of interest and importance in nanomedicine.

Professor James F. Leary is Professor Emeritus of Nanomedicine, Biomedical Engineering, and Basic Medical Sciences at Purdue University and President and Founder of Aurora Life Technologies, LLC.

Fundamentals of Nanomedicine

JAMES F. LEARY

Purdue University

CAMBRIDGE
UNIVERSITY PRESS

CAMBRIDGE
UNIVERSITY PRESS

University Printing House, Cambridge CB2 8BS, United Kingdom

One Liberty Plaza, 20th Floor, New York, NY 10006, USA

477 Williamstown Road, Port Melbourne, VIC 3207, Australia

314–321, 3rd Floor, Plot 3, Splendor Forum, Jasola District Centre, New Delhi – 110025, India

103 Penang Road, #05–06/07, Visioncrest Commercial, Singapore 238467

Cambridge University Press is part of the University of Cambridge.

It furthers the University's mission by disseminating knowledge in the pursuit of
education, learning, and research at the highest international levels of excellence.

www.cambridge.org
Information on this title: www.cambridge.org/9781107013971
DOI: 10.1017/9781139012898

First published 2022

Printed in the United Kingdom by TJ Books Limited, Padstow Cornwall

A catalogue record for this publication is available from the British Library.

Library of Congress Cataloging-in-Publication Data
Names: Leary, James F., 1948– author.
Title: Fundamentals of nanomedicine / James F. Leary.
Description: Cambridge, United Kingdom ; New York, NY : Cambridge University Press, 2022. |
    Includes bibliographical references and index. | Summary: "The first introductory book on the subject,
    this book will provide a complete grounding to this pioneering field for students and professionals across
    biomedical engineering, biology and medicine. It features a comprehensive overview of original work in
    this revolutionary field. Topics discussed include drug delivery, cell-material interaction and gene
    therapy, accompanied by real-world examples and over 140 illustrations. The book teaches readers how to
    design and test their own nanomedical systems for real-world applications in biomedical engineering,
    medicine and pharmacy"– Provided by publisher.
Identifiers: LCCN 2021048067 (print) | LCCN 2021048068 (ebook) | ISBN 9781107013971 (hardback) |
    ISBN 9781139012898 (epub)
Subjects: MESH: Nanomedicine–methods | Drug Delivery Systems–methods
Classification: LCC R857.N34 (print) | LCC R857.N34 (ebook) | NLM QT 36.5 | DDC 610.28–dc23
LC record available at https://lccn.loc.gov/2021048067
LC ebook record available at https://lccn.loc.gov/2021048068

ISBN 978-1-107-01397-1 Hardback

To my wife, Jill Norton, whose encouragement and support were essential to the completion of this book.

Contents

Preface

A little perspective might be helpful to the reader to better understand the somewhat unusual professional circumstances that led to the writing of this book. This book has been a labor of love for over eight years, but the field of nanomedicine started for me in 2000 when I and others were invited by NASA (John Hines was my excellent program officer) to invent some nanomedicine for the Mars mission. I was then a professor at the University of Texas Medical Branch (UTMB) in Galveston, Texas, just down the road from Johnson Space Center. I became one of the first researchers in the United States to be funded for a grant on nanomedicine. The book's nanomedicine origins come from two funding sources – NASA and the National Cancer Institute – in what was a bold joint initiative at the time by those two funding entities.

While pursuing my PhD in biophysics at Penn State University, I was one of the first two graduate students in the USA to train in the field of flow cytometry. I knew all of the original inventors of this field personally. My dissertation work led to my postdoctoral fellowship at Los Alamos National Laboratory in Los Alamos, New Mexico, where I was immersed in one of the birthplaces of the field of flow cytometry. After I left Los Alamos and started my career as an assistant professor of pathology at the University of Rochester, I quickly became known as the "father of high-speed flow cytometry and rare-event analysis" for my early work in these areas, and many of my more than 100 flow cytometry publications in this field are cited in this book. My first research as a young assistant professor at the University of Rochester Medical School in Rochester, New York, was in kinetics of viral binding to single cells. For that work, I fluorescently labeled viruses to measure their interactions with single cells using flow cytometry. Viruses were my first nanoparticles – from nature! Viewing nanomedical drug delivery devices was inspired by my early work with virus–cell interactions. There are many allusions in this book to biomimicry designs of nanomedical devices based on viruses in nature.

Importantly, my background in flow cytometry and single-cell analysis prepared me for being receptive to approaching nanomedicine as "single-cell medicine," which was definitely not the emphasis of medicine at the time! I was a medical school professor and researcher for the first 27 years, including 10 years as a professor of internal medicine within the Division of Infectious Diseases while at UTMB, among my many roles as professor there in five different departments and four research centers. When I moved to Purdue University in 2005, I became an endowed professor of nanomedicine where I first had my office and then my laboratories within the Birck

Nanotechnology Center. This allowed me to immerse myself in the diverse multi-disciplinary research areas of nanotechnology as well as enabling possibilities for interactions with many nanotechnology collaborators. I mainly focused on nanomedicine teaching and research, resulting in more than 40 scientific publications, thus far, in the field of nanomedicine.

At Purdue University, I instantly acquired four graduate students in the Weldon School of Biomedical Engineering. Those students needed to be quickly trained in nanomedicine, so within the next year I formulated what was presumably the world's first graduate-level nanomedicine course, Engineering Nanomedical Systems. There were no textbooks or other courses elsewhere in existence. I starting teaching this course in 2007 at Purdue University as part of the graduate program in the Weldon School of Biomedical Engineering, the source of most of my graduate students in nanomedicine. By 2012, it had become sufficiently popular at Purdue that it went worldwide on the Internet; my lectures were videotaped and archived on Purdue's nanoHUB, as well as distributed as podcasts at Apple U. I taught the course every year from 2007 through 2014, my last full academic year at Purdue University before I retired in August 2015.

This actual book grew out of earlier interactions with Cambridge University Press starting in 2010. I was asked to review another nanomedicine book proposal for Cambridge University Press. It was an edited book on aspects of nanomedicine. I evaluated the book and told Cambridge University Press that I thought that at least 2 of the 12 or so chapters would be good because of the expertise of the authors of those chapters, but that the book (as is frequently the case for edited compilations involving diverse and unconnected chapters by different authors not communicating with each other) was disjointed since it was written entirely separately by different authors without any serious interaction. I had contributed over 20 chapters to other people's books before writing my own book. I made the offhand remark to Cambridge University Press (not realizing where it would lead!) that I thought I could write a better book, as a textbook, all by myself on the subject. Cambridge University Press quickly replied, asking me what would be the title of *my* book! After some serious thought, I finally accepted their challenge.

I had already realized, after teaching my course for over four years, that there was still an unmet need for a textbook on the subject, and that the course could serve as a template for a textbook. I then responded affirmatively to Cambridge University Press's request for a book proposal. My book proposal was reviewed by several experts in the field and very quickly accepted.

My original 2010 proposal to Cambridge University Press consisted of these same 14 chapters of this book. For several years, I proceeded to flesh out more details of the book in outline form, in many cases down to individual sections, but I had very little time to actually write. Although many sections needed updating over the years and new sections were added to satisfy new developments in this rapidly changing field, it is surprising how well that original vision for the book stood up! That is particularly the case since there were no other nanomedicine courses out there. I had to figure out on my own what topics belonged in a textbook and how all of those many diverse topics should be linked together into something coherent.

Many friends, colleagues, and students have wondered what took me so long to finish the book. There were many reasons, some good and others not, but there were two major reasons. First, it is hard to find the time and energy to write a book with this level of breadth and depth across multiple scientific fields and medicine in the hours *after* working 65+ hours per week on other job-related activities. Second, I was constantly engaged in a steep learning curve to keep up with the work in nanomedicine fields of science and engineering that had technical half-lives of three to four years. Technical half-life means half of everything you learned, and thought you knew, would be superseded by new advances in the field. The science and engineering, in addition to the molecular biology and medicine, underlying nanomedicine is literally shifting under your feet, so that sections of the book needed to be updated or edited many times to reflect these advances. I like to say I have written this one book several times already!

Finally, I began writing again in earnest during 2020. The COVID-19 pandemic's forced isolation, as well as "semi-retirement", gave me an opportunity to partially escape the chaos of 2020. Throughout 2020, I frequently spent five to six (or more!) hours per day writing. Despite the pandemic, 2020 was a very good year for book writing – a silver lining in an otherwise dark year! I like to joke that when Isaac Newton escaped the plague to the countryside, he spent the year developing much of the field of physics. All I did with my year escaping to the countryside during the COVID-19 pandemic was to write this book!

This book was produced electronically – in every sense of the word! My lectures were digitally recorded at Purdue. I used those recordings and Dragon Naturally Speaking digital speech recognition software (originally invented by an MIT classmate of mine) to transcribe the video-recorded lectures to give partial content to the book. I made generous use of the lifetime digital library access given to me by Purdue upon my becoming an emeritus professor to give me access to many hundreds of scientific papers used in this book. I downloaded pdf versions of all papers cited and then organized all of the references within EndNote, a reference database software system. EndNote was then linked directly to Microsoft Word to allow insertion of all references into the text as citations. Microsoft Word was used for all writing, formatting, indexing, and automated generation of a table of contents for each chapter at the front of the book as well as a bibliography of all references used. Figures, those not taken from published literature, are all my own, mostly generated while teaching my course at Purdue ("Source: Leary teaching"). Many started out as PowerPoint slides used in my lectures and other scientific presentations at national and international meetings where I was a frequently invited speaker. I used a variety of graphic software programs, including PowerPoint, Adobe Photoshop, and Adobe Photoshop Elements. The book is complex not only due to the multidisciplinary nature of the field of nanomedicine, but also due to the complexity in terms of the many electronic pieces and how they all fit together!

Acknowledgments

I would like to acknowledge the help and support of many people during the writing of this book. First, I am grateful to my first undergraduate, graduate, and postdoc students (Jacob Smith, Tarl Prow, Peter Szaniszlo, and Nan Wang) and colleagues (Stephen Lloyd, David Gorenstein, Bruce Luxon, and Rene Rijnbrand, from the University of Texas Medical Branch, where my work in nanomedicine began) and to my colleagues and collaborators, Yuri Lvov at Louisiana Tech University, and Nicholas Kotov, now at the University of Michigan.

Second, I would like to thank the many students of my Purdue biomedical engineering graduate-level course, Engineering Nanomedical Systems, now scattered across the country and around the world, who gave me feedback that helped to improve the book. Third, and very importantly, I would like to acknowledge the support from my own very talented graduate students at Purdue (Emily Haglund, Mary-Margaret Seale-Goldsmith, Desiree White-Schenk, Lisa Reece, Jaehong Key, Michael Zordan, Meggie Grafton, Christy Cooper, and Trisha Eustaquio), who have coauthored many scientific papers with me on a number of this book's topics. It is true that the many apparently naïve questions from students teach the teacher! They were very good students and they made seminal contributions not only to this book but to the nanomedicine field at large. Fourth, thanks to other Purdue faculty colleagues (Deborah Knapp, Deepika Dhawan, Donald Bergstrom, Helen McNally, Kinam Park, Riyi Shi, Sophie Lelieve, Pierre-Alexandre Vidi, and Dmitry Zemlyanov) who were a critical part of this book, including providing some of their own guest lectures in my course.

Fifth, my South Korean colleagues at the Korean Institute for Science and Technology [KIST]; Kwangmeyung Kim, Ick Chan Kwon, and Kuiwon Choi) were an integral part of these efforts. I served as "Foreign PI" for nine years, from 2007 to 2016. We were funded by the South Korean government to conduct collaborative research on a project entitled "Nanomedicine and Molecular Imaging." This research with KIST over that nine-year period led to many international trips for our collaborations and resulted in many scientific journal publications, many of which are cited in this book.

1 The Need for New Perspectives in Medicine

Nanomedicine is the intersection of the field of nanotechnology with the field of medicine. In order to understand the basis for this intersection, it is first important to learn a little bit about nanotechnology in general as well as a few fundamentals of medicine and specific fields of science and engineering. In this book, I will use an approach to teaching championed by Dr. Albert Baez, a Harvard professor (and father to singer Joan Baez), called the "spiral approach" to teaching (Baez, 1967). The idea is to first introduce a concept at a very simple level and then gradually peel away the layers to go into greater depth and level of understanding, like peeling back the layers of an onion. I will introduce basic concepts in earlier chapters and then go into much more detail in later chapters.

1.1 Nanotechnology: Why Is Something So Small So Big?

Nanomedicine was first popularized as a future form of medicine by futurist Robert Freitas in his pioneering initial book (Freitas, 1999), which continued in a multi-volume format. He has continued to write, not only more books on different aspects of the subject but, also, articles in scientific journals (Freitas, 2005). The idea of nanomedicine was popularized by Freitas over 20 years ago. His work helped inspire this author. While his artwork seems more in the realm of science fiction than reality, the actual writing raises many of the promises of nanomedicine and is well worth reading (Freitas, 1999, 2005). Although depicted more from the viewpoint of nanorobotics, with engineering nanomanufactured devices rather than biomimetic self-assembled nanodelivery systems, as emphasized by this author in an earlier work (Leary, 2010), his work has enthused and inspired many, including this author.

The size of the scalpel determines the precision of the surgery. Nanotechnology affords us the chance to construct nanotools that are on the size scale of molecules, allowing us to treat each cell of the human body as a patient. Human disease, while frequently described in terms of either a collection of symptoms or the organ affected, is ultimately a disease of individual cells. To understand the underlying mechanisms of human disease, it is necessary to study the disease changes at the level of the single cell. Nanomedicine ultimately will allow for eradication or amelioration of human disease at the single-cell level using large numbers of self-assembling nanomedical devices that effectively parallel-process disease at the level of millions of cells simultaneously. Like nanotechnology itself, which is an atom-by-atom approach to the construction of nanodevices, nanomedicine is a cell-by-cell approach to human disease. Recently a new gene-editing technology, CRISPR (Doudna and Sternberg, 2018), has allowed us to use nanomedical delivery techniques not only to perform "nanosurgery" on a single cell but even to edit a single DNA base pair (or more bases) that causes genetic diseases. In these cases, first applied to human patients in 2020, we can now practice nanosurgery on defective molecules within a single cell. We will see more and more of this in the future as well as clever uses of nanodelivery systems to target CRISPR technology to specific single cells within the human body. This will be discussed more in Chapter 9.

Since these nanotools are self-assembling, nanomedicine has the potential to perform parallel-processing single-cell medicine on a massive scale. These nanotools can be made of biocompatible and biodegradable nanomaterials. They can be "smart" by using sophisticated targeting strategies that can perform error checking to prevent harm even if a very small fraction of them is mistargeted. Built-in molecular biosensors can provide controlled drug delivery with feedback control for individual cell dosing.

If designed to repair existing cells, rather than to just destroy diseased cells, these nanomedical devices can perform in situ regenerative medicine, reprogramming cells along less dangerous cell pathways, allowing tissues and organs to not be destroyed by the treatments, and providing an attractive alternative to allogeneic organ transplants.

Nanomedical tools, while tiny in size, can have a huge impact on medicine and healthcare. Earlier and more sensitive diagnosis will lead to presymptomatic diagnosis and treatment of disease, before permanent damage to tissues and organs. This should result in the delivery of better medicine at lower costs with better outcomes.

1.1.1   Definitions of Nanotechnology Based on Size

Attempts to simply define *nanotechnology* in terms of a particular size range are well intentioned but overly simplistic. The National Nanotechnology Initiative included not only nanomedicine applications of nanotechnology but also its more general impact (Sargent, 2010). That size limit is often set at about 100 nm, but that is not a highly agreed-upon value. How does a nanostructure differ from conventional large proteins or other chemical molecules or polymers of a similar size?

Size alone does not make this distinction. Nanostructures are fundamentally different forms of matter than simple chemicals. Their size and organization frequently take advantage of the quantum mechanical properties of these structures to have unique properties. A simple example is the extraordinary fluorescence properties and photostability of quantum dots or nanocrystals that cannot be explained in terms of the elemental composition alone.

1.1.2   Bottom-Up Rather Than Top-Down Approach

Nanotechnology is not just based on its nanoscale dimensions. For most of human history, manufacturing came from sculpting bigger objects down into smaller objects.

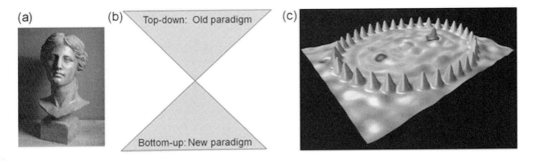

Figure 1.1 The old paradigm for most of the past few thousand years of human endeavors has been to "sculpt" larger objects into smaller objects: sculpture of Aphrodite (a) (source: www .ancientsculpture.net/images/products/small/252.jpg). The new paradigm of nanotechnology inverts this top-down process to a bottom-up approach (b), assembling nanomaterials atom by atom, as shown in (c), whereby 36 cobalt atoms are arranged in a "quantum corral" (source: www.aip.org/png/html/mirage.html).
Source: Leary (2010)

Nanotechnology represents a bottom-up, atom-by-atom assembly (Leary, 2010) (Figure 1.1).

1.1.3   Nanoscale Systems Are on the Right Scale for Nanomedicine

It is important to have tools on the proper scale for the job. Nanomedicine, as single-cell medicine, requires that the tools be smaller than the objects (i.e., single cells) they are dealing with, as shown by this classic nanomedicine size scale (Figure 1.2).

Atom-by-atom assembly, or "nanomanufacturing," still has some issues in terms of speed and practicality, but many nanostructures in biology use the laws of thermo-dynamics for "self-assembly." The concept of nanotechnology was first mentioned by Nobel Laureate Richard Feynman, who in his famous 1959 "nanotechnology" lecture "Plenty of Room at the Bottom" proposed that atom-by-atom assembly of materials might someday be possible (Feynman, 1960). Many people think that "molecular manufacturing" proposed by futurist Eric Drexler in his 1991 MIT PhD dissertation is science fiction. Unlike conventional chemical batch synthesis of finished products from raw materials, molecular manufacturing would create products in an atom-by-atom assembly in a form of desktop manufacturing (Drexler, 1991). As of 2021, this has not yet been accomplished to the degree of real molecular manufacturing, as envisioned by Drexler and others (Figure 1.3).

Due to its controversial status, Drexler's molecular manufacturing was overtly removed, perhaps unfairly, from the National Nanotechnology Initiative passed by Congress and signed into law in 2001. The original law sunset in 2008 and has not been formally renewed (Sargent, 2010). But Drexler's molecular manufacturing is a topic that will probably not go away, nor should it. If it were indeed possible, it would revolutionize manufacturing as we know it, giving us a form of "3D atomic printing" that would be the nanotechnology equivalent of current 3D printing (Figure 1.4).

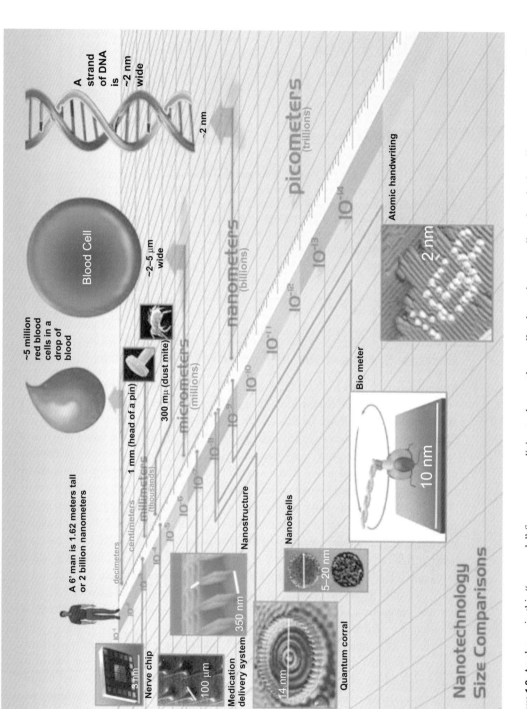

Figure 1.2 As shown in this "nanoscale" figure, nanomedicine tools must be smaller than a human cell so that single-cell treatments are possible

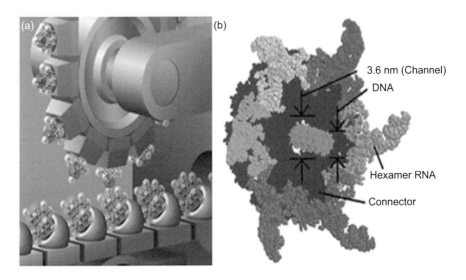

Figure 1.3 (a) The original concept of nanomanufacturing, as envisioned by Eric Drexler, is that one would need tiny machines capable of positioning atoms one at a time (source: http://metamodern.com/2009/02/27/high-throughput-nanomanufacturing). This challenging paradigm has yet to be fully realized, although there are a number of researchers attempting it. (b) The reality is that most nanomanufacturing will be done by thermodynamically driven self-assembly, in the same way that nature makes nanostructures such as Phi-29 RNA nanomotor structures (source: http://nihroadmap.nih.gov/nanomedicine/devcenters/phi29dnapackagingmotor.asp).

An Ugly Debate about the Feasibility of "Molecular Manufacturing"

Eric Drexler   Richard Smalley

Figure 1.4 In a rather heated series of back-and-forth debates, Eric Drexler and Richard Smalley argued about the feasibility of molecular manufacturing. Smalley contended that it was impossible due to the inability to pick up and position atoms quickly enough, whereas Drexler countered that smaller components could self-assemble and then be positioned as larger pieces later in the process.
Source: Leary teaching

However, quite independently of science fiction, nanomanufacturing has been going on in the biological world for many millions or even billions of years. There are numerous nanomanufactured structures in biological organisms that perform nanomanufacturing of self-assembling structures quite naturally through the laws of thermodynamics. Self-assembling nanostructures, including a wide variety of nanoparticles of diverse compositions, are being nanomanufactured by thermodynamically driven self-assembly in many laboratories, including those of this author, around the world these days (cf. reviews: Haglund, Seale, and Leary, 2009; Leary, 2019; Seale and Leary, 2009). Indeed, our own bodies are full of self-assembling, nanomanufactured components, and viruses are truly one of nature's self-manufacturing and self-assembling nanoparticle structures. Interestingly, these naturally occurring viral nanoparticles have also evolved over the millennia into a form of "nanomachines" that quite efficiently use host cell machinery to manufacture nanoparticle subcomponents for later self-assembly during the replication process. Through biomimicry (Benyus, 1997), nature inspires some of us to learn how to design synthetic nanoparticles, mimicking viruses, that do not replicate for their own purposes but rather to manufacture molecules therapeutic against a wide variety of diseases, as described in more detail later in this book. These "smart nanoparticles" are an important part of a new drug-device strategy (Figure 1.5).

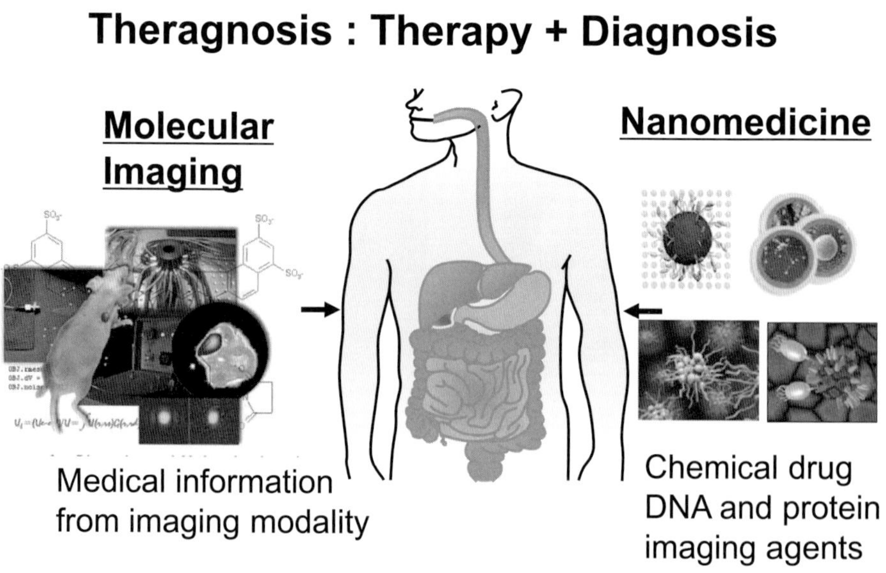

Theragnosis : Therapy + Diagnosis

Molecular Imaging

Nanomedicine

Medical information from imaging modality

Chemical drug DNA and protein imaging agents

Figure 1.5 "Smart" nanoparticles capable of both diagnostics and therapeutics will allow a new stage of "theranostics" in modern medicine.
Source: Leary teaching

1.2   The Progression of Medicine

The field of medicine has steadily progressed, particularly over the past 150 years with the discovery of disease pathogens, the importance of sterility in surgical interventions, the development of antimicrobials and vaccines, in vivo imaging, and a general embrace of new technologies. Nanotechnology just represents the next major technology to be applied to medicine. But its application will represent a fundamental paradigm shift from organ-level disease treatments to single-cell treatments. One area of medicine that has embraced nanomedicine concepts is ophthalmology (Zarbin et al., 2010a, 2010b, 2011, 2012).

1.2.1   Human Disease Really Happens at a Single-Cell Level

Human disease happens at a single-cell level, but we typically only define and treat disease when it affects the organ. We need a paradigm shift in our thinking if we are to fully understand and appreciate the power and promise of nanomedicine. We need to start thinking about human disease at a more fundamental and presymptomatic level. Indeed, if we are to be more forward-thinking and start understanding human disease in terms of regenerative medicine, or treating disease at an earlier presymptomatic single-cell level, we will need to move our thinking to abnormal changes happening within individual cells rather than macroscopic changes at the organ level. The purpose of medicine must become the way to keep people healthy and not allow them to become diseased. That will require fundamental changes in the ways we train doctors – to get them to think in terms of single-cell molecular biology and how to intervene at the single-cell level to keep patients healthy. By the time a patient manifests disease at the organ level, the disease may have already caused irreversible damage. We must also change our thinking about aging and start thinking about aging as not natural but rather the accumulation of unrepaired damage at the single-cell level. While that may appear too expensive a process, it is cheaper to repair damage at the early stages rather than waiting until extensive organ damage occurs. However, for this to happen – namely, medical interdiction at very early stages of disease – there must be molecular diagnostics at a presymptomatic stage of disease. Current healthcare practice and economics preclude this approach due to a failure to take into account the total bottom line in healthcare costs. Until that attitude changes, we will continue to ignore disease at the early stage when it is comparatively inexpensive to treat and then spend huge amounts of money later when the disease has progressed to a serious or even life-threatening stage – a case of bad healthcare economics!

1.2.2   Conventional "Modern" Medicine

Conventional medicine has progressed from fairly primitive surgical methods including "exploratory surgery" to today's more sophisticated and noninvasive in vivo imaging by simple X-rays, CT scans, MRI scans, and PET scans followed by minimally invasive laparoscopic surgery and even robotic-aided microsurgery. But with the exception of a limited number of targeted antibody therapies for specific cancers, drug therapy remains mostly untargeted drugs by either intravenous or oral administration. This has led to a crisis in the pharmaceutical industry whereby the number of drugs approved by the FDA each year has gone down drastically in the past 15 years while the cost of development has more than quadrupled. It is an unsustainable model. An easy initial partial fix to this problem is to repackage existing FDA-approved drugs into nano packages with targeted delivery and increased circulation times not only to enhance effectiveness but also to reduce side effects that can seriously affect patient health and well-being. We also need to incorporate individual genomics to know which drugs should not be administered to specific persons. This would be a huge advance for the pharmaceutical industry and lower their risk in development of new drugs. If you can exclude the people who would have severely adverse effects, beyond unpleasant side effects, many more drugs would be available for the rest of us!

1.2.3   "Personalized" or "Molecular" Medicine

There are essentially two main problems with our current approach to medicine. First, there are many very good drugs that cannot be used due to the extreme side effects, including death, in a subset of patients. Without patient-specific predictive information to tell us which patients should *not* receive specific drugs, we end up excluding these drugs from their effective use in the majority of patients. There is no way to win this game based on "population medicine." The only way to win the game is to practice individual or personalized medicine. Now that we have the genomes of tens of thousands of individual human beings sequenced, we can start to use that information to decide which drugs to give, or not to give, to individual human beings. This will allow us to include many drugs that are good and appropriate for most people, but perhaps deleterious or even lethal to some people. That advance will totally change the drug industry and make drugs more affordable by avoiding the very large and expensive clinical trials which will become largely unnecessary. These large clinical study sizes are a result of us playing a game of Russian roulette with enough people sampled to start seeing the outlier patients. That is ultimately a poor way to test new drugs. We need a greater understanding of the causes of adverse side effects based on information from individual human genomes and also modeling of the differing biochemistry and metabolism of individual humans.

This problem will be gradually overcome when we have enough specific genome information for each patient to allow us to distinguish which patients should, or should not, receive specific drugs. Some of this can be done through rapid testing of specific portions of the individual patient genome through single nucleotide polymorphism (SNP) chips that examine the DNA variations in specific genes from individual to individual. But some of this work will be difficult and slow going, particularly with multigene disorders, and we will ultimately need to use whole genome information for each individual.

1.2.4 Nanomedicine "Single-Cell" Medicine

Personalized medicine will solve only part of the problem. That still leaves the serious problem of untargeted drug delivery whereby those drugs go everywhere in a patient's body and cause unintended damage to other tissues and organs as part of the processes we refer to as side effects or, medically speaking, contraindications.

Nanomedicine is the process of treating the body, especially at a single-cell level, using nanodelivery systems to improve the targeting of drug/gene therapies to specific desired cells and to allow much smaller amounts of drugs to be put into the body in the first place. Clearly, by lowering the amount of drug needed by tenfold or a hundredfold, we are going to make a lot of progress in lessening side effects and dangerous patient outcomes. Paying attention to the design of nanomedical systems that have long circulation times in vivo is perhaps as important as targeting. The future of medicine lies in the combination of both personalized medicine *and* nanomedicine (Leary, 2010) (Figure 1.6).

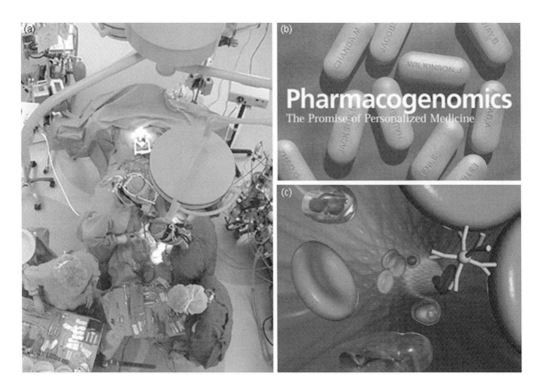

Figure 1.6 (a) Conventional medicine still mostly uses hand-guided surgery (source: www .texasheartinstitute.org/HIC/Topics/images/ordome.jpg), with some robot-guided microsurgery. (b) The future of medicine will combine the capabilities of "personalized medicine" using pharmacogenomics based on the individual patient's genome (source: http://ehp.niehs.nih.gov/ txg/members/2003/111-11/focus/focus.html). (c) Nanomedicine will work in close relation with "personalized medicine to provide superb targeting and drug delivery capabilities" or provide for real-time fluorescence-guided surgery to better see tumor margins.
Source: Leary (2010)

1.3   How Conventional Medicine Works for Diagnosis of Disease

Conventional medicine works in a series of well-defined steps initiated upon the development of patient symptoms. The following six steps should be familiar to anyone navigating the current medical care system.

1.3.1   Identification of the "Diseased State"

The first step is the identification of a "diseased state" by a patient who "doesn't feel right or well" and then goes or is taken to a clinic or a hospital emergency room. This initial self-diagnosis by the patient is highly subjective and based on symptoms that are hard to distinguish between different medical problems.

1.3.2   Collection of Medical Data by Health Professionals

The second step begins with attempts by nurses and doctors to collect simple measurements (e.g., temperature, blood pressure, heartbeat rate, palpitation to find "where it hurts" and/or any abnormal lump). By the time these measurements are taken, the disease may have become quite advanced. In addition, measurements taken in the doctor's office at a single time point are a poor substitute for continuous measurements taken over days, weeks, or months.

1.3.3   Analysis of Initial Medical Data on Patient

This is followed by a third step whereby clinical tests (e.g., blood chemistry; urine chemical analysis; blood, urine, or sputum cultures to detect abnormal numbers or types of microbes; blood cell numbers and percentages by cell subpopulation types; biopsies of tissues and their interpretation by histopathologists) are administered to try to narrow down the disease diagnosis possibilities.

1.3.4   More Advanced Examinations of the Patient

The previous step is often followed by internal examinations using noninvasive imaging, such as standard x-rays, computerized tomography (CT) scans, magnetic resonance imaging (MRI) scans, and positron emission tomography (PET) functional imaging. Before the advent of modern noninvasive imaging methods, the patient was literally opened (i.e., exploratory surgery) by a surgeon. But due to the extreme invasiveness and the risk of infection, this is seldom done anymore. It should be seen only in cases where noninvasive diagnostics fail to diagnose the disease.

1.3.5   Comparison of Patient Data with "Normal" Ranges

In almost all cases, individual results are compared with the "normal" ranges of individuals thought not to be diseased. This step brings home the point that we are still practicing population rather than personalized medicine. We attempt to determine the status of the individual patient in terms of a mythical average patient or some population-based limits of what is considered normal. Actually, this population-based diagnostic is less serious in its implications.

1.3.6   Molecular Tests to Determine Disease State

While still in their infancy, molecular tests (e.g., gene relocations, amplified gene copies) are performed. Occasionally, tests on relatives of the patient are used to establish genetically determined diseases. We will see more and more molecular tests as we start to understand disease at the single-cell and molecular level.

1.4   How Conventional Medicine Works for Treatment of Disease

To a large extent, conventional medicine is designed to stabilize the patient so that the patient heals himself or herself. That process is aided by surgery or treatments with drugs to lower tumor or infectious disease load, but the body still must mostly overcome the disease by itself. A typical sequence of events in the course of conventional therapy consists of the following steps.

1.4.1   Stabilization of the Patient: "Heal Thyself"

The major first step is stabilization of the patient (e.g., via intravenous hydration with saline, blood transfusions, simple medicines to lower dangerous fevers) so that the patient can apply a "heal thyself" strategy. The problem is that by the time many symptoms appear, it may be difficult or even impossible to stabilize the patient to enable self-healing.

1.4.2   Surgical Repair of Injuries

If the patient is not likely to accomplish this without additional help, the next step is surgical repair of injuries (i.e., reconstructive surgery), surgical or radiation removal of diseased tissues or organs, or, in the case of nonsurgical interventions, treatment with chemical drugs that are delivered locally (e.g., ointments to skin, injection of drugs into tissues or organs). However, surgery is still a macroscopic exercise whereby many normal cells (e.g., in a tumor margin) are sacrificed to try to ensure that all diseased cells are removed. This can and frequently does lead to serious organ failures and other postsurgical complications.

1.4.3   Treatment with Drugs Locally

Whenever possible, we try to treat the medical problem locally. That can involve treatment with bandages to bind up and protect open wounds from infection, treatment with antibiotic ointments, or possibly treatment with local anesthetics to reduce pain locally. This is frequently done with skin diseases but is much harder to accomplish with internal diseases.

1.4.4   Treatment with Drugs Systemically

If the problem is more systemic and cannot be solved with localized therapies, medications are delivered systemically (e.g., chemotherapy), usually by either oral or intravenous routes. Interestingly, nanomedicine may provide a third alternative by using drug-carrying nanoparticles that are able to penetrate the skin and then distribute medications further in the body through either the blood or lymphatic systems. This would reduce the risk of infections caused by rupturing the skin, which protects us from many pathogens in the environment.

1.4.5   Treatment with Targeted Therapies

Patients for whom appropriate therapies exist might be treated with targeted therapies (e.g., currently, monoclonal antibodies targeted against diseased cells), although these are still limited to a small number of diseases. The next stage beyond these targeted antibodies is the arena of nanomedicine. The earliest forms of these nanomedical approaches are already being used in limited clinical trials and will soon start to become available. They should offer an improvement in circulation time and targeting, with subsequent decreases in side effects such as neutropenia, which is caused by destruction of neutrophils that have phagocytosed a fraction of those targeted antibodies before they could find their proper targets.

1.5   Factors Limiting the Progress of Medicine

Interestingly, the rate of progress of medicine is limited less by the pace of scientific progress than by other factors frequently beyond the control of researchers, physicians, and healthcare companies. A number of factors, including economics, politics, ethics, legalities, and government regulations, can significantly affect how quickly, or even whether at all, new advances in medicine are implemented in our healthcare systems.

1.5.1   Economics

Especially in today's economy, healthcare costs are being driven by the rising prices of drugs, devices, and procedures. If new drugs, devices, or procedures are to have any hope of being accepted, they must provide savings in terms of overall patient outcomes and total care costs. Nanomedicine – by providing better therapeutics at lower doses, and targeting these doses – will reduce costly side effects and improve patient outcomes. Individual genomics will help prevent adverse medical events by revealing which patients should not receive certain drugs. It will also provide better prognostic information as to which procedure is likely to be most effective for a given patient. The current guessing game of trying one drug or procedure at a time and proceeding until an effective one is found is not only very costly. The wrong drug or procedure may preclude choosing the better option since the patient has been adversely changed by that prior incorrect or suboptimal medical path.

1.5.2   Politics

Although we would like to think that the politicians who make our laws always have our best interests in mind from a medical point of view, that is sadly not the case at this point in time. Without citing specific examples, we see the effects of an influx of money from lobbyists or attempts to emotionally manipulate voters with rhetoric and misstatements of the facts. These political factors can impede or even completely block the implementation of new medical progress for all of the wrong reasons. Conversely, money, rather than medical necessity, may drive the use of drugs that are, at best, completely useless and a waste of money and, at worst, potentially harmful to patients.

1.5.3   Ethics

I deliberately discuss ethics before legal aspects of nanomedicine. Just because something is legal does not make it ethical. As with any new technology, the law has not yet encountered the ethical boundaries. The purpose of the law is to constrain bad behavior when people are unable to self-constrain using a proper sense of morals and ethics. It is still early in the evolution of nanomedicine, but as it is increasingly used, we will begin to see areas and instances where we need to restrain ourselves. Otherwise, we will need laws to govern our use of nanomedicine. There are still only a small number of published papers that are devoted to these issues, and I include only one example here (Juliano, 2012). There will be more discussion of this topic in Chapter 14.

Diagnostics always precedes effective treatment. This creates a moral dilemma of disease diagnosis where there is no effective treatment. Some argue that diagnosis under these conditions is in some way immoral. However, people can not only plan their lives with a diagnosis but also be prediagnosis and ready for treatment when those new treatment modalities arrive. Indeed, one strategy, while somewhat controversial, is that some doctors try to keep their patients with currently untreatable diseases alive long enough to benefit from new medical treatment advances. Those decisions should probably be left to the patient and his or her own doctor.

1.5.4   Legalities

The real or imagined threat of lawsuits in our litigious society either protects us from potentially dangerous new medical treatments or prevents us from potential benefits.

Fear of lawsuits has made us a very risk-averse society. Most humans are poor at risk assessment. Telling people that some drug or behavior puts them at 10 times the risk is misleading if that initial risk is already very low. Ten times a small number is still a small number! We need to see risk in a more realistic fashion to make proper use of risk assessment in our decision-making processes. There will be more discussion of this important topic in Chapter 14.

Hopefully, the possibility of new nanomedical advances that greatly lower the risks through huge decreases in doses and better drug targeting will allow us to more freely explore the benefits of nanomedicine. With current larger doses and lack of targeting, the consequences of incorrect or suboptimal drugs or procedures are sufficiently high that we must prevent the procedure for everyone. Even if 99 percent of people will benefit, if the effects on the remaining 1 percent are sufficiently severe, the drug will not be available for the majority of people who would benefit. In some cases, those benefits may rise to the level of saving lives. Again, a combination of individual genomics for personalized medicine and nanomedicine should greatly affect the negative legal effects. It is difficult to understand why big pharma has not more fully embraced personalized medicine that would indicate which patients should *not* receive a given treatment. This knowledge would greatly expand the use of drugs that are good for 99 percent of the population but not good for the remaining 1 percent. The problem is that 1 percent of a large number is still a large number. Why not eliminate the 1 percent problem up front with personalized medicine? The cost of whole-genome sequencing is now sufficiently low that it is worth analyzing the genomes of all patients. The argument that we don't know the importance of small differences in genetic sequences from individual to individual, while currently true, means that we will never know the importance (or not) of these individual variations unless we look at a large number of patients. In this day of big data and relatively low whole-genome sequencing costs, the benefits of analyzing this vast amount of data should far outweigh the costs.

1.5.5   Regulation

Most would agree that one proper function of government is to provide regulations that protect its citizens from dangerous substances or procedures. It also provides a level playing field so that companies do not lower their prices by cutting corners and engaging in risky business behaviors. Regulation is increasingly important in medicine, where these drugs or procedures may come from different countries that might have significantly lower standards of protection. In the USA, we have a regulatory agency, the Food and Drug Administration, that oversees the potential introduction of all new medical drugs, devices, or procedures. In the case of nanomedicine, that may also involve the introduction of new "combo drugs" (i.e., drug and device) systems. Unfortunately, there is reluctance from many pharmaceutical companies to apply for combo status due to the perceived additional hurdles in the drug approval process.

1.6   Some Specific Problems with Conventional Medicine

We frequently tolerate high barriers for the introduction of new medical drugs and procedures, but we forget that many of the current ones were either grandfathered in or approved at a time when data or information was either limited or incorrect. The bar for current approval is set much higher than in the past. Many good prescription drugs in common use today would never have made it to market in the current risk-averse climate. We also fail to consider the steep price we pay for a medical system largely driven by patient symptoms.

1.6.1   Consequences of Waiting for Patient Symptoms

While the consequences of waiting for the patient to self-identify symptoms are obvious in the case of heart attacks and similar catastrophic health events, there can nonetheless be serious consequences for delayed self-diagnoses based on patient recognition of symptoms. The sensitivity of tumor detection by "feel" is very bad. For example, by the time most patients can detect a palpable tumor (i.e., they feel an abnormal lump), that lump probably contains millions of tumor cells that may or may not have metastasized to neighboring tissues or organs. The difference in subsequent treatment may then change from relatively simple surgery to the need for systemic chemotherapy. Waiting for symptoms to appear means that many patients perceive that "something is wrong" with their bodies only after a disease is quite advanced.

1.6.2   Trained People and Modern Drugs Are Expensive

Despite the current trend to blame the rising cost of modern medicine on corporate greed, there are also some good reasons for at least some of this increased cost. A century ago, a doctor needed a good "bedside manner" because in many cases, particularly before the advent of antibiotics, a doctor could do little for a patient except comfort them and hold their hand while they died. Modern medicine has extended the average human life span more than 50 percent during the last 100 years through a combination of advanced medical procedures and new drugs. For example, we are now able to save newborn babies weighing less than 2 pounds (less than 1 kilogram). That is astounding medical progress! But we are greedy and want even more. The easier advances have already been made. New advances are more difficult and will require much more research and development and clinical trials – all of which are expensive. The training of new doctors and nurses has become even more expensive since they need to not only master standard medical protcols but also understand and properly utilize these new advanced procedures and more sophisticated medications.

Drug development has become very expensive and can take years and more than a billion dollars (USD) to bring to market. One point I will emphasize repeatedly in this book is that the current model for "blockbuster," untargeted drugs is unsustainable.

But nanomedicine, as a paradigm shift rather than an extrapolation based on exciting paradigms, can provide some of the solutions by providing sophisticated targeting of drugs to diseased cells and packaging of those drugs in nanodelivery devices that will allow a more than tenfold reduction in dose to achieve similar, or better, therapeutic effects.

1.6.3   Diagnostic Technologies Are Still Relatively Primitive and/or Expensive

While noninvasive imaging technologies such as MRI, CT, or PET imaging have provided huge advances, it is still difficult to detect disease at very early stages. Although these imaging techniques can help guide surgery, they are still at a relatively crude spatial resolution. But some nanomedical techniques currently being developed can provide as much as 100 times the spatial resolution, allowing, for example, real-time fluorescence-guided surgery, which has already been performed on the first human patients (van Dam et al., 2011).

1.6.4   Relatively Crude Targeting of Drugs

Many modern drugs are PEGylated, meaning that there has been a layer of polyethylene glycol added to minimize immune system destruction and elimination of the drug to increase overall circulation time. This allows the lowering of doses, which can reduce the number and intensity of side effects. But most drugs are still untargeted, and drug pharmacokinetics follows an exponential decay and yields only about a 1–2 percent effective delivery.

1.7   Some Ways Nanotechnologies Will Impact Healthcare

The simple answer as to how nanotechnology will impact healthcare is that earlier and more sophisticated diagnoses increase chances of survival. Nanotechnology can be used to detect disease earlier through a variety of means to be discussed later in this book. By the time some macroscale physical symptoms are evident to either the doctor or the patient, permanent and irreversible damage to the patient may already have occurred. Nanotechnology will likely have a big impact on healthcare over the next several decades.

1.7.1   Nanomedicine Will Be Proactive Rather Than Reactive Medicine

Conventional medicine is reactive to tissue-level problems that are happening at the symptomatic level. Nanomedicine promises to diagnose and treat problems at the molecular level inside single cells, prior to traditional symptoms. Not only is conventional medicine reactive to symptoms, but there is also the potential for significant irreversible tissue or organ damage by the time a patient feels symptoms and seeks medical attention. Many symptoms are late-stage indications of important medical problems. Earlier, presymptomatic diagnosis and treatment usually lead to a better prognosis and patient outcomes. Interestingly, presymptomatic diagnosis is happening with increasing frequency due to the improved resolution and sensitivity of in vivo imaging (e.g., CT, MRI, or PET scans) originally being used for totally different diagnostic reasons. For example, the patient may be in for symptoms of gall bladder problems and then be diagnosed with a cancer due to discovery of a tumor during a routine in vivo imaging scan.

Nanomedical diagnostics will provide even more opportunities to diagnose diseases at the presymptomatic stage due to improved sensitivity. These nanomedical diagnoses and follow-up treatments will provide opportunities to be proactive, rather than reactive, to human disease. While many people cite increased costs as a reason not to use early diagnostics, the true costs of waiting until major or even irreversible organ damage can be far greater. Most current cost-benefit analyses fail to include the true bottom line of total costs.

1.7.2 Possibility of "Regenerative Medicine"

By providing very early diagnostics and therapy prior to actual significant tissue or organ damage, we may also have the opportunity to repair the damage before it is irreversible. These treatments will provide a new and potent form of medicine known as *regenerative nanomedicine*. Ultimately, it will be seen as far more cost-effective than our current approaches. Keeping a patient healthy is far less expensive than waiting until they are quite ill and require costly, and often futile, treatments. Bottom-line cost-benefit analysis must be performed with the total costs both to the patient and to society; currently, this frequently ignores longer-term indirect costs that can be much larger than immediate direct costs. Very early diagnostics and treatments, including regenerative medicine, can have a huge impact on society. For big pharma, this represents a new opportunity. Whereas one-time-use drugs to cure disease self-limit their own sales, recurring treatments to prevent disease can provide long-term sales benefits. Regenerative medicine drug development is a new frontier and opportunity for drug developers.

1.7.3 Blurring of Distinctions between Prevention and Treatment

All of this will lead, as it should, to a blurring of the distinction between prevention and treatment. Disease is frequently defined either by a set of symptoms or by specific injury to a given tissue or organ. If we can diagnose potential problems at a very early stage and intervene appropriately, at what point will we actually "prevent" disease? When we prevent advanced disease, we may prevent the conventional symptoms of that disease from even appearing. An example is the early introduction of antivirals to HIV-infected individuals who no longer displayed the unusual symptoms of thrush seen in AIDS patients in the early 1980s. Whereas some may argue that such early intervention is costly or unnecessary, the overall cost to both the patient and society at large may be much less, since prevention frequently is less expensive than later-stage treatments. It also means that our concept of "disease" must change to the point where current manifestations of disease are seen as a failure to maintain wellness, which should be the goal. It does mean that if we deploy very early medical approaches to many more people, those medical treatments must be as inexpensive as possible and be highly effective. Such a combination of low cost and high effectiveness is not easy to achieve. Perhaps we need a new government–industry partnership that rewards efforts to lower healthcare costs and produce a healthier population. We must move away from a fee-for-service healthcare system and toward a system where healthcare services are rewarded for preventing the development of serious health problems. While some of this has admirably been attempted by a few healthcare maintenance organizations (HMOs), a larger and longer-term effort needs to be implemented.

1.7.4   Ophthalmology Has Embraced a Nanomedicine Approach

One field of medicine that has embraced the nanomedicine approach is ophthalmology (Zarbin et al., 2010a), from a sense of the future (Zarbin et al., 2010b) to artificial vision (Zarbin et al., 2011). Although still in the early stages, the field of ophthalmology is gradually making practical advances using nanotechnology and nanomedicine. It is also trying forms of regenerative medicine to actually fix the underlying medical problems (Zarbin et al., 2012).

1.8   "Nano Hype" versus Reality

As with any new technology, there is a tendency to "hype" (i.e., exaggerate the importance of) the apparent merits of the technology. On the other end of the spectrum, there is the tendency, particularly of people who feel threatened by the new technology, to put it down or outright dismiss it. This is a familiar pattern of acceptance illustrated by the so-called Gartner Hype Cycle (Gartner, 2020). We have already had unreasonable hype and unrealistic disillusionment, and right now we are in a period of adjusting our expectations to be more realistic. However, nanomedicine is real and will slowly become the new norm for drug delivery (Leary, 2013) (Figure 1.7).

1.9   Why Nanomedicine Will Happen: The Perfect Storm

Why will nanomedicine happen despite the naysayers? The answer is simple – there appears to be a "perfect storm" of factors driving the healthcare system toward this particular solution with respect to nanomedicine and healthcare (Junger, 1997). The perfect storm paradigm is based on the "perfect storm" that hit North America between October 28 and November 4, 1991. On October 27, Hurricane Grace formed near Bermuda and moved toward the coast of the southeastern United States. Grace continued to move north, where it encountered a massive low-pressure system moving

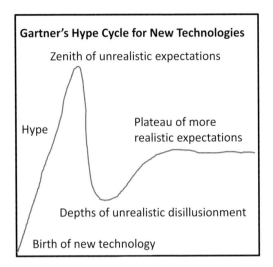

Figure 1.7 A modified Gartner Hype Cycle for new technologies starts out with hype and then plunges into "depths of unrealistic disillusionment," until it stabilizes at a "plateau of more realistic expectations."
Source: Leary teaching (www.gartner.com/en/research/methodologies/gartner-hype-cycle)

south from Canada. The clash of systems over the Atlantic Ocean caused 40–80-foot waves on October 30. It became the basis of a book (Junger, 1997) and a Hollywood movie. Beyond that, it became synonymous with the coming together of multiple forces driving things in a particular direction. In this case, I am implying that there is a perfect storm of factors in science and healthcare driving us toward the adoption of nanomedicine for targeted drug delivery combined with individual human genomics to produce personalized medicine.

Chapter 1 Study Questions

1.1 Why does size matter in bionanotechnology and nanomedicine?

1.2 Medicine has traditionally approached disease at the organ level. How is nanomedicine a fundamentally different approach?

1.3 Manufacturing nanoscale devices was not thought possible due to a variety of factors in molecular manufacturing. What are some of those factors, and how are many bio-nano devices actually assembled in the real world?

1.4 What are some of the reasons why nanomedicine will have a big impact on the future of medicine?

1.5 Why is nanotechnology (and nanomedicine) frequently called a bottom-up rather than top-down approach?

1.6 What are some of the advantages of targeted drug delivery?

1.7 Why can nanomedicine be called a presymptomatic form of medicine, and what advantages/disadvantages does this provide?

1.8 How can nanomedicine be thought of as proactive rather than reactive?

1.9 Why is symptom-based diagnosis so problematic?

1.10 Why can nanomedicine be thought of as "massively parallel processing single-cell medicine"?

1.11 Personalized medicine uses genomic information about the individual to provide some important advantages over conventional medicine. What are some of these advantages?

1.12 How is the concept of "disease state" different in nanomedicine as opposed to conventional medicine?

1.13 Why is the use of "normal" ranges not a good way to determine whether someone is ill?

1.14 What does the medical maxim "heal thyself" mean in terms of what conventional medicine provides?

1.15 Often the scientific attributes of a new technology (e.g., nanomedicine) do not wholly determine whether that technology is adopted. What are some of these nonscientific factors that may delay, or even prevent, adoption of a new technology such as nanomedicine?

1.16 When diagnosis and treatment are done presymptomatically, there is a blurring of whether we are treating disease or preventing disease. What are some of the major implications of this blurring process?

1.17 What are some of the moral/ethical dilemmas of diagnosing diseases that are not yet treatable?

1.18 Why is early diagnosis important with respect to prognosis?

2 Nanomedicine: Single-Cell Medicine

2.1 Features of Nanomedicine

A number of features of nanomedicine make it a new, unique medical discipline. In addition to its association with nanotechnology, it changes the fundamental paradigm of disease in medicine from the organ level to the single-cell level. It is paradigm shifting in a number of ways, as the following sections describe.

2.1.1 Bottom-Up Rather Than Top-Down Approach to Medicine

Befitting its nanotechnology origins, nanomedicine uses a bottom-up approach to medicine, rather than top-down. Throughout human history, most approaches have proceeded with a top-down approach, similar to sculpting smaller objects from larger starting material, as described in Chapter 1. This approach made sense when humans could only visualize and use macroscopic things visible to the eye and manipulatable with human hands. But we now live in an age when we know that things really exist on the microscopic and even atomic scale. Our efforts become indirect as we may or may not be able to visualize these things with imaging technologies that transcend human vision. We need to be able to either manipulate things on the submicroscopic or atomic level or else create devices that are self-assembling according to the laws of thermodynamics. While medicine must treat the total patient, approaches to human health and disease must begin at the single-cell level since that is where early disease begins – as molecular defects within individual cells. For example, although cancer is a disease, the best way to make progress in diagnostics and therapeutics is to detect cancer-specific biomarkers that differentiate cancer into more than 200 different diseases and then treat each subdisease specifically. This approach has been spectacularly successful in treating leukemias and lymphomas, whereas earlier, less-specific approaches failed to make significant progress.

A bottom-up approach starts with single cells both diagnostically and therapeutically. Even though single-cell technology has been around for many years now, its true promise is just starting to be appreciated. Nanomedicine is by its very nature a single-cell technology.

2.1.2   Nanotools on the Scale of Molecules

Just as human hands and chisel were on the proper scale to deal with sculptures, as described in Chapter 1, nanomedicine requires so-called nanotools that are smaller than individual human cells. Nanomedical systems containing multimodal imaging nanomaterials surrounded by layers of molecular targeting, sensing, and therapeutic molecules are on the proper scale to perform "single-cell surgery" on a massively parallel processing scale.

2.1.3   Cell-by-Cell Repair Approach: Regenerative Medicine

Since nanomedical approaches to disease at the single-cell level can intervene in the disease process much earlier than conventional macroscopic medicine approaches, there is promise of nanomedicine becoming a key component of regenerative medicine, whereby disease treatment is seen as a process of keeping the body well by treating disease sufficiently early that it prevents irreversible damage to organs and is considered a form of "single-cell medical preventive maintenance" by doing cell-by-cell repair.

2.1.4   Feedback Control System to Control Drug Dosing

In conventional medical therapeutics, there is no real control of proper dosing of therapeutic drugs. A more nanomedical engineering approach requires including a feedback control system to control proper dosing at the single-cell level. This can be accomplished by incorporating molecular sensors in the nanomedical system to control proper dosing to individual cells. In Chapter 9, specific technical approaches to construction of a feedback control system will be discussed in considerable detail.

2.2   Elements of Good Engineering Design

There are a number of elements of good engineering design that should be used in the development of effective nanomedical systems. While the general elements just involve good common sense, they are frequently ignored, resulting in ineffective nanomedical systems.

2.2.1   Begin with the End in Mind

The dictum "begin with the end in mind" (Covey, 1989) means, in this instance, that good systems should be constructed first from a holistic point of view before getting into the specific details. This avoids the problem of introducing components that either complicate or compromise the eventual product. For example, it makes little sense to use toxic subcomponents that must be subsequently removed to prevent formation of a toxic product. A better approach to avoid this type of problem would be to use so-called green chemistry that avoids the use of harsh toxic solvents that are difficult to remove.

2.2.2   Use a Previously Tested General Design, If Possible

Many problems have already been solved by others working on different issues. The trick is seeing the current problem as only a variant of those previously resolved problems. Unfortunately, science has become so divided in terms of subspecialties that many scientists are unable to see the similarities. As scientists, we frequently read too narrowly and fail to ever see the useful approaches that can be borrowed from other fields of science.

2.2.3   Use Biomimicry Whenever Possible

Biomimicry is copying the designs that nature has already given us (Benyus, 1997). In the natural evolution of organisms, nature has tried many (sometimes in the millions or more!) possible designs and permutations. The result is one or more designs that actually work, but maybe suboptimally. It would be foolish to ignore these "free design" gifts from nature. On the other hand, it would be equally foolish to assume that these natural designs of biomimicry are the best possible design. We should study what nature did to achieve the level of function represented by these designs and then see if we can improve upon them. This is an approach frequently followed in the pharmaceutics industry whereby molecules that exist in nature (e.g., natural drugs) can be improved upon to get around their limitations.

    One of the biggest uses of biomimicry in my own work has been the study of nature's viruses that have evolved to target and enter specific types of cells. My work in nanomedicine over the past 20 years has involved design of nonreplicating, synthetic virus-like structures that use the evolutionary design of natural viruses and biomimicry to target specific cells, promote entry across cell membranes, and retarget inside single cells for advanced drug delivery and even the concept of "manufacturing drugs" inside living cells for treatment of chronic diseases (Haglund et al., 2009; Leary, 2019; Seale and Leary, 2009). This idea of manufacturing molecules inside living cells has been used to create new vaccines against the coronavirus responsible for COVID-19. The mRNA templates inside liposomes get taken up by patients' cells, which then manufacture portions of the coronavirus spike proteins responsible for binding to cells. These viral spike proteins are then released by cells into the bloodstream, where they help activate the immune system against COVID-19. This overall process is yet another example where a nanomedicine approach to disease is already saving lives.

2.2.4   Use Multiple Specific Molecules to Do Multistep Tasks

If the overall nanomedical problem can be broken down into specific steps, then solutions can be viewed at the level of specific molecules doing specific tasks. A collection of different molecules, perhaps organized distinctly in different layers, can accomplish complex multistep tasks, if properly ordered (Leary, 2019). Fundamentally, the basic drug-targeting problem can be seen as a multistep process (Haglund et al., 2009) (Figure 2.1).

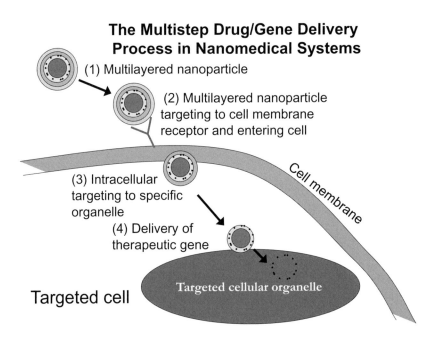

The Multistep Drug/Gene Delivery
Process in Nanomedical Systems

(1) Multilayered nanoparticle

(2) Multilayered nanoparticle
targeting to cell membrane
receptor and entering cell

Cell membrane

(3) Intracellular
targeting to specific
organelle

(4) Delivery of
therapeutic gene

Targeted cell

Targeted cellular organelle

Figure 2.1 Drug delivery is a multistep process. Good design recognizes this and breaks the overall design into multiple steps usually, but not always, characterized by a specific molecule for each step.
Source: Seale, Haglund, et al. (2007)

2.2.5   Control the Order of Molecular Assembly to Control the Order of Events

To control a multistep set of tasks, it is important to solve the tasks in the proper order. This implies controlling when specific molecules become involved in the overall process. If a nanomedical system is designed in multiple layers, each of which is controlled through molecular sensors, then we can create a "programmable" nanomedical system (Haglund et al., 2009; Leary, 2019; Seale, 2007a, 2007b, 2009). Programming is usually thought of in terms of computer software, but it is much more general than that instance. It is really a way to control the order of events. Multilayered nanoparticles, with each layer exposed subsequently through molecular sensors, can be made capable not only of controlling the subsequent order of events but also making complex decisions through the chemistry (Figure 2.2).

2.2.6   Perform the Molecular Assembly in Reverse Order to the Desired Order of Events

Although it should be obvious from the previous section, the multiple layers should be ordered in the multilayered nanoparticle in the reverse order from the desired order of events (Leary, 2019). This is due to the fact that usually only the outermost current

Result: A Multicomponent, Multifunctional "Programmable" Nanodevice or Nanosystem

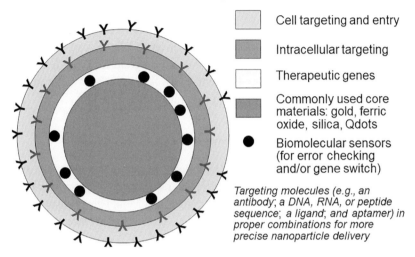

Cell targeting and entry

Intracellular targeting

Therapeutic genes

Commonly used core materials: gold, ferric oxide, silica, Qdots

● Biomolecular sensors (for error checking and/or gene switch)

Targeting molecules (e.g., an antibody; a DNA, RNA, or peptide sequence; a ligand; and aptamer) in proper combinations for more precise nanoparticle delivery

Figure 2.2 The multistep drug-targeting process can perhaps best be accomplished by a multilayered strategy whereby each layer contains the molecule(s) necessary to accomplish that step.
Source: Seale, Zemlyanov, et al. (2007)

layer is able to be exposed to interact with cells and molecules in the human body. If a layer is peeled off due to its molecular sensing system, the next layer is exposed and able to perform its function. If not, then the nanoparticle will not do that subsequent step, which itself is an important decision at the molecular level. The result of constructing a multilayered approach is generation of a smart nanodevice with a decision tree capable of autonomous decision making, including advanced targeting for drug delivery in the human body.

2.3   Building a Nanomedical Device

Building a nanomedical device is a multistep process that starts with the innermost layer and gradually works itself out, layer by layer, to the outside. When the device is functional, its outermost layer controls stealth and initial targeting while inner layers concern the intracellular secondary targeting and drug delivery. The core not only provides a scaffold for building up layers, but also often serves as part of the in vivo diagnostic system giving location information based on noninvasive in vivo imaging technologies such as CT, PET, and MRI.

2.3.1   Choice of Core Materials

Since we are designing a nanomedical system in reverse order, the logical place to start is to choose appropriate core materials upon which to build subsequent layers of the total system. This is where it becomes interesting. The core material can be thought of as "active" or "passive" core material in the total process. An active core might be magnetic such that external magnetic fields can be used to control its subsequent location or concentration or even permit localized heat by an oscillating magnetic field for single-cell hyperthermia. A passive core might involve use of that core material as a contrast agent for X-ray, MRI, or PET.

2.3.2   Add Drug or Therapeutic Gene

Since drug delivery should usually occur at the end of a multistep targeting process, it is logical to put the drug in a place where it will only be released if it has gone through this decision tree and error-checking process. It should also be a good place to protect the drug from degradation in the blood and release it once it is safely inside the targeted diseased cell.

2.3.3   Add Molecular Biosensors to Control Drug/Gene Delivery

Molecular biosensors should ideally control the dosing and release of drugs at the correct therapeutic dose for an individual diseased cell. Individual cells may require different drug doses. Single-cell nanomedicine approaches allow for slightly different treatments for individual cells. This usually means that the molecular biosensors, as part of a feedback control system, should be tightly coupled to the drug or therapeutic gene and sensitive to the treatment needs of individual cells.

2.3.4   Add Intracellular Targeting Molecules

Delivering a nanomedical system to an individual cell is only the first step of the total multistep targeting process. The interior of a cell is huge compared to the size of a nanoparticle. For this reason, there should be additional intracellular targeting molecules to help guide the nanomedical system to a subcellular organelle or to the aberrant molecules that cause disease at the single-cell level.

2.3.5   Result: A Multicomponent, Multifunctional Nanomedical Device

The result of combining the above atoms and molecules in a holistic design is a multicomponent, multifunctional nanomedical device capable of targeting and delivering therapeutic molecules to the diseased cell without significantly injuring many nearby healthy cells. Nanosurgery at the single-cell level should have as its goal the elimination, or at least minimization, of the traditional tumor margin whereby nearby healthy cells are removed in conventional surgery in order to be as certain as possible that cancer cells are not left behind. A significant number of cells removed in conventional surgery are normal, healthy cells. At the macroscopic level, even using microsurgery techniques, surgeons cannot see where the diseased cells end and the normal cells start. Nanomedical approaches' capability for "real-time" fluorescent-guided surgery based on the targeting of fluorescently labeled nanoparticles has recently been demonstrated (Low, Singhal, and Srinivasarao, 2018).

2.3.6   For Use, Design to Delayer, One Layer at a Time

Usually, we think in terms of delayering the multilayered assembly one layer at a time, from outermost to innermost, to properly control the order and location of subsequent events. Controlling the order of events is the most basic part of any programming language and applies equally as well to nanomedicine as it does to computer programming. You just need to wrap your head around the fundamental concept of programming!

2.3.7   The Multistep Drug/Gene Delivery Process in Nanomedical Systems

Overall, we need to think of therapeutic drug/gene delivery in terms of a multistep process rather than the conventional approach of just delivering the drug/gene system into the blood and hoping that it will reach the right cells before being degraded in the bloodstream. That hopeful, but naïve, and nontargeted process requires that drug dosing be massive in order to fight rapid elimination from the body. This results in injury to healthy cells, which can be seen in patients at the total body level in terms of unfortunate and sometimes deadly side effects. A properly targeted nanomedical system should be capable of circulating in the body for many hours or days and be given withat least 10 times and perhaps 100 times less drug than conventional methods. This will result in elimination of most side effects and a much more effective overall therapy. This is the basic concept behind time-release formulations of drugs. While it is difficult to maintain such nanodelivery systems in the circulation for long periods of time, it is possible, by using good engineering techniques and paying attention to surface properties of nanodevices, to increase the circulation time enough to greatly decrease the overall dose given to the patient. Early-stage efforts in the area of increasing circulation times of drugs currently use pegylation techniques to escape opsonification and immune-response early elimination of drugs from the human body.

2.4   The Challenge of Drug/Gene Dosing to Single Cells

The main challenge of drug/gene dosing to specific single cells lies in the targeting/mistargeting process described below. Once the targeting problem is satisfactorily resolved, the next challenge is delivering the right dose for treatment efficacy.

2.4.1   Precise Targeting of Drug Delivery System Only to Diseased Cells

The essence of nanomedicine is the precise targeting of drugs to diseased cells while sparing, or at least minimizing, nearby normal cells' exposure to the drug. This requires a good cell surface biomarker molecule that distinguishes a normal cell from a diseased one. If there is such a biomarker present on diseased cells, there is a good probability of a successful nanomedical approach to that disease treatment. A more difficult situation occurs when the "biomarker" is just increased or decreased in number compared to that on a normal cell. Then targeting becomes a concentration-driven process, which is what we would hope to avoid. An example, nonetheless still useful, is that of targeting folate receptors, overexpressed on many cancer cell types but also expressed on normal cells (Frigerio et al., 2019). This is not a simple case of mistargeting to the wrong cells but failure to be able to fully distinguish between normal cells and cancer cells. It is a topic that will be discussed in more detail in a Chapters 3 and 13.

2.4.2   How to Minimize Mistargeting

We can minimize incorrect targeting, or mistargeting, by a variety of means. First, we can make sure that the targeting mechanism (e.g., antibody, peptide, aptamer) has high specificity and avidity. Second, if an antibody is against a specific epitope (a problem to be discussed in Chapter 3), it is probably a good idea to choose a different antibody against a different epitope that may not be shared by both normal and diseased cells.

2.4.3   How to Deliver the Right Dose per Cell

Delivering the right dose to a single cell can be challenging. If a certain number of nanoparticles need to be delivered to a single cell to achieve therapeutic efficacy, then the problem is driven by Poisson statistics and probabilities, including ratio of nanoparticles per nearby diseased cells, probability for attachment and uptake and intracellular targeting, and a variety of other factors that become part of the overall efficiency.

A better way to approach the problem is to try to use a drug that will induce apoptosis at the level of a single nanoparticle taken up by a single cell. Then, once this threshold is reached, it does not matter how many targeted nanoparticles reach each cell. But if a single nanoparticle can trigger apoptosis, we must be very careful not to have instances of random, nontargeted nanoparticles entering healthy cells. A better idea might be to require a few nanoparticles to bind to a cell before inducing apoptosis. That might eliminate the problem of a single nanoparticle mistargeting and binding to a normal cell.

2.4.4   One Possible Solution: In Situ Manufacture of Therapeutic Genes

Another possible solution is to bypass the problem of delivering the right number of nanoparticles per cell and use nanoparticles to introduce a drug- or gene-producing template complete with a feedback control biosensor into a cell whereby one achieves

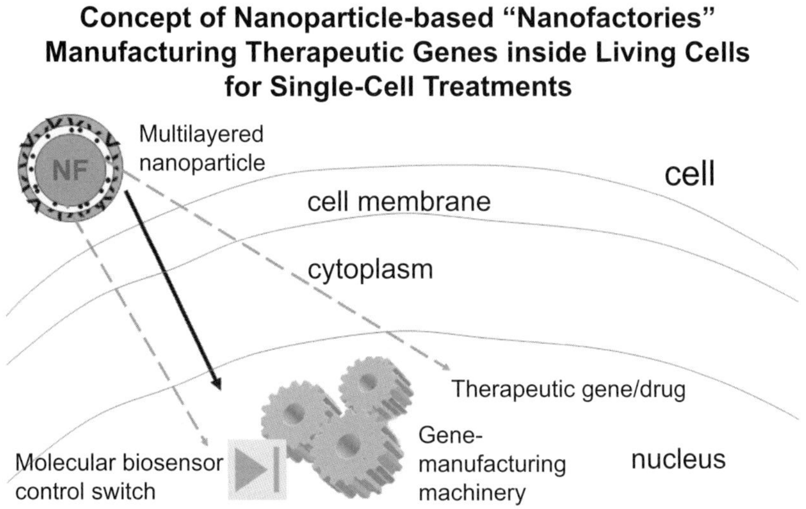

**Concept of Nanoparticle-based "Nanofactories"
Manufacturing Therapeutic Genes inside Living Cells
for Single-Cell Treatments**

Figure 2.3 An elegant solution to the problem of delivering the proper drug dose to a single cell is to embed a drug/gene manufacturing assembly into the cytoplasm of a cell and have it deliver a drug dose using a biosensor feedback control system.
Source: Prow et al. (2005)

an in-situ manufacture of therapeutic drugs (Prow et al., 2005). Then it does not matter if the right number of nanoparticles per cell is reached. The whole system will self-regulate to the right dose to achieve the therapeutic result (Figure 2.3).

An additional advantage of this approach is that it can be used to treat chronic diseases where the goal may not be to eliminate a diseased cell but rather to reregulate the expression of specific gene products of that cell to be in the range of production of a normal cell. This is much harder, but possible, to achieve and will be driven by high financial incentives to use nanomedicine to treat chronic diseases rather than diseases for which there is a one-time use for a cure.

2.5   Bridging the Gap between Diagnostics and Therapeutics

An important feature of nanomedicine is that it bridges the traditional gap between diagnostics and therapeutics. If designed properly, it can have both properties and become a "theranostic" device. This deserves a more detailed discussion of the nature and merits of theranostic devices in the following sections.

2.5.1   How Conventional Medicine Is Practiced in Terms of Diagnostics and Therapeutics

Conventional medicine usually starts with the patient being given a test in response to patient complaints about symptoms. The doctor plays detective by basing his or her diagnosis on a collection of symptoms that are unfortunately frequently shared by a number of possible diseases. The various diagnostic tests are given to narrow down the possible underlying diseases until a diagnosis is declared. The problems with this approach are numerous. First, and perhaps most importantly, symptoms are usually a late-stage occurrence after irreversible tissue damage has already occurred. Second, the assumption is made that the symptoms are caused by a single disease. Sometimes there are several diseases occurring simultaneously within a given patient, which can lead to conflicting or even contradictory results. Third, the process of reaching a diagnosis by process of elimination can be very costly and time-consuming. Our current healthcare system usually requires that the patient be given the cheapest tests first before proceeding to more expensive tests. There is nothing wrong with that approach provided the severity of the disease progression will not compromise the ultimate therapeutic outcome and that incremental efforts and incorrect therapies do not preclude the use of better alternatives in the future. A patient is not necessarily the same after treatment 1, which might actually preclude a more efficacious treatment 2.

The other problem with conventional medicine is that diagnostics and therapeutics are carried out by different doctors who frequently do not communicate very well with each other, frequently caused by subspecialty training and a different medical vocabulary. There is also the so-called hammer-and-nail problem." A doctor whose specialty is a particular "nail" will frequently recommend a therapy of the specific "hammer" (i.e., therapeutic) for that particular "nail" (i.e., diagnosis). For example, a surgeon will almost always want to cut. A radiation oncologist will want to use radiation treatment rather than a chemotherapeutic approach. This is not to say that the doctors are promoting their own approach. It is just that doctors are more comfortable staying within the bounds of their own medical training.

2.5.2   The Consequences of Separating Diagnostics and Therapeutics

Conventional medicine separates diagnostics from therapeutics. These two functions are usually carried out in separate places with little or no communication between the two. It can also lead to additional treatment delays, which may compromise the patient's overall prognosis. This can lead to serious negative consequences since there is no tight linkage between the patient responses to therapeutics. One of the goals of nanomedicine should be to integrate as much as possible initial diagnostics with initial therapeutics in a process known as theragnostics or theranostics.

2.5.3   A New Approach: "Theragnostics" (or "Theranostics")

Another way that nanomedicine is fundamentally different from conventional medicine is that diagnostics and therapeutics can be carried out simultaneously, so-called theragnostics or theranostics, even possibly linked through nanosensor feedback loop systems connecting therapeutics delivered per single cell to diagnostics in those same single cells. Diagnostics then monitors the real-time response to therapeutics at the single-cell level.

2.6   How Theragnostics Relates to Molecular Imaging

Theragnostics often relates to noninvasive and molecular imaging techniques because these technologies make it possible to know the location of the nanomedical systems within the human body. Location is important in medicine. Even though the emphasis of nanomedicine is at the single-cell level, it helps to know the location of those single cells. Organs still do tell us something about disease because single cells are organized into tissues and organs. The choice of core material determines what types of molecular and noninvasive in vivo imaging can be used to determine the location of these nanomedical systems.

2.6.1   Conventional Imaging Is Not Very Specific

Conventional imaging is usually done at the organ level. It is usually macroscopic and can have spatial limits that allow a view of disease only at the level of millions of cells. At that point, there is frequently irreversible tissue damage. There is a need for higher-resolution, noninvasive, in vivo imaging to get down to the resolution of at least small numbers of cells. For example. PET imaging can detect much smaller tumors than MRI and is considered the current standard for determining whether someone is "cancer-free," meaning only that the cancer is below the detection limit, not that the tumor is totally absent. Advances in these technologies, such as the use of T1 and T2 contrast agents (e.g., using specific materials sensitive to T1 and T2 MRI in the core material) can considerably enhance the resolution of MRI. This will be discussed in more detail in the following sections and in Chapter 5, where the use of core materials as a contrast agent is discussed.

2.6.2   Types of In Vivo Imaging

There are now many types of in vivo imaging (e.g., CAT scans, MRI scans, PET imaging), both singly and in combination. Many of these modalities are minimally invasive and represent a tremendous advance for medicine. These imaging modalities are differentiated by their spatial resolution and their cost. As such, conventional medicine usually starts with the least expensive and lowest resolution. This makes sense as long as lower-resolution imaging scans do not miss disease at early stages when they can be treated most effectively. But false negative results can become very expensive when disease is allowed to progress rapidly after the initial diagnostic imaging!

2.6.2.1   X-Rays and Computed Axial Tomography Scans

X-ray imaging is sensitive to electron-dense bodily materials. As such, it is most effective at detecting bone breaks because bone contains a lot of electron-dense elements. It is not particularly sensitive to soft tissue diagnostics unless an X-ray contrast agent (typically a heavy metal that is "electron dense") is added. X-ray imaging can be performed at two levels. In many cases it is sufficient to detect the

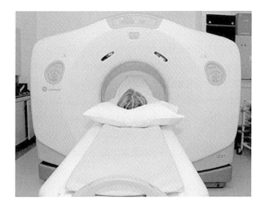 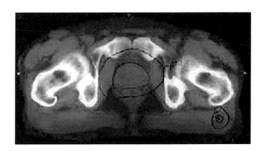

Figure 2.4 Although two-dimensional X-ray imaging is simple and fast, it suffers from the problem of parts of the image being hidden behind other structures. By creating a three-dimensional X-ray image, the image can be rotated to reveal structures that would be hidden in a two-dimensional image.
Source: www.cancerhelp.org.uk/help/default.asp?page=148

medical problem. However, it is susceptible to missing medical problems if the medical problem is hidden from view due to the angle of irradiation. This is one of the many reasons why 3D X-ray imaging, known as computerized axial tomography (CAT) was first invented. Since CAT imaging requires the construction of a 3D image from multiple angles of detected X-rays, it is computationally intensive and correspondingly much more expensive. Frequently, it is not done unless there is a perceived need for higher-resolution imaging than a simple two-dimensional X-ray, which is very fast and inexpensive, particularly now that even 2D imaging employs reusable imaging devices rather than old-style photographic film materials (Figure 2.4).

Many nanomedical systems contain cores that are composed of electron-dense materials. Since these nanoparticles can be targeted to disease, it represents an important new frontier for X-ray imaging diagnostics that can use targeted, electron-dense nanoparticles to identify diseased cells in soft tissues that would not normally be amenable to detection by X-rays. This would allow X-rays, and particularly CAT scan systems, to provide many of the advantages of more expensive modalities such as MRI and PET. Many of these core materials are both X-ray and MRI contrast agents, so if a diagnosis is equivocal with one imaging technology, the patient can quickly be rescanned using the same targeted nanoparticles with the other imaging technology. It is also sometimes beneficial to use a combined MRI/PET or X-ray/MRI image to help localize tumors to specific organs.

2.6.2.2 Magnetic Resonance Imaging Scans

Magnetic resonance imaging (MRI) uses the ability of pulsed magnetic fields to change the magnetic polarization of atoms and molecules, either naturally occurring

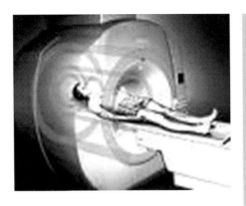
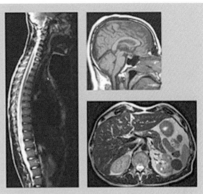

Figure 2.5 MRI can deeply penetrate soft tissue to noninvasively image internal organs and to visualize abnormalities such as tumors.
Source: http://nobelprize.org/nobel_prizes/medicine/laureates/2003/press.html

or introduced through contrast agents, to perform noninvasive 3D imaging of soft tissues. Since the core materials of nanomedical systems can contain MRI contrast agents, targeted nanoparticles can significantly improve the resolution of MRI imaging, allowing detection of disease at an earlier stage. MRI uses intense magnetic fields to detect the relaxation of flipped atoms (e.g., usually hydrogen atoms within water molecules) back to their original positions. MRI is good for deep visualization of soft tissues (Figure 2.5).

2.6.2.3 Positron Emission Tomography Scans

Positron emission tomography (PET) is a much more expensive imaging modality due to its use of contrast agents containing positrons (anti-matter electrons), which are expensive to produce and have short lifetimes (sometimes hours or minutes). But since they are distinct from naturally occurring atomic particles in the body, they provide a powerful contrast with a signal with essentially a zero background (Czernin and Phelps, 2002) (Figure 2.6).

To date there has been limited incorporation of PET probes in nanoparticles due to vastly increased costs and lack of nearby and available cyclotrons capable of generating these PET probes, which then need to be rapidly incorporated into appropriate nanoparticles – a complex and costly set of linked procedures. As such, this is still mostly in the realm of research-level centers rather than common clinical settings.

2.6.2.4 Optical In Vivo Imaging

If deep penetration is not needed, then slightly modified optical imaging methods can be used to image subregions below the skin of an animal or a human. Usually, these

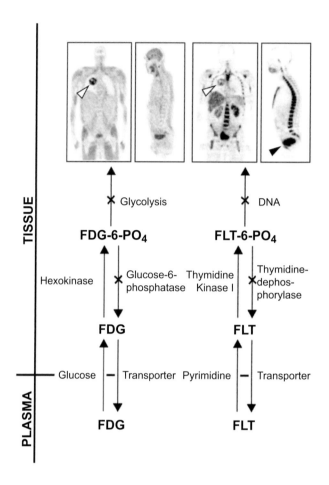

Figure 2.6 Top: FDG is phosphorylated to FDG-6-PO₄ by hexokinase. Since the activity of glucose-6-phosphatase is negligible, FDG-6-PO₄ is essentially trapped in tumor cells (arrow). Bottom: Fluorothymidine is phosphorylated by thymidine kinase to FLT-6-PO₄ and accumulates in tumor cells.
Source: Czernin and Phelps (2002)

techniques use longer wavelengths of irradiating light to achieve deeper tissue penetrations (Figure 2.7).

2.6.2.5   Molecular Imaging

Since nanomedical systems operate at the molecular and supramolecular level, they are at the level needed to provide molecular subcellular-level imaging. Molecular imaging has the promise to provide for diagnostics almost at the single-cell level. This is important since disease occurs at the single-cell level and prognosis is much better if it is diagnosed and treated at the earliest possible stage (Betzig et al., 2006). But currently, most molecular imaging techniques are difficult to perform in vivo (Figure 2.8).

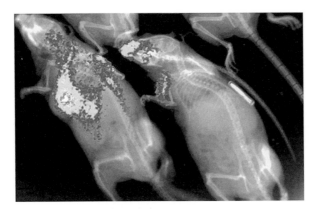

Figure 2.7 Optical imaging, in many cases using fluorescent probes targeted to organs or tumors, allows their visualization if close to the surface. In this instance, the white patches of the gray-level image are really red fluorescence labeling sites of tumors.
Source: www.kodak.com/US/en/health/s2/products/imgStationXRayImagingModule/index.jhtml

2.7   Engineering Nanomedical Systems for Simultaneous Molecular Imaging

While nanoparticles will naturally congregate at sites in the body with leaky vasculature via enhanced permeability retention (EPR), a topic to be discussed in more detail in Chapter 6, they will be easily washed away and not retained at the actual site of the diseased cell unless the nanomedical system binds with sufficient avidity to the diseased cells of interest.

2.7.1   Using Cell-Specific Probes for Molecular Imaging of Nanomedical Devices

A variety of molecules (e.g., antibodies, peptides, aptamers) that bind to cells can be attached to nanomedical systems to retain the imaging contrast agents sufficiently to permit molecular imaging. These types of targeting molecules will be discussed in much more detail in Chapter 3. The size of the retained nanoparticles determines the limits of resolution of the molecular imaging. But the most important requirement is that the imaging contrast agents not drift away to the point of destroying diseased cellular resolution.

2.7.2   Breaking the "Diffraction Limit": Nano-Level Imaging

The limits of spatial resolution in imaging are normally determined by the so-called diffraction limit, but these are not normal times! The diffraction limit in physical optics says that due to diffraction, the limit of optical instruments to distinguish between two objects is approximately half the wavelength of the light used to examine

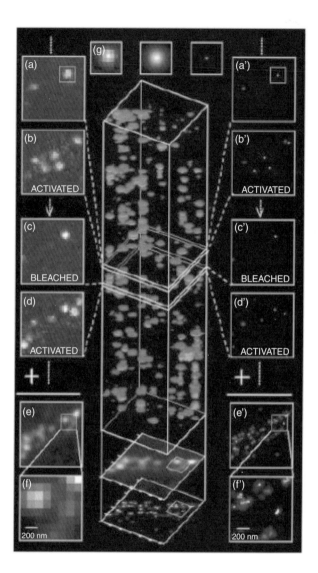

Figure 2.8 Displayed in this example of molecular imaging is "the principle behind PALM. A sparse subset of PA-FP molecules that are attached to proteins of interest and then fixed within a cell are activated (a and b) with a brief laser pulse at 405 mm and then imaged at 561 mm until most are bleached (c). This process is repeated many times (c and d) until the population of inactivated, unbleached molecules is depleted. Summing the molecular images across all frames results in a diffraction-limited image (e and f). However, if the location of each molecule is first determined by fitting the expected molecular image given by the PSF of the microscope [(g), center] to the actual molecular image [(g), left], the molecule can be plotted [(g), right] as a Gaussian that has a standard deviation equal to the uncertainty of x, y in the fitted position. Repeating with all molecules across all frames (a–d) and summing the results yields a super-resolution image (e and f) in which resolution is dictated by the uncertainties of x, y as well as by the density of localized molecules. Scale: 1 mm in (f) and 4 mm elsewhere." Source: Betzig et al. (2006, 1643)

those two objects. Since near UV irradiation has smaller wavelengths than blue or red irradiation, the smallest diffraction limit occurs with visible wavelengths in the near UV. Due to this normal optical diffraction limit, nanoparticles in the most useful size range (e.g. below 250 nm) cannot normally be individually imaged. This means that most nanoparticles cannot normally be imaged through a conventional optical microscope. Electron microscopic analyses, with much shorter wavelengths than visible light, have much higher resolution but remove nanoparticles used for nanomedicine from their normal aqueous environment, leading to many potential artifacts.

Scientists have discovered a variety of tricks to defy, if not overcome, the diffraction limit. Super-resolution microscopy (Zeiss, 2019) allows resolutions much below 250 nm and even as low as 10 nm, offering the possibility for visualizing all but the smallest nanoparticles (gold and quantum dot nanoparticles can be as small as 2 nm). It works by calculation and therefore does not defy the normal physical diffraction limits, as described in more detail in Chapter 4.

2.8   Theragnostic Nanomedical Devices

"Theragnostic" nanomedical devices combine diagnostic capabilities and the ability to guide therapeutic approaches to the site of the diseased cells. If the nanomedical devices are properly designed, there is the possibility of activating the therapeutic treatment immediately after diagnostic information concludes that therapeutic treatment is warranted.

2.8.1   Using Nanomedical Devices to Guide Separate Therapeutic Devices

These dual-capability nanomedical devices offer the promise of uniting diagnostics with therapeutics, particularly if there is an activatable switch that can be used if the diagnostic information dictates subsequent therapeutics. That signal to an activatable switch on a nanoparticle can be delivered by a variety of means, including light activation, DC or AC magnetic fields, heat, and ultrasound. By making the activation switch external to the body, the dangers of a switch self-actuating can be minimized.

2.8.2   When Might We Want to Combine Diagnostics and Therapeutics?

There are many reasons to combine diagnostics and therapeutics in a dual-use device. First, the diagnostics and therapeutics can be done immediately and without the necessity of a patient going to another care site. As all doctors know, you never want the patient to leave until the entire process is complete. Lack of follow-ups by patients, the most common problem, occurs for many reasons, including scheduling problems, patient anxiety during the interim period between appointments, and economic hardships.

Chapter 2 Study Questions

2.1 What are some elements of good engineering design to consider when designing nanomedical systems?

2.2 What is biomimicry, and why is it important in the design of nanomedical devices?

2.3 What is a multicomponent design of nanomedical systems, and how does it tend to improve overall performance?

2.4 How does the choice of core material affect the utility of nanomedical devices? Give some examples.

2.5 How does a multilayered system lead to programmability, that is, the ability to control multistep processes important to good nanomedical systems? How does this affect your order of assembly?

2.6 What are some of the advantages of targeted nanomedical systems over untargeted ones?

2.7 Why is intracellular targeting important and sometimes critically important?

2.8 How does a good nanomedical system reduce side effects of drugs?

2.9 What is directed therapy? Give some examples.

2.10 How do nanomedical systems improve noninvasive medical imaging?

2.11 What are four types of noninvasive imaging? What are their differences in terms of spatial resolution and cost?

2.12 Molecular imaging provides theranostic information by using what types of properties of the nanoparticles?

2.13 Compare the main forms of molecular imaging (MRI, CT, PET, optical/fluorescence) in terms of (a) spatial resolution, (b) sensitivity, and (c) cost.

2.14 What are some of the core materials used for imaging? How does each type of core material dictate the form(s) of molecular imaging?

2.15 How is molecular imaging different from traditional imaging methods?

2.16 Most forms of molecular imaging are structural, meaning they respond to structure rather than function. But PET imaging can also act as a functional imaging agent. What function does it measure?

2.17 What challenges does PET imaging present in terms of more general use?

2.18 Can nanomaterials provide for more than one form of molecular imaging?

2.19 How do some nanomaterials serve as MRI contrast agents? What do they do to the signal to improve contrast?

2.20 Why is near-infrared fluorescence in the near-infrared part of the spectrum?

2.21 What is real-time fluorescence-guided surgery? How does it help a surgeon do a better job?

2.22 Why does theragnostics represent a different paradigm in terms of the way traditional medicine is practiced? Why might combining diagnostics and therapeutics in a single device be a good idea?

2.23 Is it possible to develop a theranostic nanomedical device whose therapeutics function can be "turned on" only if its diagnostic information tells us that therapy is needed?

2.24 What are the essential elements of a nanomedical system?

2.25 Why does the multistep process of a nanomedical system in vivo tend to drive a design that is multilayered?

2.26 What are the requirements of a good targeting strategy?

2.27 What are some of the consequences of mistargeting?

2.28 What is a side effect? Give an example of a side effect. Why is it important to minimize side effects?

2.29 Why does a Boolean targeting strategy tend to be more specific than a non-Boolean strategy?

2.30 Why is a multiplex targeting system difficult to achieve?

2.31 What are some of the ways we can design a nanomedical system to minimize side effects?

3   Targeted Drug Delivery

3.1   Overview: Targeting Nanosystems to Cells

A number of different targeting molecules can be used to target nanoparticles to specific cells for diagnostics or therapeutics. The main categories are antibodies, peptides, and aptamers. Each targeting molecule type has its advantages and disadvantages, as described in the following sections.

3.1.1   Antibodies as Targeting Molecules

Antibodies as targeting molecules for nanoparticles containing diagnostic and/or therapeutic molecules seems an obvious choice. There are many antibodies commercially available. The specificity of antibodies is typically higher than that of other targeting molecules such as peptides or aptamers. While they are often a good choice as a targeting molecule, they are quite large and suffer from limitations due to steric hindrance. Sometimes it is a good strategy to start the overall design process with an antibody and then, if steric hindrance problems occur, move to a biomimetic peptide that mimics the antibody. A biomimetic peptide has targeting properties similar to a given antibody but is smaller in size and often cheaper and faster to produce.

3.1.2   Structure of Antibodies

To use antibodies properly, it is important to have some basic understanding of antibody structure. Even if all you ever use is commercially available antibodies, you need to know the many options available to you and which one is best for your application. Figure 3.1 shows the basic subunits of an antibody that led to these options. An antibody is fairly large, weighing approximately 200 kD depending on the antibody structure type, which varies slightly depending on whether it is IgG, IgM, or another subtype. Its two heavy chains can be dissociated if the disulfide bonds holding them together are broken by treatment with appropriate enzymes such as papain. More often, the Fc tail portion of the antibody can be removed from the main structure by pepsin enzymatic cleavage forming an F(ab)$_2$ bivalent binding fragment. If that fragment is further treated with papain, then a univalent F(ab) fragment is formed. These fragments are frequently offered as options by antibody vendors. Full Ig has the advantage of providing more sites for binding fluorescent probes far away from the

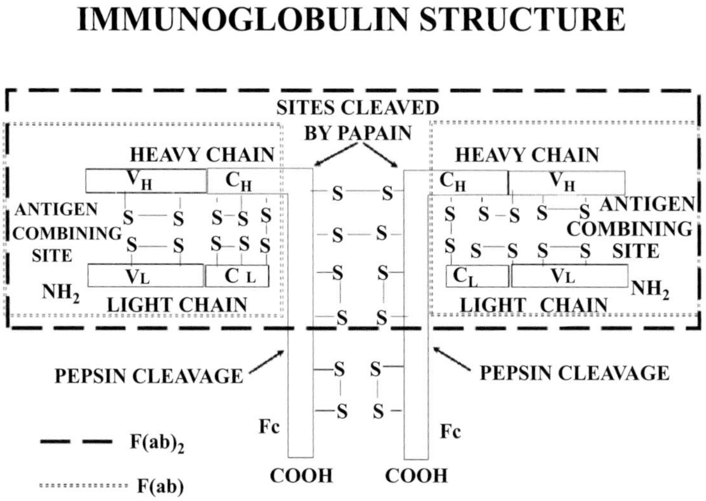

IMMUNOGLOBULIN STRUCTURE

Figure 3.1 An antibody is an immunoglobulin that has molecular shape recognition of a portion of another molecule. It consists of two basic "large chain" backbones joined by disulfide bonds and two "light chains," each joined to a heavy chain to form antigen combining sites.
Source: Leary teaching

antigen combining site, leading to a brighter fluorescent antibody while retaining binding specificity. But the more probes that are attached, the higher the probability that they will be close enough to the antigen combining site to affect that antibody binding specificity. Some cells that have Fc receptors, meaning that they bind antibodies by the Fc portion of the antibody rather than by its antigen combining site, result in nonspecific antibody labeling. In that case, a prudent strategy might be to use an F(ab)$_2$ fragment, which may have fewer fluorescent probes attached but still be reasonably bright. If the bivalent structure of the F(ab)$_2$ antibody causes undesired receptor clustering on the targeted cell, then it may be wise to use an F(ab) fragment antibody. It will probably be less fluorescently bright, but it may be a very good choice when fluorescent antibodies are not really needed and a fluorescent nanoparticle is used instead (Figure 3.1).

3.1.3   Labeling of Antibodies with Fluorescent Probes

Antibodies can be labeled with a wide variety of fluorescent probes of different colors. Then complex heterogeneous cell mixtures can be subdivided by multicolor immunofluorescence into different cell subpopulations. This allows us to assess the specificity of our targeting of nanoparticles to specific cell subpopulations and to minimize mistargeting.

The process of attaching fluorescent molecules to antibodies (F/P ratio) is inherently a combinatorial labeling process driven by the relative frequencies of molecules and fluorescent probes. This means that there is always a collection of antibodies with

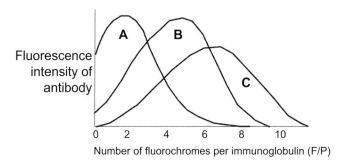

Figure 3.2 The number of fluorochromes binding per immunoglobulin is a Poisson distribution. Unlabeled Ig will outcompete labeled Ig, and Ig with a large number of fluorochromes will become increasingly nonspecific. To obtain the highest specificity brightness ratio, one should stay in the region shown. This can be done by HPLC purifying the labeled antibody to remove unlabeled Ig and overlabeled Ig. It is best to have an F/P ratio between 1 and 6.
Source: Leary teaching

different numbers of fluorescent probes attached that is in a Poisson distribution, as shown in Figure 3.2.

Antibody vendors typically characterize the fluorescent probe labeling of their antibodies in terms of the F/P ratio, which, unless purified by high-performance liquid chromatography (HPLC), corresponds to the average (mean) of the Poisson labeling distribution shown in Figure 3.2. If the F/P ratio is very low (e.g., 1.2), then there will be a mixture of labeled and unlabeled antibodies, and the unlabeled antibody will unfavorably outcompete the labeled antibody for the antigenic sites on the cells. If the F/P ratio is too high (e.g., >9), then the fluorescent probes may have bound to portions of the antibody that affect its specificity. In most cases, one should not use antibodies with F/P ratios greater than about 6 unless they have been HPLC purified, which provides a windowed subset of the total Poisson distribution in a narrower range. This can help address the preceding two problems. HPLC-purified antibodies can provide a narrow window of F/P ratio, but this process can add expense to the manufacturing of the antibodies.

3.1.4   Lock–Key Models of Antibody Shape Recognition of Antigens

Antibodies recognize antigens (or, more specifically, one or more epitopes of an antigen) in a manner that has usually been called the antibody lock–key model, as shown in Figure 3.3.

Polyclonal antibodies typically contain multiple different antibodies, each targeting a different epitope of an antigen. This means that many polyclonal antibodies need to be adsorbed against cells or tissues to remove the undesired antibody clones. Monoclonal antibodies are specific against a single epitope of an antigen. This can be good unless that epitope is shared with other antigens. When staining cells, it is possible for the antibody to see many different epitopes on different antigens. That is

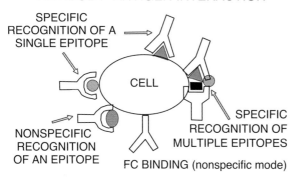

LOCK–KEY MODEL OF
ANTIBODY–ANTIGEN INTERACTION

SPECIFIC
RECOGNITION OF A
SINGLE EPITOPE

CELL

NONSPECIFIC
RECOGNITION
OF AN EPITOPE

SPECIFIC
RECOGNITION OF
MULTIPLE EPITOPES

FC BINDING (nonspecific mode)

Figure 3.3 Antibodies recognize different epitope regions of antigens, explaining why different antibodies can recognize different epitopes of the same antigen.
Source: Leary teaching

particularly true when staining chemically fixed cells whereby the antibody is exposed to the entire proteome of the cell. In that case, one should use flow cytometry–certified antibodies, which are specific for a specific epitope on a specific antigen and will not label epitopes on other antigens. For this reason, antibodies used on Western blot gels will often not be good enough for flow cytometric analysis. Nonspecific staining is obvious on gels only because antibodies cross-react at other gel locations and are therefore spatially distinct.

For the above reasons, it is important to distinguish which epitope(s) is being recognized by a specific antibody. Sometimes an epitope may not be accessible to an antibody, which explains why not all antibodies are equally good at recognizing an antigen. Since polyclonal antibodies are really a cocktail of several different antibodies, this explains why not all monoclonal antibodies (which recognize a single epitope) may be effective in targeting an antigen depending on whether that epitope is accessible to the antibody.

The bioconjugation of antibodies to most nanoparticles is a fairly straightforward process. Indeed, most of the nanoparticle targeting in the literature uses antibodies. It is certainly the easiest and fastest way to rapidly prototype a system in vitro. But use of antibodies as targeting molecules for nanoparticles in vivo has many caveats. First (and foremost) is the problem of immunogenicity. Monoclonal antibodies produced in mice can cause severe, and possibly fatal, immune reactions in vivo in many people. The way to alleviate this problem is to "humanize" the mouse monoclonal antibody by mutagenizing or otherwise changing the amino acid sequences in a portion of the overall antibody to reduce undesired human immune reactions to mouse proteins. This is exactly what is done to solve this problem for mouse monoclonal antibodies used as anti-cancer agents in humans. The process, however, is not perfect. Some humans continue to experience reactions such as anaphylactic shock, and this adverse reaction is hard to predict. Second, the process of "humanizing" the amino acid sequences can make the humanized antibodies expensive. Third, the antibody itself, regardless of final amino acid sequences, can stimulate undesired immune reactions. Fourth, antibodies are really too big to be used for targeting of small nanoparticles. With a molecular weight of 180–240 kD (depending on Ig subclass), the antibody is sufficiently big to cause steric hindrance problems (e.g., the surface area available to bind antibodies on the surface of nanoparticles can be very limiting). Finally, not all potential receptor targets are immunogenic enough to elicit a response in animals (e.g., mice) necessary to produce an antibody. It is not possible to make antibodies against every desired epitope. Fortunately, there are a number of possible alternatives (e.g., peptides and aptamers) that can be produced synthetically and do not require either being immunogenic or being produced in animals.

3.1.5   Peptide Targeting

One of the most obvious and popular alternatives to antibodies are biomimetic peptides. These are much smaller peptide sequences that mimic the antigen recognition of antibodies. These peptides fold in ways that recognize the shape of the desired target in a manner similar to the conventional lock–key model of antigen–antibody binding. Increasingly, these biomimetic peptides are becoming commercially available. Like their much larger cousins (antibodies), these peptides are also relatively easy to bioconjugate to the surface of nanoparticles. Since they are much smaller, it is easier to bind more of these molecules to the limited surface area of nanoparticles. Another advantage of these targeting peptides is that they can be made totally in vitro and do not require being antigenic to any animal's immune system. The challenge is finding the correct amino acid sequence. Usually this is done by panning or other high-throughput screening techniques that simply find the sequences that actually bind to the desired target. Increasingly, these sequences are becoming known and are being published in the literature or are becoming commercially available. These peptides can be produced relatively easily and inexpensively in large quantities using conventional peptide sequence machines. It is also easier to slightly change the amino acid sequences of these peptides to "humanize" them to avoid undesired immune reactions within humans.

3.1.6   Aptamer Targeting

An interesting alternative to peptides as targeting molecules is the use of aptamers made from oligonucleotide precursors. These can be made from either DNA or RNA oligonucleotides. Aptamers also have shape recognition similar to antibodies and peptides. They can be easily and inexpensively made – provided that the correct sequence is known. The number of commercially available aptamers is much smaller than those of antibodies or biomimetic peptides, but the numbers are growing each year. One of the largest repositories of useful aptamer sequences (Lee et al., 2004) and their targets can be found at www.freebase.com/base/aptamer.

There are a number of unusual advantages of aptamers as targeting molecules. One of the most important is the almost complete lack of adverse immune reactions in vivo in humans. Indeed, it seems to be that the immune systems of most humans (with the exception of those suffering from relatively rare diseases caused by production of anti-nuclear antibodies in these individuals) ignore DNA sequences circulating in blood. This may also be why looking for rare mutated DNA sequences in blood from circulating tumor cells that have been lysed sometimes works diagnostically. Another unusual, and valuable, property of aptamers is that (depending on type) they can be duplicated in large numbers by polymerase chain reactios (PCRs). This allows both their manufacture and detection by PCR. In the latter case, this means that the sensitivity of diagnostics can be greatly enhanced by amplifying the number of targeting molecules directly at the target site in a process known as in situ hybridization. In addition, one can make very interesting nanobarcoding (Eustaquio and Leary, 2012a, 2012b) of nanoparticles not only to serve as potential targeting molecules but also to be included as secondary nontargeting molecule labels. This variation of in situ hybridization with in situ PCR amplification (Bagasra, 2007, 2008) of oligonucleotide sequences attached to nanoparticles (effectively "nanobarcoding" them) enables their use to allow detection of multiple differently labeled nanoparticle systems simultaneously.

There are many more important details of the above approaches that need to be discussed and understood. This more detailed information is important to allow for proper choice of targeting molecules for given applications.

3.2 Antibodies: Polyclonal, Monoclonal, and "Monoclonal Cocktails"

There are two fundamental categories of antibodies – polyclonal antibodies and monoclonal antibodies. Each type of antibody has its advantages and disadvantages, as described in the following sections. Interestingly, since monoclonal antibodies can at times be too specific, artificial mixtures of specific monoclonal antibodies, known as "monoclonal cocktails," can be constructed that combine the advantages of monoclonal and polyclonal antibodies while eliminating most of the disadvantages.

3.2.1 Where Do Antibodies Come From in Nature?

In nature, antibodies are produced in animals and humans by exposure to materials that serve as antigens and illicit an immune response in these individuals. Proteins serve as the largest class of antigenic agents, and as few as six amino acids have been shown to elicit an immune response. This is also why "humanizing" the amino acid sequences of antibodies to prevent adverse reactions in some humans is so challenging.

While antigens are sometimes administered as potential allergens to humans by doctors to patients suffering from allergies, it is considered unethical and improper to try to produce antibodies directly in humans for subsequent commercial use. Hence, antibodies are made in other species, primarily in mice, but also in other species such as rats, rabbits, sheep, and goats. For this reason, all of these antibodies are "foreign" proteins that can cause severe immune reactions if introduced into humans.

3.2.2   How Do We Make Them in the Laboratory?

Polyclonal antibodies are made in the laboratory by introducing an antigen intravenously into an animal, sometimes with adjuvants to help stimulate the animal's immune response to produce antibodies against that antigen. Since it takes only a small number of amino acids on a protein to be antigenic, there are frequently multiple sub-antigenic sites (i.e., epitopes) on a given protein antigen that can cause production against that small part of the protein. The animal often produces multiple different antibodies simultaneously (i.e., polyclonal antibodies). Blood is removed periodically from the animal to harvest these antibodies, which need to be separated from other blood proteins and cells. Since every animal is different, the polyclonal antibodies will potentially be different in both the scope of the polyclonality of the antibody and the avidity (i.e., the degree to which it attaches to the epitopes). The animal obviously has a finite lifetime and amount of antibody it can produce, so the reliability and reproducibility of the antibody is always in question.

For this reason, researchers sought out alternatives to these methods to produce larger (potentially infinite) and more reliable sources of antibody of more defined specificity. In 1984, Caesar Milstein, his postdoctoral student Georges Kohler, and Niels Jerne shared the Nobel Prize for discovery of the process by which monoclonal antibodies could be produced (Kohler and Milstein, 1975). Their discovery has revolutionized modern medicine both in terms of diagnostics and therapeutics. Production of monoclonal antibodies has become a billion-dollar industry, and many thousands of monoclonal antibodies are now commercially available. Such monoclonal antibodies used in small quantities for diagnostics are relatively inexpensive. However, their use in vivo for treatment of a variety of cancers and other diseases is relatively expensive due to both the very large quantities required and the need to "humanize" the monoclonal antibodies, as described previously.

3.2.3   Monoclonal Antibodies

The production of monoclonal antibodies begins in a manner similar to the production of polyclonal antibodies, with the raising of an immune response against an injected antigen. However, instead of merely harvesting the antibodies secreted from the animal's B-cells into their blood serum, we harvest the B-cells themselves. Since there are potentially many different monoclonal antibodies being produced from different individual B-cells, we must isolate individual clones of B-cells at the single-cell level. These individual cells are hybridized with an immortal mouse myeloma cell line to produce an immortal hybridoma cell that secretes the desired antibody, which must be chosen by a screening process to get the desired hybridoma cell. This hybridoma cell line can then be grown into larger quantities in two different ways. If injected back into the peritoneal cavity of another mouse, the tumor cells will grow into a large tumor – essentially using the mouse as a tissue culture vessel. Some of these tumors can grow very large and cause pain and suffering for these animals. For this reason, the European Union now bans this method of producing large amounts of monoclonal antibodies. A second method produces monoclonal antibodies in lesser, but still satisfactory, amounts. The hybridoma cells are grown in tissue culture, and the secreted antibodies are merely harvested from the supernatant of the culture. Since there are many other proteins and molecules in the culture supernatant, the antibodies are frequently further purified by high performance liquid chromatography (HPLC).

3.2.4   Therapy Problems with Mouse Monoclonal Antibodies

When the first monoclonal antibodies were produced in mice and used directly for human therapies, a number of problems arose, including severe immune reactions and even death of some of these patients. It was essentially injecting mouse proteins (in the form of mouse monoclonal antibodies) into humans and eliciting a large, and sometimes fatal, immune response. This problem led to the "humanizing" of mouse monoclonal antibodies, as discussed in the next section.

3.2.5   "Humanizing" Monoclonal Antibodies to Reduce Host Immune Reactions

It was subsequently found that certain amino acid sequences in the mouse monoclonal antibodies were eliciting this unfavorable immune response in the patients. Fortunately, other technologies had been developed to "mutagenize" or otherwise alter these specific amino acid sequences to ones that would produce much less adverse immune response. These "humanized" amino acid sequences produce a much lower immune response in most, but not all, patients. Residual immunogenicity appears to reside in the complementarity-determining region (CDR) of the antibody structure, which is the molecular sequences of the immunoglobulin responsible for binding the antigen. Unfortunately, some patients still go into anaphylactic shock within 15 minutes of injection with these antibodies. It is probably not possible to perfectly "humanize" monoclonal antibodies for all humans, but it may also not be necessary in all cases (Getts et al., 2014).

3.2.6   Why Antibodies May Not Be a Good Choice for Targeting Nanosystems to Cells

With all of the preceding problems, it is not hard to see that full-size antibodies may not be the best choice as targeting molecules for nanomedicine. It is possible to use specific enzymes to cleave the full-size antibody molecules into $F(ab)_2$ or $F(ab)$ fragments, which reduce both their size and adverse immune response. But it is perhaps better to consider other targeting molecule alternatives.

3.3   Peptide Targeting

Peptides can be used for targeting nanoparticles to cells. They are attractive targeting molecules for nanomedicine because they are more appropriately sized molecules compared to nanoparticles and also because they can be readily manufactured in large quantities at high purity. By recognizing other molecules, they can serve as ideal targeting molecules for nanoparticles or biosensors (de la Rica, Pejoux, and Matsui, 2011).

3.3.1   How Does a Peptide Target?

Peptides, like antibodies, can target in a "lock–key" type of model based on specific protein folding that then conforms to the shape of a specific antigen. While antibodies have components that help preserve this specific folding, care must be taken to preserve the shape recognition of the much smaller peptide sequences used for targeting, in general, by avoiding any conditions that would denature a protein.

3.3.2   Examples of Peptide Targeting

Here is an example of not only peptide targeting of a nanoparticle to a cell but also the same peptide capable of cell membrane penetrating and intracellular retargeting – all within a seven amino acid peptide sequence! In this example, the peptide sequence, LTVSPWY, based on its previous success of induction of oligonucleotide uptake in the SKBr3 human breast cancer cell line (Shadidi and Sioud, 2003), was attached to quantum dot nanoparticles (Qdots) and targeted to SKBr3 cells (Haglund et al., 2008, 2009). The Qdots with LTVSPWY targeting peptides bound to the SKBr3 cells, penetrated the cell membrane (transporting the Qdots inside), and then helped retarget these same Qdots toward the nucleus of those cells.

Clearly not all peptides combine cell targeting, uptakes, and retargeting in a single peptide sequence. In most cases, a different cell-penetrating peptide sequence would also need to be added to the surface of the nanoparticle to accomplish this second step, and a third peptide sequence for intracellular retargeting might need to be added. The only theoretical limit to the number of different peptides that can be put on a given nanoparticle is the amount of surface available on that nanoparticle.

3.3.3   Advantages and Disadvantages of Peptide Targeting

The main advantage, in addition to the reasons previously discussed, of using peptides for targeting of nanoparticles to cells is their small size. Depending on the size of the nanoparticle, antibodies may be a case of "the tail wagging the dog," meaning that the antibodies on a nanoparticle may be as large as, or larger than, the nanoparticle itself. A number of different peptides may be needed to deal with different targeting steps as well as cell membrane penetrating functions.

3.4   Aptamer Targeting

Aptamers can also be used as targeting molecules. They have a number of advantages. First, they are composed of DNA or RNA nucleotides, which are generally not likely to cause an immune response in vivo. Second, they can be easily manufactured synthetically in large quantities at high purity. Third, not commonly considered, they can be PCR amplified, which can be of great utility when trying to study the biodistribution of nanoparticles in tissues. The utility of PCR amplification of non-aptamer oligonucleotides as "nanobarcodes" for studying the biodistribution of nanoparticles has been demonstrated (Eustaquio, Cooper, and Leary 2011a; Eustaquio and Leary, 2011, 2012a, 2012b).

3.4.1   What Are Aptamers and How Do They Target?

Aptamers are oligonucleotides consisting of DNA or RNA oligonucleotides. They are capable of targeting by molecular shape recognition since they fold into three-dimensional shapes due to their tertiary structures. In addition to their ease of synthetic manufacture as long as the correct sequence is known, they tend to not be immunogenic in most humans.

3.4.2   Some Different Types of Aptamers

Some aptamers consist of oligomer strings of DNA nucleotides. These are the simplest forms of aptamers. Other aptamers can be RNA: RNA or RNA: DNA or LNA (locked nucleic acid), which is a modified RNA nucleotide. The ribose portion of an LNA nucleotide is modified with an extra bridge connecting the 2' oxygen and 4' carbon. This bridge "locks" the ribose in the 3'-endo conformation, which is often found in A-form duplexes. LNA nucleotides can be mixed with DNA or RNA residues to form more complex structures.

3.4.3   How Do You Make Aptamers?

Aptamers can be constructed using published methods (Blind and Blank, 2015) or purchased from commercial sources. Databases containing aptamer sequences that bind to known antigens should first be searched (Lee et al., 2004). If the aptamer sequence is known, then aptamer targeting molecules can be readily manufactured.

3.4.4   How Do You Screen for Useful Aptamers?

If the correct sequence to link to an antigen is not known, combinatorial libraries of aptamer sequences can be screened against the antigen of interest either by simple panning methods or by more sophisticated flow cytometry screening of aptamers on bead libraries (Yang and Gorenstein, 2011; Yang et al., 2003) (Figure 3.4).

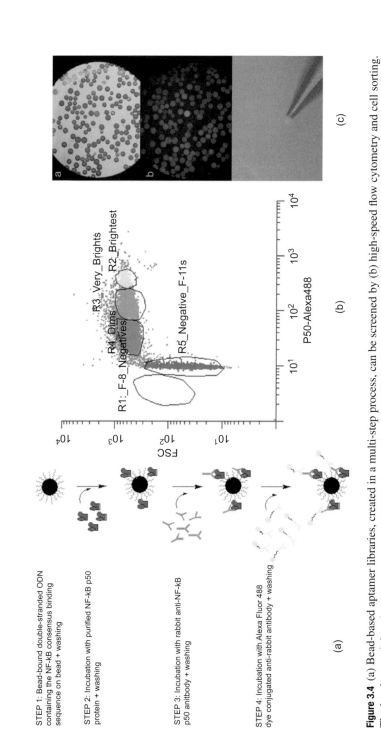

STEP 1: Bead-bound double-stranded ODN containing the NF-kB consensus binding sequence on bead + washing

STEP 2: Incubation with purified NF-kB p50 protein + washing

STEP 3: Incubation with rabbit anti-NF-kB p50 anitbody + washing

STEP 4: Incubation with Alexa Fluor 488 dye conjugated anti-rabbit antibody + washing

(a)

R3_Very_Brights
R2_Brightest
R4_Dims
R1:_F-8_Negatives
R5_Negative_F-11s

FSC

P50-Alexa488

(b)

(c)

Figure 3.4 (a) Bead-based aptamer libraries, created in a multi-step process, can be screened by (b) high-speed flow cytometry and cell sorting. The beads containing the aptamers most avidly binding to the antigens of interest can be detected and individually sorted. (c) The aptamer sequence can then be determined by direct sequencing.
Source: Yang et al. (2003)

3.4.5 How Do You Find Known Aptamers and Their Targets?

Aptamer sequences and their targets can be found from literature searches. But the most comprehensive aptamer database is maintained by the Ellington lab (Lee et al., 2004). There are also a growing number of companies that provide commercial sources of known aptamer sequences. A schematic of different ways aptamers can be conjugated to nanoparticles for targeted drug delivery (Fu and Xiang, 2020) is shown in Figure 3.5.

Whereas DNA aptamers are generally quite stable, RNA aptamers must be kept away from RNase, which can degrade it; hence, the suspending fluids are all DNase and RNase free to prevent this degradation.

3.4.6 Ligand and Small Molecule Targeting

Another way of targeting nanomedical systems to diseased cells is the use of small molecules, often as ligand–receptor binding. A particularly important one is the use of folate ligand that binds to folate receptors, which have been shown to be elevated in many types of human cancers (Frigerio et al., 2019). This has been applied to the delivery of fluorescent nanomedical systems to cancer cells to highlight surgical margins for real-time fluorescence-guided surgery, which allowed detection of tumors as small as 0.1 millimeter as opposed to the conventional surgical detection of about 3 millimeters (Low et al., 2018). The primary mechanisms of folate ligand–receptor binding and internalization are shown in Figure 3.6.

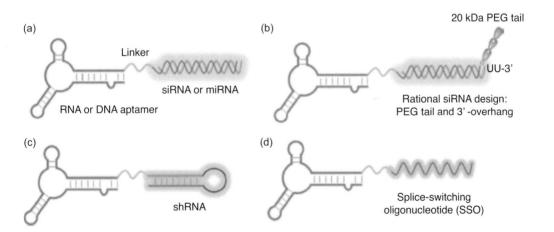

Figure 3.5 Covalent aptamer-therapeutic oligonucleotide chimeras. (a) Schematic of the first generation aptamer (RNA or DNA)-siRNA/miRNA c by an annealing step to make the chimera. (b) Schematic of the optimized second-generation chimera. The aptamer portion was truncated from 71 to 39 nt. A two nucleotide (UU) overhang and polyethylene glycol (PEG) tail were added to the 3¢-end of the guide strand (red) and the 5¢-end of passenger strand (blue), respectively. (c) Schematic of the aptamer-shRNA chime ic of the AS1411 DNA aptamer-SSO chimera. (d) SSO: splice-switching oligonucleotide.
Source: Zhou and Ross (2014)

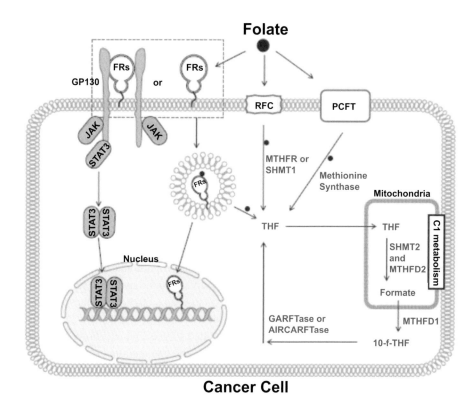

Cancer Cell

Figure 3.6 "Main aspects of folate receptor signaling and C1 metabolism. Three types of folate transporters/receptors are known to exist in humans to facilitate the uptake of folate: Folate Receptors (FRs), Reduced Folate Carrier (RFC) and Proton-Coupled Folate Transporter (PCFT). Left side: Folate binding to FRs can induce STAT3 activation via a GP130 co-receptor mediated JAK-dependent process. Folate can also bind FRs undergoing endocytosis and upon released FRs are set free to act like transcription factors. Right side: Folate, through an interlinked set of mitochondrial and cytosolic reactions, support the C1 metabolism and the main pathway reactions are depicted [1, 6]. THF: tetrahydrofolate; MTHFR, methylenetetrahydrofolate reductase; SHMT1/2, serine hydroxymethyl transferase in cytosol (1) and mitochondrial (2); MTHFD1, methylenetetrahydrofolate dehydrogenase 1; MTHFD2, methylenetetrahydrofolate dehydrogenase 2; 10-f-THF: 10-formyl- tetrahydrofolate; GARFTase: glycinamide ribonucleotide formyltransferase; AICARFTase: 5-aminoimidazole-4-carboxamide ribonucleotide formyltransferase."
Source: Frigerio et al. (2019)

3.5   Assessing Targeting

It is important to assess the specificity and sensitivity of the targeting molecules. If the targeting of nanoparticles is highly specific, then it will be highly efficient in bringing drugs to diseased cells while sparing normal cells. This will lead to fewer side effects and allow a general lowering of total drug exposure per patient. If the targeting is also sensitive, then it will be able to target smaller numbers of diseased cells and be effective in eliminating those diseases.

3.5.1 The Importance of Quantitatively Assessing Cell Targeting

In order to determine the quality of the targeting process, we must quantitatively determine the amount of specific, and nonspecific, targeting. This allows us to know whether we are improving, or making worse, the targeting process with a given modification of our targeting procedure.

3.5.2 Rare-Event Targeting of Cells In Vitro and In Vivo

If the diseased cells to be targeted are rare cells (typically less than 1 percent, and perhaps a lot less than that!), then the targeting process must be especially specific and sensitive. This section discusses methods to evaluate the specificity and sensitivity of antibodies and antibody-labeled nanomedical systems in vitro. It presupposes that an appropriate in vitro simulation of the in vivo situation has been created by adding in other cell types so the in vivo situation can be properly modeled in vitro. It must also deal with the problem that, no matter how good or careful the cell preparation, there will be dead and damaged cells in the cell mixture. These dead or damaged cells can bind antibodies (and nanomedical systems attached to them) nonspecifically. That may, or may not, be important in the overall treatment process, but the effect of dead and damaged cells is to nonspecifically bind up these nanomedical systems, making them less available to bind to the diseased cells and thereby making treatments less effective. Dye exclusion methods (e.g., trypan blue, propidium iodide, and other similar methods) are not effective in excluding all of the dead and damaged cells. Indeed, what we are measuring is the ability of antibodies to stick nonspecifically to cells, not the ability of cells to exclude dyes due to damaged membranes. For rare cells, this constitutes a serious problem since the remaining dead/damaged cells can be hundreds or even thousands of times more populous than the rare cells of interest. Nonetheless, it has been amply demonstrated by many researchers that irrelevant antibodies, which should not bind specifically to cells, can be used to exclude dead/damaged cells sufficiently to permit the detection of rare cells. For much more insight on the targeting and detection of rare cells, I have published a number of highly detailed references on the subject (e.g., Leary, 1994, 2005).

3.6 Assessing Nanomedical System (NMS) Targeting at the Single-Cell Level

It is important to assess nanoparticle targeting to diseased cells in a heterogeneous mixture of normal and diseased cells in order to quantitatively assess both targeting and mistargeting efficiency to all cell types to be encountered by the nanoparticles. Since nanoparticles are much bigger than free antibodies or peptides, their kinetics of binding to cells is much different. Due to their size, they can also represent potential problems in steric hindrance whereby their size on the cell surface prevents the binding of more nanoparticles. However, since individual nanoparticle binding may be sufficient for detection or therapeutic delivery, this does not necessarily represent a serious problem for either diagnostics or therapeutics.

3.6.1   Fluorescent Labeling of NMSs

Nanoparticles provide multiple possibilities for fluorescent labeling depending on their size and structure. If the nanoparticles have a lipid-like layer, so-called membrane-tracking dyes can be used that merely partition into these hydrophobic interior regions. Alternatively, fluorescent molecules can be chemically attached to either the interior or exterior of the nanoparticle. One important consideration frequently ignored is the quantum efficiency of the fluorescent dye that will be in the given microenvironment. For example, dyes such as FITC are pH dependent. If the microenvironment is acidic, the quantum efficiency of FITC can drastically decline, making FITC either very dim or even invisible.

Fluorescent labeling is also dependent on the size of the nanoparticle. If the nanoparticle is very small, there may be limitations based on the surface area available for fluorescent molecules. Most organic dyes have quantum efficiencies affected by dye concentration. Higher concentrations of some fluorescent dyes (e.g., some cyanine dyes) can cause the dye to self-quench due to dye stacking such that the nanoparticle can become dimmer, rather than brighter, at high fluorescent dye concentrations.

3.6.2   First Estimates of NMS Binding by Fluorescence Microscopy

It is frequently better to first assess the quality of fluorescent nanoparticle labeling by conventional fluorescence microscopy. Whereas individual fluorescent nanoparticles cannot be distinguished due to optical diffraction limits, the fluorescent nanoparticle labeling will appear similar to fluorescent molecule labeling, albeit with more of a nonuniform or "punctate" labeling. Focusing up and down on 2D monolayers of cultured cells can provide some sense of whether the nanoparticles are mostly on the surface or inside these cells. More definitive measures of the location of the nanoparticles require 3D microscopy methods such as confocal microscopy.

3.6.3   Estimates of Cell Surface or Internal Binding of Nanoparticles by Confocal Microscopy

Confocal microscopy provides a way of virtual optical sectioning of cells such that the location of fluorescent nanoparticles within individual cells can be more accurately determined. Since the fluorescent molecules within nanoparticles are often partially protected from oxidation, they are subsequently more protected from photobleaching. Again, whereas individual nanoparticles cannot be distinguished due to optical diffraction limits, the 3D location on or inside individual cells can be determined much more accurately than by conventional 2D fluorescence microscopy.

3.6.4   Flow Cytometric Quantitation of NMS Binding to Specific Cell Types

If the cells of interest can be made into a single cell suspension, then the binding of nanoparticles can be determined at the level of the number per cell, albeit without being able to determine the spatial location of these nanoparticles, and can be quantitatively determined by flow cytometry. An introduction to flow cytometry methods is discussed in Section 3.8.

3.7   Image Analysis of NMS Binding to Single Cells

If cells cannot be made into a single-cell preparation suitable for flow cytometry, the binding of nanoparticles to single cells can still be semi-quantitatively assessed using scanning image cytometry methods. The term *scanning image cytometry* means that the analysis area of the image analysis system can be confined to approximately the single-cell level. Unlike flow cytometry, this cannot be done perfectly, and both illumination and fluorescence collection cannot be perfectly confined to single cells, thereby rendering it "semi-quantitative." Nonetheless, it is still a highly useful technique to assess the binding efficiency of nanoparticles to single cells.

3.7.1   Ability to Scan/Locate Cells of Interest

The first problem in quantitative image analysis is locating the cells of interest amid a frequently large area containing many cell types. Scanning image systems, where all of the scan is in the stage movement, are slower than those with a flying spot moved optically with minimal stage movement. It should be noted that the degree of resolution needed will dictate the amount of time it takes to scan a wide area. Image analysis algorithms to find the desired cells can be quite complex. Additional image analysis algorithms can be formulated using either expert systems or self-learning neural networks.

3.7.2   Photobleaching Challenges

Most organic dyes, if unprotected from oxidation, are subject to photobleaching. This is not a problem in flow cytometry since the excitation of the cells occurs only for microseconds. Nor is it a problem for most modern imaging systems with very rapid excitations. But it is a problem with conventional fluorescence microscopy or for some confocal microscopes, the latter case being due to the wide depth of single-photon confocal microscopes. Fixed cells can be examined for many minutes under conventional fluorescence or confocal microscopes if anti-fade solutions are used to block photobleaching effects. Unfortunately, conventional anti-fade solutions do not work on live cells. One possible solution is to carry out the nanoparticle binding to live single cells and then chemically fix the cell–nanoparticle complexes afterward to protect them from photobleaching.

3.7.3   Detection of Nanoparticles by Super-Resolution Microscopy

Basic laws of physics dictate a lower limit of resolution to about 250 nm, depending on the wavelength, based on visible light. Recently, super-resolution microscopy has permitted the visualization of individual nanoparticles as small as about 40 nm in diameter (Betzig et al., 2006) and higher resolutions based on computed positions of nanoparticles and Brownian motion. Since nanoparticle–cell interactions must necessarily take place in an aqueous environment, conventional transmission or scanning electron microscopy (since they require vacuum microenvironmental conditions) could not be used to visualize these interactions at the single-nanoparticle level. Atomic force microscopy could be used, but it is difficult and arduous. The advent of super-resolution microscopy has opened a new era for studying and visualizing nanoparticle–cell interactions on live cells in aqueous environments.

3.8   A Quick Overview of Single-Cell Targeting Assessment by Flow Cytometry

Cytometry is the measurement of single cells. This means that it has a natural and important relationship with nanomedicine. I spent 20 years of my research career using flow cytometry before applying it to nanomedicine starting in the year 2000. Flow cytometry allows the rapid (1,000–100,000 cells/second) and highly quantitative and reproducible measurement of nanoparticle–cell interactions, including their differential interactions on different cell subpopulations. It is a natural marriage of single-cell measuring technology with single-cell medicine (i.e., nanomedicine) that was immediately obvious to me when I began my work in the field of nanomedicine more than two decades ago.

3.8.1   Basic Principles of Flow Cytometry

A flow cytometer can be considered a type of virtual fluorescence microscope (Figure 3.7).

Instead of putting cells on a slide, we flow them past various light excitation sources one cell at a time in a form of a "virtual slide." Typically, cells flow through the focal point of lenses that are part of the fluorescence collection optics in an attempt to gather as much of the fluorescence light associated with the individual cell being examined and to minimize fluorescence background not associated with the cell.

There are a number of advantages of a flow cytometer over a fluorescence microscope. First, the cell is only in the light source for a few microseconds, so photobleaching does not occur. Second, highly quantitative measurements on single cells can be performed at rates in excess of 10,000 cells/second and even higher rates for analysis of very rare cells that would likely never be seen by conventional fluorescence microscopy. Third, the measurements are highly quantitative and reproducible, both typically within a coefficient of variation (CV; standard deviation

A Flow Cytometer Is a "Virtual Fluorescence Microscope"

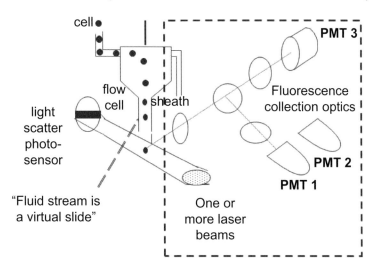

Figure 3.7 A flow cytometer is really just another form of a fluorescence microscope, but instead of putting cells on a slide, we flow them individually through the focal point of the fluorescence collection optics.

Source: Leary teaching

divided by the mean) of less than 2 percent. Fourth, the sensitivity for fluorescence detection on a modern flow cytometer can be as low as about 30 fluorescent molecules per cell due to the fact that fluorescent molecules can be excited and re-excited thousands of times while traversing the light source. This leads to a sensitivity amplification factor of thousands of times for every fluorescent molecule associated with the cell without photobleaching losses. Indeed, it is possible to claim that flow cytometry, especially when combined with cell sorting, is one of the most powerful biological instrumentation technologies invented in the twentieth century! It is now found in virtually all major hospitals and biological research centers around the world in more than 65 countries.

A disadvantage of flow cytometry is typically a zero-resolution measurement whereby fluorescence is collected on a whole single-cell basis such that the distribution of fluorescence within the cell is typically not measured. A second disadvantage is that the morphology of the cell is typically not measured. A third disadvantage is that the tissue architecture is destroyed when single cells are made into a single-cell suspension suitable for flow cytometric analysis.

However, most of these disadvantages can be overcome by using a flow cytometer–cell sorter, which is a first-stage flow cytometer followed by a second-stage cell sorter that typically sorts single cells within saline-based droplets in a variation of inkjet printing, as shown in Figure 3.8.

Then cell subpopulations identified by flow cytometry can be physically separated by cell-sorting technology so that other tests and observations can be performed on

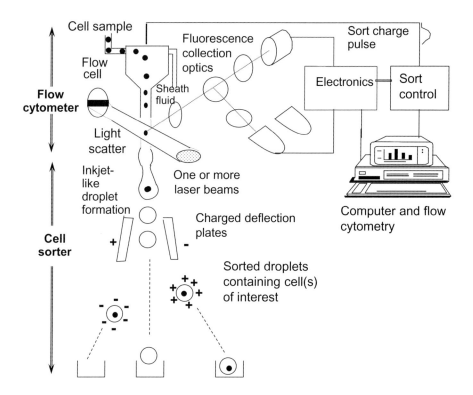

Figure 3.8 A flow cytometer–cell sorter combines the powerful analytic advantages of a flow cytometer with the ability to separate single cells for subsequent analyses by other technologies, including image analysis, confocal microscopy, and various molecular characterization technologies such as PCR amplification of DNA or RNA for gene expression and genomic analyses.
Source: Leary teaching

these isolated cell subpopulations. Flow cytometry combined with cell sorting is indeed a powerful merger of technologies, as you will see in further chapters of this book.

Various fluorescence probes on a single cell can be rapidly excited by one or more lasers (some commercial sorters can have five or more lasers), and fluorescence colors from the probes can be detected on multiple sensors (currently, more than 25 different fluorescent colors can now be measured on commercially available cell sorters). When a cell with desired properties is selected by the sort control electronics, a radio-frequency charge pulse is sent to the saline-like fluid stream, which is also being shaken by piezoelectric transducers to form well-ordered and spaced droplets. Some of these droplets contain cells, and those cells can be collected by putting positive or negative electrostatic charges on the surface of those droplets, which can then be electrostatically deflected in a variation of conventional inkjet technology found in many computer printers.

3.8.2   Use of Flow Cytometry for Assessing Specificity and Sensitivity

Since flow cytometry can make multiple measurements per cell, it is capable in many cases of distinguishing different cell types on the basis of cell surface biomarkers, which identify cell subpopulations on the basis of specific cell surface molecules. Those capabilities allow us to assess the targeting specificity of our nanomedical system, and those targeting systems can be systematically improved and reevaluated. The sensitivity of the targeting system can assess what fraction of the diseased cells are actually targeted, rather than missed.

3.9   Rare-Event Analysis of NMS Targeting to Desired Cells

It should be obvious that in vivo targeting of nanoparticles to diseased cells is inherently a rare-event targeting and analysis situation. This is particularly so for drugs introduced intravenously. Because the probability of finding the right cells is so low, as drugs circulate, we must take steps to increase their circulating lifetime. This dictates strategies for protecting drugs in blood and preventing their elimination. This will be the subject of numerous sections throughout this book and was a research area focus of mine for more than 20 years.

3.9.1   Strategies for Rare-Cell Detection

A detailed review of the basics of rare-cell detection (Leary, 1994) discusses not only the cell-labeling strategies to reduce nonspecific background but also the best methods of detection by high-speed flow cytometry and subsequent data analysis techniques.

3.9.2   More Advanced Flow Cytometry for Ultra-Rare-Cell Detection

A good review of advanced flow cytometry techniques for detection of "ultra-rare-cell" subpopulations (Leary, 2005) points out that it is equally important to specify negative selection parameters in order to exclude major-cell subpopulations and dead or damaged cells that bind antibodies and other targeting molecules nonspecifically. The best antibodies in the world rarely detect to a specificity of 1 in 1,000 cells, and detection levels for nanomedicine need to be several orders of magnitude better than that. The use of two or more positive targeting molecules and a cocktail of fluorescently labeled (in a different color) negative targeting molecules (which should *not* bind!) helps ensure specificity to a level of one cell in a million, even when there are many dead or damaged cells present (Leary, 1994, 2005, 2009). This is really the major secret to good rare-event detection because it is otherwise impossible to sift through levels of background to find the true-positive rare cells amid a much higher level of false-positive cells.

3.9.3   Examples of Nanoparticle Targeting to Rare Cells

Targeting of nanoparticles to rare cells is relatively new. A discussion of the techniques and challenges is provided in recent publications (Cooper and Leary, 2014, 2015). The binding kinetics of nanoparticles to cells is quite different from the binding kinetics of molecules. This means that the amount of time needed to properly label cells of interest depends on many additional factors, such a nanoparticle size.

As the rare leukemic-cell subpopulation goes lower and lower in frequency, it requires more flow cytometric parameters to distinguish the rare cells from the background cells in venous peripheral human blood. This requires a sophisticated multicolor immunofluorescence immunophenotyping strategy (Cooper and Leary, 2014), as shown in Figure 3.9.

In addition to the increased complications of multicolor immunophenotyping, the data become high-dimensional. This means that multidimensional, bioinformatic

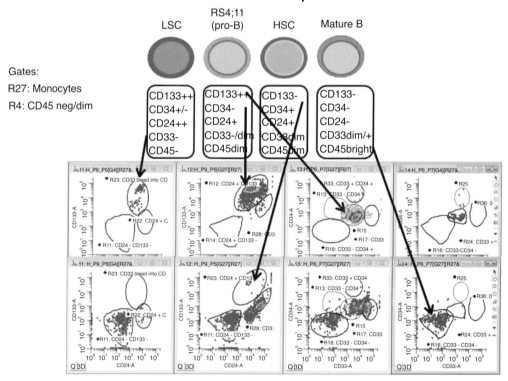

Figure 3.9 Flow cytometric scatterplots for subsets of LSC, RS4;11, HSC, and mature B cells reveal subpopulations of cells delineated on the basis of Boolean combinations of cluster designation (CD), the international system for classifying biomarkers.
Source: Cooper and Leary (2014)

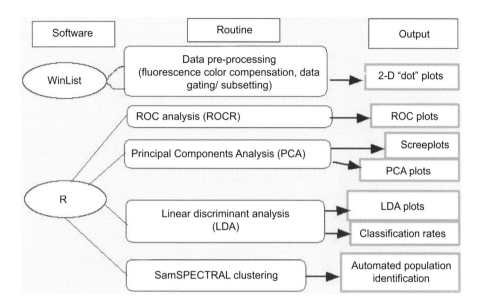

Figure 3.10 Multiparameter flow cytometric analysis of data can use data mining techniques in R bioinformatics software to find cell subpopulations in higher dimensional data space that cannot be directly visualized.
Source: Cooper and Leary (2015)

approaches to the analysis of this flow cytometric data must be used. Some strategies for bioinformatic analysis of multidimensional flow cytometric data (Cooper and Leary, 2015) are shown in Figure 3.10.

3.9.4 Rare-Cell Sampling Statistics

It is important to note that rare-cell sampling statistics should be calculated when determining how many cells must be sampled to have statistical confidence (typically at the 95 percent confidence limit) in your results. While most people think that rare-cell sampling involves Poisson statistics, which is merely a convenient approximation, the true statistics are based on combinatorial statistics. Rare-cell sampling statistics are inherently difficult to calculate because they involve combinatorial statistics, which are difficult to calculate. But a simplified method, using a log transformation, was developed by us that permits their simple calculation (Rosenblatt et al., 1997).

Chapter 3 Study Questions

3.1 What different types of molecules can be used for targeting nanoparticles? What are the advantages and disadvantages of each?

3.2   What are the differences between monoclonal and polyclonal antibodies? How is each made? What are the advantages and disadvantages of each in terms of targeting?

3.3   What does it mean when we "humanize" monoclonal antibodies? Why is that important in terms of immune system interactions?

3.4   Describe the lock–key model of antibody–antigen targeting.

3.5   What is an epitope? How small can an epitope be? Why is that important?

3.6   Compare polyclonal and monoclonal antibodies in terms of their targeting to epitopes. How can a monoclonal antibody sometimes be "too specific"?

3.7   What do we mean by Ig subtype? Fc fragment? F(ab)$_2$ fragment? F(ab) fragment? Why is it important to understand what these mean?

3.8   What do we mean when we say that adding fluorochrome molecules to antibodies is a "Poisson process"? What is an F/P ratio? Why is it bad to have an F/P value that is too low or too high? How can HPLC purification be used to reduce these problems?

3.9   What is an antibody–antigen titration curve, and why is it important to know where you are on the curve in your experiment?

3.10   What is a cluster designation in terms of antibodies? Why is it important to understand cluster designations?

3.11   What do we mean by immunophenotyping? How can immunophenotyping be used to distinguish between different types of cancers or between two different developmental stages during differentiation?

3.12   Antibodies are commonly used as targeting molecules for in vitro applications. Why might antibodies *not* be the best molecule for targeting molecules to cells in vivo?

3.13   What are some of the complications of using antibodies to target cells?

3.14   What do we mean by direct immunofluorescence? Indirect immunofluorescence? Biotin–avidin immunofluorescence? What are the advantages and disadvantages of each?

3.15   Quantum dots have been used to replace the role of the fluorochrome for immunofluorescence. What are some of the advantages and disadvantages of quantum dots for immunofluorescence?

3.16   What is a peptide, and how can it be used to target nanoparticles to cells?

3.17   How are peptide targeting molecules identified and produced? What is a phage library? How can it be screened to find good peptide targeting molecules?

3.18   What is an aptamer? What are some of the advantages of using aptamers to target nanoparticles to cells in vivo?

3.19 What is combinatorial chemistry? How can it be used to produce aptamer libraries? How can these aptamer libraries be screened?

3.20 Why are high-throughput screening and isolation strategies important for both peptide and aptamer targeting molecules? Describe some of these high-throughput technologies.

3.21 Ligand–receptor targeting can also be used to target nanoparticles to cells. Describe a ligand receptor targeting scheme. What are its advantages and disadvantages?

3.22 Describe some technologies that can be used to evaluate targeting.

3.23 What is the difference between targeting sensitivity and targeting specificity?

3.24 Why is it important to evaluate targeting at the single-cell level? What is wrong with "bulk-cell" targeting methods?

3.25 Why is evaluation of mistargeting also very important for nanomedicine? How does mistargeting relate to side symptoms?

3.26 Why is multiparameter flow cytometry an important technology for targeting assessments?

3.27 How is a flow cytometer really just a different type of microscope? Describe how the general subsystems of a flow cytometer are analogous to a microscope. How are they different?

3.28 What are some of the cellular parameters that can be measured with a flow cytometer? Why is it important to simultaneously measure these cellular parameters when evaluating targeting?

3.29 What are some of the fields of basic science and medicine that have been significantly impacted by flow cytometry?

3.30 What are the main advantages of flow cytometry over other similar technologies? What are its main disadvantages?

3.31 What is a flow cytometer–cell sorter? Why is it such a powerful cell separation technology? What are its major limitations?

3.32 What are the main advantages of fluorescence confocal microscopy over conventional fluorescence microscopy? What are the main disadvantages?

3.33 Nanoparticle targeting in vivo can be said to be a "rare-cell targeting" situation. Why is that true, and why is it important when devising nanoparticle targeting strategies?

3.34 What are some of the challenges for targeting when facing rare-cell targeting situations?

4 Drug Delivery Cell Entry Mechanisms

4.1 Introduction

It should be recognized that trafficking of molecules both into and out of a cell is a normal activity necessary for cell survival. Cells have evolved elaborate strategies for these transport processes while ensuring the integrity of the cell membrane. Some of these transport processes are thermodynamically driven, whereas others involve active transport mechanisms (Figure 4.1).

Pinocytosis and phagocytosis are processes whereby a cell can ingest smaller or larger objects, respectively. Nanoparticles are typically in the size range to be taken up by these processes. Nanoparticle aggregates are particularly susceptible to uptake by phagocytic cell types, in particular by neutrophils in the peripheral blood that evolved to ingest bacteria in blood to prevent septicemia. That is a common problem for existing monoclonal antibody treatments for more than 25 different human cancers. This may result in the death of neutrophils, leading to patient neutropenia; its effects on nanomedical treatments would be to reduce the effective delivered dose by absorbing nanomedical systems in blood. As will be discussed in Chapter 7, most of the nanoparticle aggregation is caused by bad zeta potentials, which can be largely avoided through proper nanomedicine design techniques that are discussed in detail in that chapter. In the case of nanoparticles, the surface coatings can very much help the process of uptake. Other more specific uptake mechanisms include receptor-mediated uptake whereby a nanoparticle can be targeted to that receptor, which then, upon binding the nanoparticle, actively transports the nanoparticle across the cell membrane. Another important mechanism of nanoparticle uptake is through the use of membrane-permeating peptides, which, if coated to a nanoparticle, can help transport that nanoparticle across the cell membrane by mechanisms other than pinocytosis or phagocytosis (Figure 4.2).

Studies of quantum dot (Qdot) uptake show that Qdot surface coatings can have a profound impact on their cellular uptake. Polyethylene glycol (PEG) modification essentially blocked nonspecific Qdot delivery into the cells. On the other hand, Qdots coated with COOH (COOH is a carboxyl group containing both hydroxyl and carbonyl functional groups, allowing it to participate in hydrogen bonding as both hydrogen acceptors and hydrogen donors in help forming salts and other chemical compounds) were internalized quickly and in large amounts by both cancerous and noncancerous cells (Xiao et al., 2010). Qdot cellular uptake involves three major

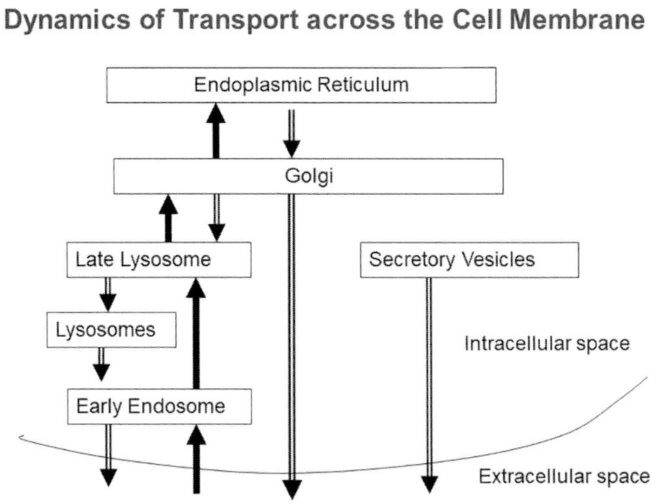

Dynamics of Transport across the Cell Membrane

Endoplasmic Reticulum

Golgi

Late Lysosome

Secretory Vesicles

Lysosomes

Intracellular space

Early Endosome

Extracellular space

Figure 4.1 Cells have adapted to find ways to safely transport molecules across the cell membrane without violating the integrity of that cell membrane.
Source: Leary teaching

Some Mechanisms for Transport of Nanoparticles across the Cell Membrane for Drug Delivery

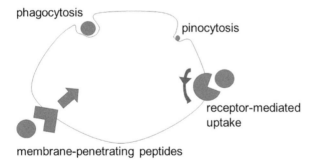

phagocytosis

pinocytosis

receptor-mediated uptake

membrane-penetrating peptides

Figure 4.2 Mechanisms by which a nanoparticle can enter a living cell include pinocytosis, phagocytosis, receptor-mediated uptake, and transport using membrane-permeating peptides.
Source: Leary teaching

stages, including endocytosis, sequestration in early endosomes, and translocation to later endosomes or lysosomes (Figure 4.3).

4.1.1   General Problem of Cell Entry

Much of the literature regarding targeting to cells involves only targeting to cell surface molecules for drug delivery. Far fewer publications are devoted to the

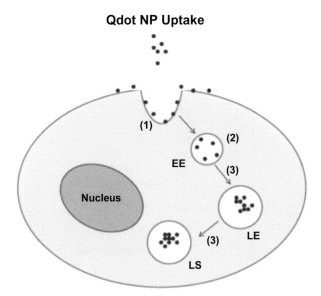

Figure 4.3 Postulated Qdot cellular uptake pathway. The process comprises three major stages: (1) endocytosis, (2) sequestration in early endosomes (EE), and (3) translocation to later endosomes (LE) or lysosomes (LS).
Source: Xiao et al., (2010)

subsequent critical intracellular targeting steps as well as the mechanisms of cell entry. Compared to a cell, a nanoparticle is like a small planet with respect to a very large star! There is still a long way to go to get to the proper place for effective drug delivery. Delivering drugs to the doorstep of a cell often does little to achieve effective therapeutics. One must deliver it to the proper subregion of a cell's interior to achieve effective drug dosing for therapeutic effects. There are many mechanisms by which a cell prevents entry of foreign molecules and far fewer mechanisms by which it promotes uptake.

Drug entry can be either passive-entry or active-entry transport. Passive-entry mechanisms are mainly due to thermodynamics and partitioning based on hydrophilic and hydrophobic mechanisms as well as phagocytosis and pinocytosis. Active-entry mechanisms included receptor-mediated uptake as well as facilitated entry by membrane-permeating peptides.

4.1.2   Choosing Modes of Cell Entry

When designing a nanomedical system, one must either count on the molecular composition of the outer layers of the nanomedical system and their potential interaction with the cell membrane (i.e., passive uptake), add specific molecules such as molecules that bind to receptors for receptor-mediated uptake (i.e., active uptake), or use so-called membrane-permeating peptides that promote pulling the rest of the nanomedical system across the cell membrane in the absence of receptor-mediated uptake.

4.1.3   How Does Nature Do It? (Biomimicry)

Luckily, nature has given us many biomimicry examples of nanosized objects (e.g., viruses) that are able to gain entry to human cells. Some of these viruses gain entry to the cell by thermodynamics or active partitioning into the cell membrane. Other viruses contain membrane-permeating peptides (e.g., Tat protein on human immuno-deficiency viruses). In fact, Tat is a commonly used membrane-permeating peptide in the design of many nanomedical systems.

4.2   Nonspecific Uptake Mechanisms

There are a number of processes by which particles gain entry to cells based totally on their size, shape, and general composition, not requiring receptor-mediated uptake. Some of the mechanisms of cell uptake involve natural processes for taking in small (pinocytosis) or larger (phagocytosis) particulates (Bergtrom, 2020), as shown in Figure 4.4.

4.2.1   Pinocytosis by All Cells

Pinocytosis is a thermodynamic process whereby small particles gain entry to cells (Kruth, 2011). Pinocytosis is a nonspecific pinching off of small vesicles that engulf extracellular fluid containing solutes and small particulates. Receptor-mediated endo-cytosis relies on the affinity of receptors for specific extracellular substances. Upon binding their ligands, the receptors aggregate in differentiated regions of the cell

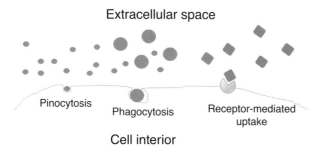

Types of Endocytosis

Extracellular space

Pinocytosis   Phagocytosis   Receptor-mediated uptake

Cell interior

Figure 4.4 Three main kinds of endocytosis are phagocytosis, pinocytosis, and receptor-mediated endocytosis, which bring small, dissolved molecules or particulates of various sizes from the extracellular fluid to the interior of the cell.
Source: Leary teaching

membrane called coated pits. The coated pits then invaginate and pinch off, forming a coated vesicle, thereby bringing their extracellular contents into the cell. After the coated vesicles deliver their contents to their cellular destinations, the vesicle membranes are recycled to the plasma membrane.

4.2.2   Phagocytosis by Some Cells

Phagocytosis is a process by which larger particles gain entry to human cells (Aderem and Underhill, 1999). In phagocytosis, phagocytic cells extend pseudopodia by membrane evagination. The pseudopodia of cells engulf particles of food that end up in digestive vesicles (i.e., phagosomes) inside the cytosol. A lysosome fuses with the phagosome, after which stored hydrolytic enzymes are activated. The result is the digestion of the engulfed particles. Phagocytosis begins upon contact between the outer cell surface and those particles.

4.3   Receptor-Mediated Uptake

Receptor-mediated endocytosis is perhaps the best understood mechanism for bringing larger substances into cells and a highly desirable way to bring nanomedical systems into the diseased cell. For many reasons, outlined in the following sections, it is a highly preferred method of cell entry.

4.3.1   Receptor-Mediated Transport of Desired Molecules

Receptor-mediated uptake is perhaps the most useful method because it can be highly specific. Some molecules on the cell surface are designed by nature to transport specific molecules that bind to them across the cell membrane. This is a necessary life process by which cells transport molecules that they need for basic metabolism and cell reproduction across their membranes. The cell knows how to accomplish receptor-mediated uptake very efficiently!

4.3.2   Example: Transferrin Receptor Transport of Iron for Metabolism

A specific example of receptor-mediated uptake of molecules across the cell membrane is the case of cell surface transferrin molecules, which help transport iron across the cell membrane. Transferrin transports iron across the cell membrane and is essential to human health at the single-cell level.

4.4   Nanoparticle Uptake

Nanosized objects also are naturally taken up by cells due to their composition, size, and shape. The specific reason for this natural uptake is less well understood than receptor-mediated uptake of molecules needed for basic cell metabolism. Nonetheless, we can use cells' ability to take up nanosized objects by designing nanomedical systems of the appropriate size and shape.

4.4.1   Surface Coatings on Nanoparticles

Surface coatings can have a profound effect on nanoparticle uptake, as well as perceived nanoparticle toxicity, which will be discussed more in Chapter 11. For example, coating of gold nanoparticles with acrylic greatly enhanced their uptake into cells (Krpetic et al., 2011). In another published example, surface coating had a profound impact on the cellular uptake of Qdots. PEG modification essentially blocked nonspecific Qdot delivery into the cells. On the other hand, Qdots coated with COOH were internalized quickly and in large amounts by both cancerous and noncancerous cells. Qdot cellular uptake involves three major stages, including endocytosis, sequestration in early endosomes, and translocation to later endosomes or lysosomes (Xiao et al., 2010), as shown in Figure 4.5.

4.4.2   Size and Shape Matter

The dictum that "size matters" applies to the proper size and shape of nanomedical systems, which can take advantage of natural mechanisms such as pinocytosis and phagocytosis (Chithrani, Ghazani, and Chan, 2006) (Figure 4.6).

Interestingly, shape also matters (Chithrani et al., 2006) (Figure 4.7).

A variety of different shapes have differing efficiencies of uptake by cells. Although the exact mechanisms by which this uptake occurs are not well understood, it is probably due to some combination of electrical charge distribution and general

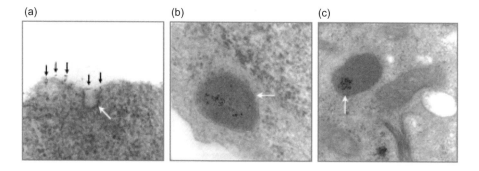

(a)   (b)   (c)

Figure 4.5 TEM images illustrating the process of Qdot cellular uptake and intracellular translocation. (a) Qdot endocytosis through plasma membrane invagination. Black arrows point to Qdots attached to the cell surface; the white arrow denotes the membrane pit engulfing Qdots. (b) Qdots sequestered and dispersed in early endosomes (white arrow). (c) Qdots condensed in late endosomes/lysosomes (white arrow).
Source: Xiao et al. (2010)

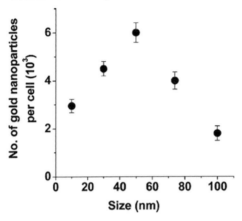

Gold NP Cellular Uptake as a Function of Size

Figure 4.6 Dependence of cellular uptake of gold nanoparticles as a function of size. Source: Chithrani et al. (2006)

thermodynamic factors. Some of this uptake potential appears dependent on the local radius of curvature, probably due to a combination of electrostatic interactions between the nanoparticle and the cell surface and perhaps thermodynamically driven partitions of lipid-rich materials on nanomedical systems and on any cell membrane.

4.4.3   Inhibitors Help Explore the Mechanisms of Nanoparticle Uptake

In order to explore the cellular mechanisms for nanoparticle uptake, it is necessary to use inhibitors to dissect which pathway(s) are important. One experimental strategy is to use drug inhibitors that block specific pathways of uptake. A conclusion from one of these highly detailed studies (Zhang and Monteiro-Riviere, 2009) for these Qdot nanoparticles and these specific cells was that Qdot nanoparticles with carboxylic acid surface coatings were recognized by lipid rafts but not by clathrin or caveole in human epidermal keratinocytes. The Qdots get internalized into early endosomes and then transferred to late endosomes or lysosomes. An excellent example of the detailed work in this publication is shown in Table 4.1.

4.4.4   Agglomeration Reduces Uptake

Agglomeration, the clumping of nanoparticles, generally reduces cell entry, except perhaps in the case of phagocytic cells, which are designed to swallow up larger particulates by phagocytosis. In general, it is a good idea to design nanomedical systems that are not only monodispersed in the bloodstream but also resistant to agglomeration. Agglomeration is a sign that the zeta potential of the nanomedical system is incorrect and must be altered. Large proportions of Chapter 7 are devoted to the intricacies of zeta potential design and testing.

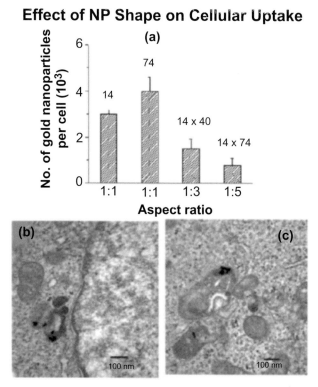

Figure 4.7 The effect of the shape of nanoparticles on cellular uptake and transmission electron microscopy images of rod-shaped gold nanoparticles internalized within the cells. (a) Comparison of uptake of rod-shaped nanoparticles (with aspect ratios of 1:3 and 1:5) and spherical nanoparticles. Transmission electron microscopy images of rod-shaped gold nanoparticles with aspect ratio (b) 1:3 and (c) 1:5 internalized inside vesicles of HeLa cells. Source: Chithrani et al. (2006)

4.4.5   Nanoparticles Tend to Agglomerate Inside Cells as They Are Taken Up

While not fully understood, nanoparticles, once taken up by cells, tend to agglomerate in clusters within subcompartments of the cell. This is an important phenomenon that needs further study since agglomerated nanoparticle tend to behave differently in terms of targeting and drug delivery.

4.4.6   Agglomerated Qdots Inside Cells

If you see nanoparticles inside cells visualized at the light microscopy level, you are not seeing individual nanoparticles, which are below the optical diffraction limit and therefore cannot be seen directly. Instead, you are seeing agglomerated nanoparticles (in this case, Qdots) inside cells, as shown in Figure 4.8.

Table 4.1. Inhibitors Used to Explore the Relevant Cell Mechanisms for Nanoparticle Uptake

Name	Function	Final Conc.
Nocodazole	Disrupts microtubule (cytoskeleton)	10.0 lg/ml
CytD	Inhibit F-actin polymerization (cytoskeleton)	10.0 lg/ml
MbCD	Cholesterol-depletion reagent (caveolae/lipid rafts)	5.0 mg/ml
Lovastatin	Cholesterol depletion (caveolae/lipid rafts)	10.0 lg/ml
Genistein	Inhibit F-actin recruitment to clathrin pits [a](clathrin)	10.0 lg/ml
CPM	Inhibitor of Rho GTPase (clathrin)	10.0 lg/ml
Y-27632	Related to cytoskeleton and melanosome transfer	10.0 lg/ml
NaN$_3$	ATP inhibitor (ATP)	3.0 mg/ml
DMA	Naþ/Hþ exchanger inhibitor (macropinocytosis)	10.0 lg/ml
WMN	PI3K inhibitor (macropinocytosis)	100.0 ng/ml
LY	PI3K inhibitor (macropinocytosis)	10.0 lg/ml
PTX	Inhibitor of a Gi subunit (GPCR)	100.0 ng/ml
CTX	Activator of a Gs subunit (GPCR)	1.0 lg/ml
U-73122	PLC inhibitor (downstream of GPCR)	4.0 lg/ml
SRP	PKC inhibitor (downstream of GPCR)	1.0 lg/ml
TrpI	Inhibitor of PAR-2 pathway (melanosome)	1.0 mg/ml
NCM	Melanosome inhibition (melanosome)	122.1 lg/ml
BMA1	Inhibits endosome acidification (endosome)	100.0 ng/ml
CRQ	Inhibits endosome acidification (endosome)	125.0 lg/ml
BFA	Interferes with Golgi, endosome, and lysossome	10.0 lg/ml
PolyI	Scavenger receptor inhibitor	10.0 lg/ml
FCD	Scavenger receptor inhibitor	10.0 lg/ml
LDL	Ligands for LDL receptor	10.0 lg/ml
AcLDL	Ligands for scavenger receptor	10.0 lg/ml

Source: Zhang and Monteiro-Riviere (2009).

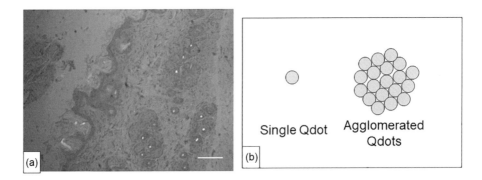

Figure 4.8 (a) Bright spots within the tissue correspond to clusters of agglomerated Qdots.
(b) Since these clusters are visible by optical microscopy, they are much larger than
individual Qdots. Indeed, the size of these Qdot clusters indicates that they must contain many
thousands of Qdots.
Source: Adapted from Haglund et al. (2009)

4.5   Drug Delivery by "Shedding"

One mechanism of drug delivery is to try to get the nanomedical systems as close as possible to the diseased cells either by targeting or by simple enhanced permeability and retention (EPR). Then a layer of the nanomedical system is simply shed, either releasing the drug near the outside of the cell or shedding inside the cell after uptake. This bypasses the problems of normal dilution of drug molecules in the blood, delivering a much higher local concentration of drug to the vicinity of the diseased cells. This means that a therapeutically effective drug dose can be delivered to diseased cells without exposing the patient to high systemic concentrations of the drug. Drug uptake in conventional medicine is largely driven by concentration gradients. This is a better way to introduce drugs at much higher concentration to the diseased cells of interest.

4.5.1   Extracellular Drug Delivery by Shedding

One way to deliver drugs is to just increase the local, as opposed to systemic, drug concentration by shedding nanomedical system layers and releasing a drug near a diseased cell that has been successfully targeted. The drug release might be triggered, for example, by low pH in the extracellular environment, which may be an indication of a nearby cancer cell. Generally, this method is fairly nonspecific as it may release drugs not only around diseased cells but also around normal cells; however, it provides some therapeutic advantage to conventional drug delivery methods.

4.5.2   Intracellular Drug Release by Shedding

It is usually better to release a drug once it is inside a cell. If the drug is properly designed in terms of active groups, it will be resistant to leakage back out of the cell. The advantages of releasing a drug inside a cell are numerous, but in large part they relate to the high specificity of drug targeting as well as a very high local drug concentration in the cell interior.

4.6   Technologies for Measuring Nanomedical System Interaction with Cells

It is important to have quantitative, or at least semi-quantitative, measures of the interactions of nanomedical systems with single cells. Evaluation of the effectiveness of targeting these cells, as well as the response, to nanomedical therapeutics is a critical factor in evaluating the proper design of these systems.

4.6.1   Single-Cell Measurements to Detect the Location of Nanomedical Systems

Without quantitative measures, the effectiveness of nanomedical system design is impossible to evaluate. Such quantitative measures need to evaluate not only the proper targeting of diseased cells, and the minimization of mistargeting to neighboring normal cells, but also the diseased cell's responses to the targeted drug. Flow cytometry can reveal the association of nanoparticles with cells, but it has difficulty determining whether the nanoparticles are outside or inside a cell because it is a so-called zero-resolution device. Two-dimensional microscopy can provide better information about whether the nanoparticles are inside or outside a cell by focusing up or down and doing human pattern recognition. This kind of determination is subjective. More definitive information about the location of nanoparticles can be determined by three-dimensional confocal microscopy. But again, individual nanoparticles are below the optical diffraction limit in size, so you are mostly seeing agglomerated nanoparticles inside a cell, as discussed previously.

4.6.2   Below "Optical Limit" Imaging

One of the challenges of imaging, as a way to evaluate targeting and cellular responses, is that cells are very small and nanoparticles smaller yet. Nanomedical systems are typically below the normal optical limits (0.25–0.5 μm), depending on the wavelength of light used. To improve on this to single-nanoparticle level requires a new form of confocal microscopy, namely, super-resolution microscopy (Zeiss, 2019), which actually calculates the location of a particle using Brownian motion. This calculated image allows super-resolution microscopy to see below the diffraction limit. Although there are many new "optical tricks" to obtain "super-resolution" (meaning detection below normal optical limits), these methods are difficult to employ in practical in vivo imaging of patients, so they are mainly used for in vitro or ex vivo analyses.

4.6.3   Nanomedical Systems with X-Ray Dense, Fluorescent, Metallic, or Magnetic Cores

Since optical imaging is difficult or impossible to perform at a high-enough resolution for in vivo imaging, a number of other imaging modalities with spectral wavelengths below the normal visible wavelengths can be used. While it is difficult to obtain spatial resolution at the single-cell level with these imaging modalities, they are nonetheless useful for whole-body imaging.

4.6.4   Study Live Cells with Minimally Invasive In Vivo Imaging?

Usually, pathologists perform diagnostics on biopsied cell material whereupon the cells have been chemically fixed. As such, the cells are no longer alive. In the ideal case, we would like to study intact, live cells within patients by minimally invasive in vivo imaging. MRI and CAT scans enable clinicians to look inside patients noninvasively. MRI scans are more sensitive to water in muscles and soft tissue, whereas CAT scans are more sensitive to bone, which contains electron-dense compounds.

4.7   Nanomedical Systems Evaluation Technologies

There are a number of different technologies that have been used to quantitatively, or semi-quantitatively, examine nanomedical systems on or within a cell. Many of these technologies can only be used by ex vivo imaging, meaning cells and/or tissue needs to first be biopsied from the patient.

4.7.1   Flow Cytometry: A Zeroth-Order Imaging Device

Flow cytometry, a zeroth-order imaging device, is one of the great medical technologies of the twentieth century (Givan, 2001, Goetz, Hammerbeck, and Bonnevier, 2018; Shapiro, 2001), and it has been critical to the diagnosis and treatment of many diseases. Basically, a flow cytometer is a virtual fluorescence microscope whereby cells in suspension, rather than on a slide, are examined one by one as they flow past a light source (Figure 4.9).

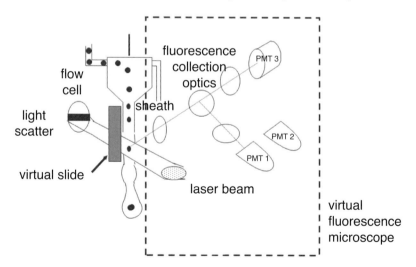

Flow Cytometer: Just Another Form of a Fluorescence Microscope
(Tip the picture [not the flow cytometer!] on its side)

Figure 4.9 A flow cytometer is really just another form of a fluorescence microscope, but instead of putting cells on a slide, we flow them individually through the focal point of the fluorescence collection optics.
Source: Leary teaching

Mostly it has been a technology that can be used on live or fixed human cells on biopsied material. The most common way to biopsy material has been liquid biopsies (Biswas, Ganeshalingam, and Wan, 2020), a fancy current term for examining blood samples obtained from patients by venous (or pin prick) blood sampling. Biopsied tissue from other body organs first needs to be dissociated into a single-cell suspension using proteolytic enzymes appropriate for that tissue. Unfortunately, these proteolytic enzymes may remove some types of biomarker molecules from the cell surface. Different biomarkers have different sensitivities to particular enzymes, which is why the choice of these enzymes is very important. More recently, there have been attempts to perform single-cell flow cytometry in vivo (Tuchin, Tarnok, and Zharov, 2009; Yu et al., 2020) by deeper optical sampling of cells flowing through arteries, veins, and even capillaries. This is particularly important as a means of studying circulating tumors in vivo. Importantly, it can sample far more blood than could be processed by trying to obtain liquid biopsies.

Flow cytometry is mostly a "zero-order imaging" modality whereby signals from the whole cell (e.g., fluorescence or light scatter) are quantified. While some attempts have been made to localize fluorescence measurements within subregions (e.g., the nucleus) of a cell (Leary et al., 1979), mainly done by time-of-flight measurements, they have also been used to look at receptor densities on cells by providing more reliable measurements of cell surface area and dividing the total fluorescence per cell by this surface area (Jett et al., 1980). Receptor density may be more important than absolute receptor number in terms of initiating action on a cell by receptor internalization (e.g., insulin or transferrin receptors; Rudolph et al., 1985).

4.7.2   Scanning and Transmission Electron Microscopy

Scanning and transmission electron microscopy can be performed only on biopsied material and usually on cells that have been chemically fixed since they usually require these measurements to be performed in a vacuum (Figure 4.10).

Recently, there have been attempts to perform measurements on live cells whereby measurements are made in a brief stretch of time between cells being in a liquid medium before they are quickly exposed to vacuum conditions (Putnam and Yanik, 2009).

4.7.3   One- and Two-Photon Confocal Microscopy

Confocal microscopy has been employed for some time to obtain three-dimensional reconstruction of cells within biopsied tissues. Confocal microscopy carries out virtual optical sectioning of cells and tissue to capture the 3D structure of cells and tissues (Figure 4.11).

One-photon confocal microscopy irradiates most of the cell or tissue and is of lesser resolution than two-photon confocal microscopy. As such, it is quite subject to photobleaching and usually requires fixed cells or tissues plus antifade solutions. Two-photon confocal microscopy can form small spots of energy for each optical

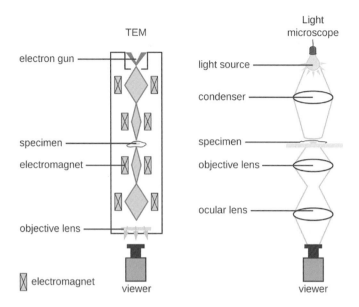

Figure 4.10 Electron microscopy is analogous to light microscopy, but with magnetics instead of lenses to guide the electrons to their targets. Transmission electron microscopy (TEM) can only penetrate relatively thin sections of tissue, but, unlike light microscopy, it has the resolution to see individual nanoparticles inside cells (or at least inside thin sections of cells) since its wavelengths are much smaller than the nanoparticles. Nanoparticles usually, but not always, have electron-dense cores that are capable of being visualized by TEM. Otherwise, the cells and nanoparticles need to be coated with some electron-dense fixatives. In any case, the cells must generally be fixed to survive under the vacuum requirements of TEM. Scanning electron microscopy (SEM) is slightly different from TEM, whereby surfaces of cells and tissues are coated with electron-dense materials and electrons bounce off these coatings into a secondary detector to give images of the outside of cells and tissues. SEM would be used to visualize the presence of nanoparticles on the outside of cells.
Source: https://courses.lumenlearning.com/microbiology/chapter/instruments-of-microscopy/

section to minimize photobleaching, but care must be taken to not literally burn the particular section of the cell or tissue!

A problem with confocal microscopy in performing 3D imaging of fluorescent objects inside cells is that normal fluorescent probes are subject to photobleaching such that the fluorescence fades while the cell is being scanned. Anti-fading solution can partially counteract this fading on fixed, but not living, cells.

One advantage, in addition to seeing the 3D architecture of the cells in the tissue, is also to see the spatial relationship between cells of different types within that tissue. Before confocal microscopy, one had to take thin sections of tissue, typically chemically fixed, due to depth-of-field focusing problems. With confocal microscopy, virtual, rather than physically real, thin sections of live tissue can be imaged without the need to physically slice that real tissue into thin sections. This was accomplished first by one-photon confocal microscopy, but this method suffered from photobleaching effects through the entire tissue. With the advent of two-photon confocal

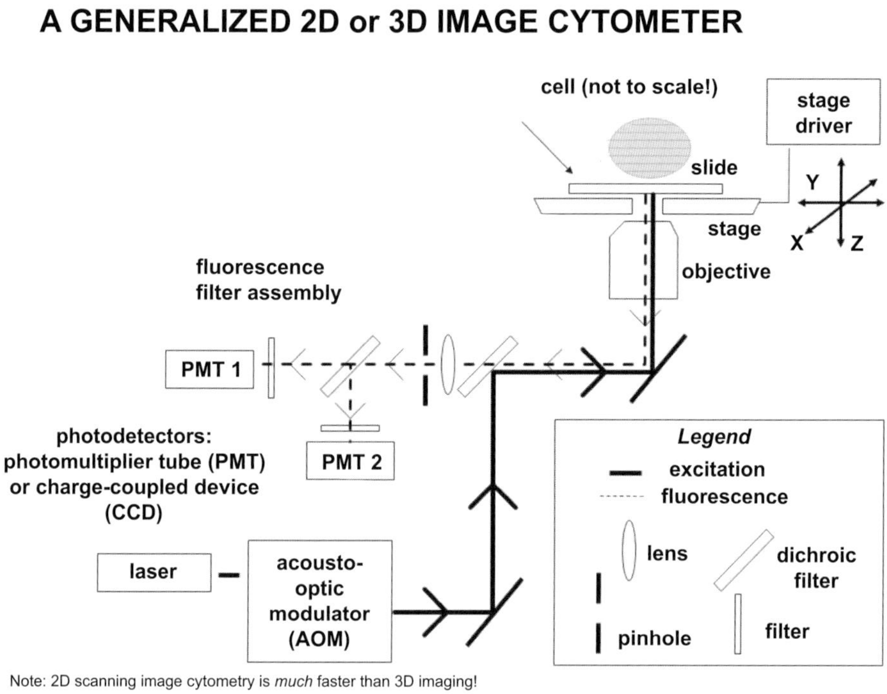

Figure 4.11 An image cytometer can provide 2D imaging of a cell to give spatial information about the intracellular location of nanoparticles. If it "optically sections" a cell by forming narrow layers of images in the third dimension ("confocal microscopy"), it becomes a 3D image. Source: Leary teaching

microscopy, there is considerably less photobleaching at the cost of much more expensive lasers and the need to be careful not to physically burn subregions of the tissue due to the high local intensities of excitation energy.

4.7.4   Hyperspectral Imaging

Hyperspectral imaging has become a powerful addition to image analysis (Johnson, Wilson, and Bearman, 2006). Unlike the standard three colors of RGB imaging, hyperspectral imaging collects images at many different wavelengths. Sometimes much more detail can be seen inside cells and tissue with hyperspectral imaging techniques (Johnson et al., 2006; Lu and Fei, 2014). It can also sometimes do a better job of distinguishing nanoparticles both on the outside and inside of cells (Figure 4.12).

A tissue sample illuminated by a light source is projected through a front lens into an entrance slit, which only passes light from a narrow line. After collimation, a dispersive device (e.g., a prism, grating) splits the light into a series of narrow spectral

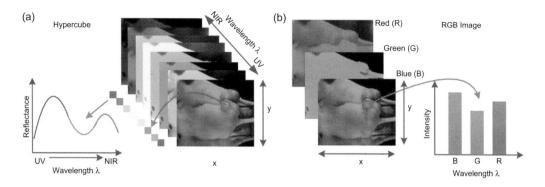

Figure 4.12 Comparison between hypercube and RGB image. Hypercube is a three-dimensional data set of a two-dimensional image on each wavelength. (a) Reflectance curve (spectral signature) of a pixel in the image. RGB color image has only three image bands on red, green, and blue wavelengths, respectively. (b) Intensity curve of a pixel in the RGB image.
Source: Lu and Fei (2014)

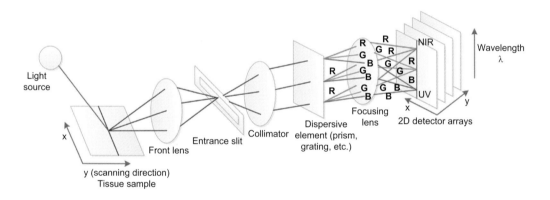

Figure 4.13 Schematic diagram of a "pushbroom" (line-scanning) hyperspectral imaging system where R(Red), G(Green) and B(Blue) samples are imaged to different planes of the multiple 2D detector arrays.
Source: Lu and Fei (2014)

bands that are then focused onto a detector array. Slit width controls the amount of light entering the spectrograph. In this way, for each pixel interval along the line defined by the slit, a corresponding spectrum is projected on a column of the detector array. Thus, each line of the targeted area on a tissue sample is projected as a 2D image onto the detector, with one spatial dimension and one spectral dimension. By scanning over the tissue specimen or moving the camera across the tissue sample in a "pushbroom" or line-scanning fashion, a hyperspectral imaging camera collects 2D images for adjacent lines, creating a hypercube with two spatial dimensions and one spectral dimension (Lu and Fei, 2014), as shown in Figure 4.13.

4.7.5   Surface Plasmon Resonance Imaging

Recently, surface plasmon resonance (SPR) imaging, typically with gold nanoparticles, has been used to much advantage in disease diagnostics (Duval et al., 2007). It does require the excitation and signal detection to be accomplished by exposing tissue either by surgical resection or by use of endoscopy in internal body cavities.

SPR is a label-free optical detection technique used to monitor and analyze biomolecular interactions in real time. The imaging capability enables users to visualize the entire working area at one time. SPR is the resonant oscillation of conduction electrons at the interface between negative and positive permittivity material stimulated by incident light. SPR is the basis of many standard tools for measuring adsorption of material onto planar metal surfaces or onto the surface of metal nanoparticles (Duval et al., 2007), as shown in Figure 4.14.

4.7.6   Atomic Force Microscopy

Atomic force microscopy (AFM) has been used to visualize nanoparticles within or on the surface of single cells (Jalili and Laxminarayana, 2004). Usually, this is done on

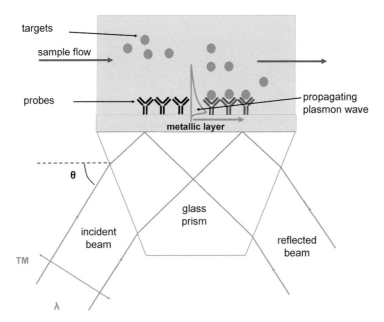

Figure 4.14 Surface plasmon resonance occurs when light is coupled to a thin layer of metal—usually silver or gold—under certain conditions (specific angle of incidence, wavelength, polarization, and metal thickness). An evanescent plasmon wave propagates along the metal-dielectric interface. Changes in either the refractive index or biomolecular layer thickness, usually from molecules bound to that surface, in the vicinity of the metallic interface can be monitored as a shift in the resonance curve.
Source: Duvall et al. (2007)

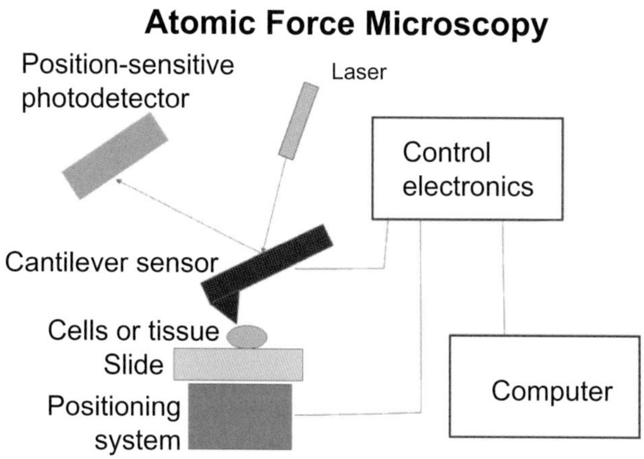

Atomic Force Microscopy

Position-sensitive
photodetector

Laser

Control
electronics

Cantilever sensor

Cells or tissue
Slide
Positioning
system

Computer

Figure 4.15 The specimen is "tapped" to sense its presence. Since the movements are on a tiny scale, they are amplified by bouncing a laser beam off a cantilever into a position-sensitive detector.
Source: Leary teaching

chemically fixed biopsied material, but recent efforts have included performing these measurements on live cells within fluids by bio-AFM (Figure 4.15).

4.8   Some Ways to Enhance Imaging

Sometimes image analysis needs a little help in achieving sensitivity within the required limits of image resolution. One way to accomplish this added contrast resolution is to use contrast agents.

4.8.1   Electron-Dense Contrast Agents to Enhance Scanning and TEM

Both transmission electron microscopy (TEM) and scanning electron microscopy (SEM) have problems visualizing cells and tissues because normally cells and tissues do not contain any electron-dense materials. This lack of contrast can be partially compensated for by using electron-dense contrast agents.

4.8.2   MRI Contrast Agents for In Vivo Imaging

Nanomedical systems can use core materials that contain magnetic resonance imaging (MRI) contrast agents. So-called T1 contrast agents were initially employed to great advantage. But one of the most used agents, gadolinium, has caused some safety concerns (Malikova and Holesta, 2017) due to potentially serious side effects of long-term gadolinium deposition in tissues, particularly kidneys, until they were chemically modified into macrocyclic forms. However, there remain concerns about both linear

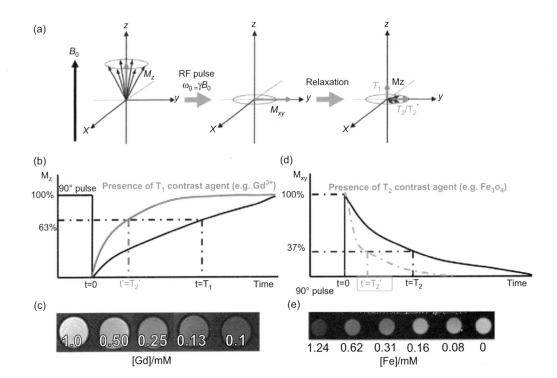

Figure 4.16 (a) Principle of MRI: Proton spins align parallel, or antiparallel, to the magnetic field and precess under the Larmor frequency; at the magnetic moment of the proton around the external magnetic field after induction of the radio frequency pulse, the magnetization of the spin changes; excited spins undergo both T1 and T2 relaxation processes. (b and c) T1 relaxation of protons is shortened under the presence of T1 contrast agents (e.g., Gd3+), which will generate a brighter image. (d and e) T2 relaxation of protons is shortened under the presence of T2 contrast agents (e.g., Fe$_3$O$_4$ nanoparticles), which will generate a darker image. Source: Xu et al. (2011)

and macrocyclic gadolinium (Cowling and Frey, 2019). But there is no reason we need to limit ourselves to T1 contrast agents. Although T1 positive MRI contrast agents are easier to visualize, T2 negative contrast agents can work equally well and, thus far, have exhibited fewer safety concerns, particularly in the case of ferric oxide, which is naturally reabsorbed into the body. More recently, less inherently toxic T2 contrast agents such as iron oxide have been used and recently improved when the iron nanoparticles were in an octopod conformation (Zhao et al., 2013). While T2 contrast agents are more challenging to measure than T1 contrast agents they are ultimately very sensitive. The basic concepts behind T1 and T2 MRI contrast agents is shown in Figure 4.16.

T1 contrast agents make images brighter, whereas T2 contrast agents make images darker. Gadolinium is a common T1 contrast agent, but concerns about its toxicity to patients led to development of macrocyclic gadolinium compounds that reduced these effects. Ferric oxide is a common T2 contrast agent that is not toxic to the patient under the doses commonly used.

Chapter 4 Study Questions

4.1   What are two nonspecific uptake mechanisms? How do they differ?

4.2   What is the general pathway of transport through the cell's internal machinery (e.g., endosomes)?

4.3   How can surface coatings on nanoparticles affect their uptake?

4.4   How do size, shape, and agglomerated clusters of nanoparticles affect uptake of nanoparticles by cells?

4.5   What is receptor-mediated uptake, and why is that an important way to provide for specific uptake by cells?

4.6   How can we use inhibitors to study specific uptake pathways?

4.7   The uptake of nanoparticles is frequently studied using fluorescence confocal microscopy. What are the advantages and disadvantages of this approach?

4.8   Transmission electron microscopy (TEM) is ultimately the "gold standard" for measuring nanoparticle uptake in cells. What are some of the problems with conventional TEM in terms of determining the 3D location of nanoparticles inside cells?

4.9   Acrylate coatings have been used to enhance the uptake of gold nanoparticles (Krpetic et al., 2011). How does preaggregation of these nanoparticles change the uptake pattern?

4.10   Both the microenvironment and external stimuli can affect nanoparticle uptake and subsequent therapeutic response. What are some of these external stimuli? What kinds of nanoparticle responses can you get with these stimuli? Give some examples.

4.11   Superparamagnetic iron oxide (SPIO) nanoparticles can do many things in terms of responsiveness to external triggers, particularly magnetic fields. What are some of these responses, and what important diagnostic and therapeutic goals can be achieved with SPIO nanoparticles?

5   Nanomaterial Cores for Noninvasive Imaging

5.1   Introduction

A wide variety of nanomaterials have been used as core materials for the subsequent construction of nanomedical systems. The core material is important in terms of the detection technologies that can be used for diagnostics, as shown in Table 5.1.

An extensive review of the many types of nanoparticles and shapes explores their chemical and physical properties (Burda et al., 2005).

5.1.1   Core Building Blocks

The core building blocks of a nanomedical system can have both structural and functional attributes. Structurally, it is a starting point for building multilayered nanomedical systems with a combination of genes or drugs, targeting molecules, and stealth molecules to evade the immune system in vivo. It is the starting place in the manufacturing process. In the philosophy of beginning with the end in mind, it is also the starting place for the overall design in terms of determining how subsequent layers will be attached.

The core can also have functional attributes. For example, it might be composed of nanomaterials that give it X-ray or MRI contrast agent capabilities. It might be magnetic, which makes many subsequent manufacturing steps easier as well as giving it the capability of serving as a source of hyperthermic therapy if placed in an oscillating magnetic field.

5.1.2   Functional Cores

Giving the core functional capabilities greatly expands the overall effectiveness of the nanomedical system both for diagnostics and therapeutics. For example, by constructing the core out of electron-dense materials, it will subsequently serve as an X-ray contrast agent for noninvasive CAT scan imaging. If that electron-dense material happens to be something like iron oxide, the core can simultaneously serve as an X-ray contrast agent as well as a T2 MRI contrast agent. This allows the nanomedical systems to be used for two-step diagnostics whereby a patient might first undergo a CAT scan to determine whether additional MRI imaging is needed. In this example,

Table 5.1. Nanomaterials for Nanomedical Systems

Core Material	Detection
Iron oxide	X-ray and MRI, if fluorescent probe added
C60 and carbon nanotubes	Fluorescent probe added
Gold	Surface plasmon resonance
Silver	Fluorescent probe added
Quantum dots	Size intrinsic fluorescence
"Next-generation" quantum dots	Size intrinsic fluorescence
Hybrid materials	Mixture of detectable properties

Note: Core materials should be chosen at the beginning of the nanomedical system overall design process to account for how they are to be detected in vivo. Source: Leary teaching.

the iron oxide can be superparamagnetic, allowing it to be used to produce hyperthermia at the resolution of a single cell if placed in an oscillating magnetic field that causes the particle to shake and produce heat at the single-cell level as a form of single-cell hyperthermia. Such a magnetic core also aids during the manufacturing process by allowing simple recovery of a nanomedical system with a magnet rather than having to recover them by ultracentrifugation. This is extremely practical in the manufacture of nanoparticles because it provides for multistep manufacturing without the use of ultracentrifugation.

A gold nanoparticle core not only serves as a convenient starting place for chemical attachment of subsequent molecules by cysteine residues. It can subsequently also be used with a light source to produce a surface plasmonic resonance signal that can be used for diagnostics.

5.1.3   Functionalizing the Core Surface

"Functionalizing" the core surface means preparing it chemically for attachment of subsequent molecules. This is particularly important if the core material itself is not very convenient chemically in terms of attaching subsequent molecules. Kumar provides an excellent source of information about biofunctionalization, meaning attachment of biomolecules to nanoparticles (Kumar, 2005).

5.2   Ferric Oxide Cores

Ferric oxide nanoparticles are inexpensive, biocompatible, and of low toxicity, making them ideal candidates for construction of nanomedical systems. Depending on the nature of the ferric oxide nanomaterials, they can also provide many useful nanomedicine functions and can be made in a wide range of sizes and shapes (Park et al., 2000) (Figure 5.1).

Ferric Oxide Nanospheres and Nanorods

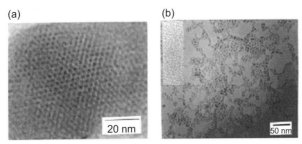

Figure 5.1 Transmission electron micrographs (TEM) of (a) spherical iron nanoparticles with diameters of 2 nm and (b) rod-shaped iron nanoparticles with dimensions of 2 nm × 11 nm. (Inset: High-resolution electron micrograph of a single nanorod.) The images were obtained with a JEOL JEM-2000EX II instrument.
Source: Park et al. (2000)

5.2.1   Paramagnetic Cores

Paramagnetic cores are permanently magnetized ferric oxide nanomaterials (Toy and Karathanasis, 2016). In general, using paramagnetic nanoparticles is a bad idea for nanomedicine because permanently magnetized nanoparticles tend to clump together. Inside the human body, this may lead to potentially dangerous embolisms. A much better idea is to use superparamagnetic nanoparticles that are only magnetized in the presence of a magnetic field and will not tend to spontaneously agglomerate.

5.2.2   Superparamagnetic Cores

Superparamagnetic ferric oxide cores are magnetic only in the presence of a magnetic field. They also do not tend to clump, thereby reducing the dangers of causing an embolism in vivo. This allows nanomedical systems to be magnetically "harvested" at different manufacturing steps in the overall synthesis process. It also allows them to be guided and concentrated locally at specific sites within the body using external magnetic fields. Superparamagnetic nanoparticles do not tend to aggregate due to residual magnetization, but they can still agglomerate due to poor zeta potential properties. Superparamagnetic nanoparticles do have size restrictions. If larger ones are needed, there is a way of creating a "raisins-in-a-bun" type of structure that preserves the overall property of superparamagnetism (Key, Cooper, et al., 2012).

5.2.3   Ferric Oxide Nanorods

Ferric oxide nanorods have all of the advantages of spherical magnetic nanoparticles with the additional advantages associated with the lack of particle symmetry. They can be magnetically rotated in vivo to effectively stir microenvironments and also to produce highly localized hyperthermia.

5.2.4   Overall Advantages and Disadvantages of Each Form of Magnetism and Geometry

As always, the best choice of magnetism and geometry depends on the overall design goals. It is possible to easily construct water-soluble ferric oxide nanoparticles to begin the process of producing nanomedical drug delivery systems (Cooper et al., 2008).

Ferric oxide nanoparticles can be formulated in a wide variety of sizes and shapes, but to keep them superparamagnetic to prevent agglomeration, it is necessary to make a "raisins-in-a-bun" structure whereby many smaller superparamagnetic nanoparticles, either spherical (Key, Cooper, et al., 2012), or cubic (Key et al., 2016), are imbedded in a larger overall structure. They can also be made into nanoflakes, which have interesting properties (Cervadoro et al., 2014) that help increase their T2 contrast ratio for MRI. A recent review shows how a top-down fabrication design can lead to production of nanoparticles with interesting properties (Aryal et al., 2019).

5.3   C60, Carbon Nanotubes, and Graphene

Carbon is typically used in two very different formats for core material. In a closed-structure C60, Buckminsterfullerene (named after Buckminster Fuller, the architect of geodesic dome fame) is a type of fullerene with the formula C60. It has a cage-like, fused ring that can contain drugs. But the volume of C60 is extremely small, so only tiny amounts of relatively small molecules can reside inside. In a more extended, open form, carbon atoms can form a "nanotube" of indefinite length that can contain larger amounts of still fairly small drug molecules. These nanotubes are also rod-shaped, leading to different possible modes of cell entry (Susi, Pichler, and Ayala, 2015) (Figure 5.2).

5.3.1   Size and Structure of C60

C60 is a truncated icosahedron with 60 vertices and 32 faces (20 hexagons and 12 pentagons, where no pentagons share a vertex) and with a carbon atom at the vertices of each polygon and a bond along each polygon edge. The van der Waals diameter of a C60 molecule is about 1.01 nanometers (nm). This means that the internal volume of C60 is VERY small, allowing it to carry only a small amount of a drug that is itself small in size. For this reason, it is not a very practical vehicle for drug delivery.

5.3.2   Elongation of C60 into Carbon Nanotubes

Under proper thermodynamic conditions, C60 can be elongated into carbon nanotubes. These nanotubes have very small internal diameters, so trying to load drugs inside these nanotubes can still be problematic. The drugs must be small. But the elongated structure of carbon nanotubes can have considerable external

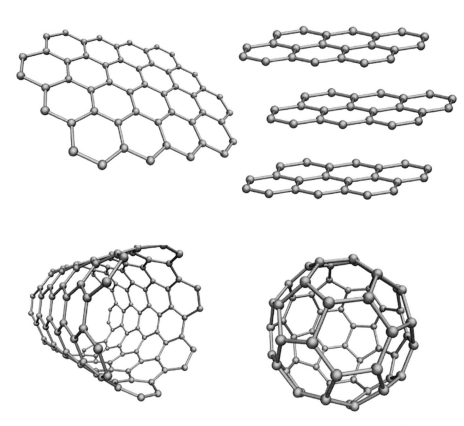

Figure 5.2 Carbon nanoparticles with widely varying sizes and shapes can be made. C60 Buckminsterfullerene particles in the shape of geodesic domes can contain a small amount of a small-molecule drug. Graphene sheets and carbon nanotubes offer very different geometries for bio-eluting stents and other structures.
Source: Susi et al. (2015)

surface area and can be labeled on the outside with drug or sensing molecules forming carbon nanowires.

5.3.3   Graphene

Another form of repeated carbon structure is a two-dimensional sheet of linked carbons known as graphene. While not very useful for normal drug delivery, these structures can be useful for sensing molecules in tissues and blood by using graphene drug-eluting stents.

5.3.4   Advantages and Disadvantages

One of the chief advantages of C60 and carbon nanotubes is that they are both fairly biologically inert. The elongated, rod-shaped structure of carbon nanotubes makes them favorable for cell uptake and drug delivery.

5.4   Gold and Silver Cores

Gold (Au) and silver (Ag) core materials are easy to make and have many useful properties for their detection (Rosi and Mirkin, 2005). Gold is literally the "stuff of stars," meaning that it is made by second-generation stars. It is also particularly easy to bioconjugate biomolecules to gold and silver through cysteine amino acid residues, making them a popular nanomaterial (Figure 5.3).

5.4.1   Gold Cores

Gold is a frequently used core material for construction of nanomedical systems. Gold nanoparticles are easy to manufacture and can be formulated into a wide variety of different shapes, including spheres, rods, and stars. One of the chief attractions is that the initial bioconjugation of subsequent layers can be done by a simple conjugation of proteins through their cysteine residues. Detection can be accomplished through surface plasmon resonance.

5.4.2   SERS Molecular Imaging and Flow Cytometry

These surface plasmon resonance properties of gold and silver nanoparticles can be used as surface-enhanced Raman spectroscopy (SERS) probes. The SERS effect can

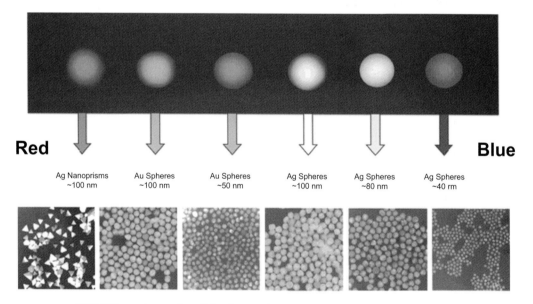

Figure 5.3 Gold and silver nanoparticles are easy to construct in a wide range of sizes. Their color depends on size, with the larger ones red and the smaller ones blue. Their synthesis leads to uniformly sized nanostructures, as seen from the electron micrographs.
Source: Rosi and Mirkin (2005)

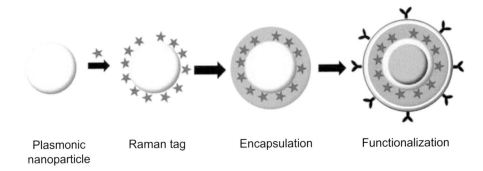

Plasmonic   Raman tag   Encapsulation   Functionalization
nanoparticle

Figure 5.4 Anatomy of a SERS tag. A SERS tag generally consists of a plasmonic nanoparticle with the desired resonant wavelength, a Raman active compound that confers a particular spectral signature, a coating to stabilize and protect the SERS tag, and a targeting molecule such as an antibody or a nucleic acid that confers molecular specificity to the SERS tag. Source: Nolan and Sebba (2011)

increase the Raman signal intensities by up to 10^{14}–10^{15} times, resulting in a detection sensitivity comparable to fluorescence (Nie and Emory, 1997). SERS tags can also be made for flow cytometry probes (Nolan and Sebba, 2011), as shown in Figure 5.4.

5.4.3   Spherical Gold Nanoparticles

It is comparatively easy to produce highly spherical gold nanoparticles of any desired size (De Souza et al., 2019). One of the additional reasons for the popularity of gold nanoparticles is the ease of attaching proteins by cysteine residues.

5.4.4   Gold Nanorods

If the processes for generating gold nanoparticles are slightly altered, these spherical structures can be fairly easily elongated into gold nanorods (Perezjuste et al., 2005). These gold nanorods have interesting and important properties, including tuning of surface plasmon resonance signals (Zweifel and Wei, 2005) (Figure 5.5).

5.4.5   Gold Nanostars

Gold nanoparticles can be made into other interesting and potentially important shapes, including gold nanostars (Mousavi et al., 2020). Nanoparticles with plasmon resonance capabilities have optical scattering in the near infrared (NIR) and are useful for biophotonic imaging. The nanostars exhibit polarization-sensitive NIR scattering and can produce "twinkling nanostars" caused by frequency-modulated signals in response to a rotating magnetic field gradient since they have ferromagnetic cores sensitive to gyromagnetic magnetic fields. This periodic "twinkling" can be converted

Gold Nanorods for Optical Imaging

(a)   (b)

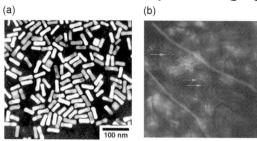

Figure 5.5 (a) Gold nanorods of high uniformity (Zweifel and Wei, 2005). (b) Gold nanorods, which fluoresce red, were photographed inside the blood vessels of a live mouse by researchers in Purdue's Weldon School of Biomedical Engineering and Department of Chemistry. The researchers have taken a step toward developing a new type of ultrasensitive medical imaging technique that works by shining a laser through the skin to detect the tiny rods injected into the bloodstream. In tests with mice, the nanorods yielded images nearly 60 times brighter than conventional fluorescent dyes, including rhodamine, commonly used for a wide range of biological imaging to study the workings of cells and molecules.
Source: Purdue University photo courtesy of Weldon School of Biomedical Engineering and Department of Chemistry (Dr. Alex Wei)

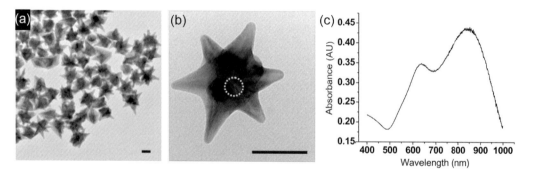

Figure 5.6 (a and b) TEM images of gold nanostars with superparamagnetic cores (bar = 50 nm), with approximate position of Fe_3O_4 core outlined by dashed circle; (c) optical extinction spectrum.
Source: Wei et al. (2009)

into Fourier-domain images with a dramatic reduction in background (Wei et al., 2009) (Figure 5.6).

5.4.6   Gold Nanoshells

Thin gold layers can be added to nanoparticles of other compositions, creating gold nanoshells. These nanoshells have interesting and important tuning properties, depending on the shell thickness, making their detection in tissues more sensitive. It

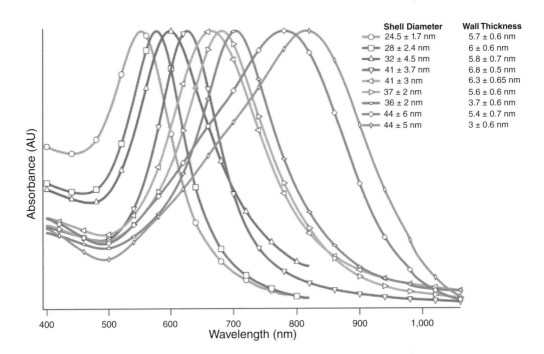

Figure 5.7 Optical resonances of gold shell–silica core nanoshells as a function of their core/shell ratio. Metal nanoshells are a new type of nanoparticle composed of a dielectric core such as silica coated with an ultrathin metallic layer, which is typically gold. The optical response of gold nanoshells depends dramatically on the relative size of the nanoparticle core and the thickness of the gold shell. By varying the relative core and shell thicknesses, the color of gold nanoshells can be varied across a broad range of the optical spectrum that spans the visible and the near infrared spectral regions. Gold nanoshells can be used in surgeries to highlight specific tumors. They can also be used in expensive point-of-care diagnostic tests.
Source: Schwartzberg et al. (2006)

is also possible to pump heat into gold nanoshells for localized hyperthermia (Loo et al., 2004; Rastinehad et al., 2019; Schwartzberg et al., 2006) (Figure 5.7).

The size and shape of gold nanoshells can be controlled experimentally to yield "tunable" nanoparticles (Prevo et al., 2008).

5.5   Silver Nanoparticles

Silver is also a good core material. Silver nanoparticles have antimicrobial properties, as evidenced by silver's use in everyday items, such as food containers, touting their antimicrobial properties (de Lima, Seabra, and Duran, 2012; Kim et al., 2007). Although the antimicrobial properties of silver nanoparticles tend to dominate discussions of silver nanoparticles, silver nanoparticles, like gold nanoparticles, can be made in a wide range of sizes and geometries that can provide unique plasmonic properties. They can also be made into complex shapes (González et al., 2014) (Figure 5.8).

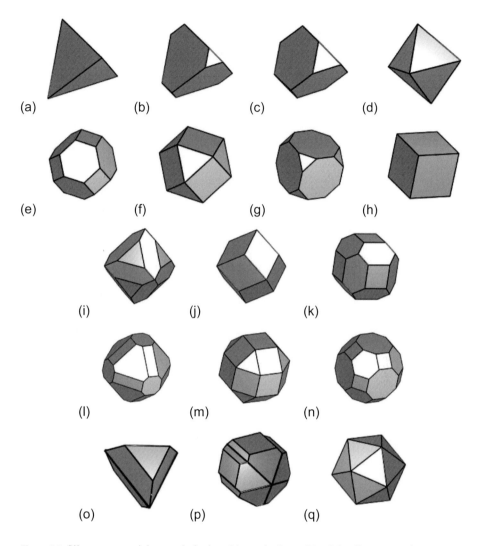

Figure 5.8 Silver nanoparticle morphologies: (a) tetrahedron, (b) minimally truncated tetrahedron, (c) truncated tetrahedron, (d) octahedron, (e) truncated octahedron, (f) cuboctahedron, (g) truncated cube, (h) cube, (i) truncated rhombic dodecahedron, (j) rhombic dodecahedron, (k) cubo-rhombic dodecahedron, (l) doubly truncated octahedron, (m) small rhombicuboctahedron, (n) great rhombicuboctahedron, (o) triangular prism, (p) Marks decahedron, and (q) regular icosahedron.
Source: Gonzalez et al. (2014)

5.6   Quantum Dot Nanoparticles

Quantum dots are "size/wavelength tunable," meaning that their spectral properties of fluorescence depend on their size. The smaller quantum dots are bluish, whereas the larger ones are reddish, with all of the colors of the spectrum in between. Although the actual diameters range from about 4 nm to 7 nm, most quantum dots become

10–14 nm in diameter once hydrophilic surface coatings are added, allowing these highly hydrophobic nanomaterials to exist in aqueous environments. They are commercially available in a wide variety of colors. While they can be excited at a variety of different wavelengths, an interesting property is that they all have very strong near-UV excitation spectra that allow one excitation frequency light source to be used to excite multiple quantum dots of different fluorescent colors. They are extremely resistant to photobleaching, which makes long-term fluorescence microscopy possible. One misunderstood property of quantum dots worth remembering is that their fluorescent lifetimes are about 10 times longer than typical organic fluorescent dyes. In a typical flow cytometer, a fluorescent dye is excited and re-excited thousands to tens of thousands of times, depending on the size of the excitation spot, while they are traversing the excitation light source. This means that whereas quantum dots are very bright with the fluorescence microscope, they are not nearly as bright when they are used in a flow cytometer since they take 10 times longer to re-excite. They do work in a flow cytometer, but they are better for fluorescence imaging applications due to their much longer fluorescence lifetimes (Figure 5.9).

Quantum Dot Nanoparticles

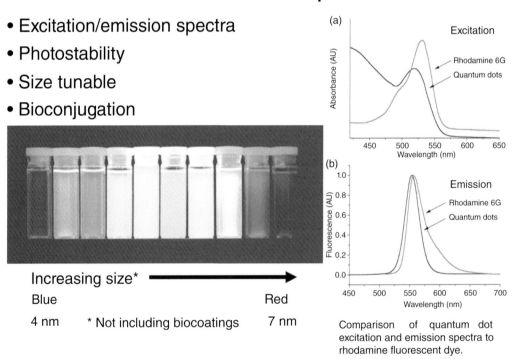

- Excitation/emission spectra
- Photostability
- Size tunable
- Bioconjugation

Increasing size* ⟶

Blue   Red

4 nm   * Not including biocoatings   7 nm

Comparison of quantum dot excitation and emission spectra to rhodamine fluorescent dye.

Figure 5.9 The emission spectra of Qdots varies according to the size of the Qdots from about 4 nm diameter for blue Qdots to about 7 nm diameter for red Qdots, not including any biocoating, which usually adds another 4–5 nm. (a) Absorption spectrum and (b) emission spectrum.

Source: Leary teaching

Quantum dots are very bright and do not tend to photobleach, making them excellent candidates for diagnostics using image analysis that requires longer irradiation times. But their fluorescence lifetimes are roughly 10 times longer than conventional fluorophores. This means that they are excited and then re-excited roughly 10 times less often during flow cytometry analyses. Quantum dots work for flow cytometric analyses, but they are not as useful as initially thought. They also run into steric hindrance problems because they are much larger than normal fluorescent probes.

5.6.1   Size Determines Spectral Qualities

Quantum dots are usually made of semiconductor nanomaterials. Their size varies typically from about 4 nm to 7 nm in diameter, not counting any biocoatings. After coatings to make them water soluble are added, they are usually about 12–14 nm in diameter. They are very small and cannot be individually imaged by conventional fluorescence microscopy. One sees only clusters of quantum dots large enough to be visible (Haglund et al., 2008).

5.6.2   Quantum Dots Might Be Cytotoxic

Since many quantum dots are made of semiconductor materials that are potentially cytotoxic, it was initially surprising that these cytotoxic effects were not seen. However, upon a more thorough analysis, it was revealed that the biocoatings on these quantum dots were temporarily shielding cytotoxic effects (Haglund et al., 2008). Later studies showed that whereas most single-cell measures of cytotoxicity were negative, the quantum dots were inducing apoptosis, a more sensitive measure of cytotoxicity, in those cells (Eustaquio and Leary, 2012b; Leary, 2014). Typical biocoatings are shown in Figure 5.10.

5.6.3   Next-Generation Quantum Dots May Be Better!

Despite these seemingly negative criticisms, there remains a use for quantum dots in ex vivo imaging of tissues for diagnostic purposes. Better still is the promise of organic quantum dots. These new developments toward less potentially toxic organic quantum dots ("next-generation" Qdots) may benefit from developments in the organic LED (OLED) technology used in some televisions.

5.7   Silica Cores

Silica nanoparticles can be easily synthesized in a variety of sizes and can embed conventional fluorescent molecules and prevent photobleaching (Lai et al., 2003). Mesoporous silica nanoparticles can be a good material for acrolein drug release, which is useful for treating spinal cord injuries (White-Schenk, Shi, and Leary, 2015). There are many factors governing their influence on cells (Pednekar et al., 2017) (Figure 5.11).

Quantum Dot Nanoparticles

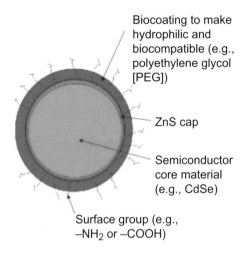

Biocoating to make hydrophilic and biocompatible (e.g., polyethylene glycol [PEG])

ZnS cap

Semiconductor core material (e.g., CdSe)

Surface group (e.g., –NH$_2$ or –COOH)

Transmission electron microscopy (TEM) image of amino-functionalized Qdots. Size was determined to be ~10 nm.

Figure 5.10 Biocoatings enable the Qdots to exist in hydrophilic environments as well as provide convenient surface group "hooks" (NH$_2$ and COOH groups) for bioconjugating additional molecules to the overall structure.
Source: Haglund et al. (2008)

Mesoporous Silica Nanoparticles (MSNs) for Drug Release

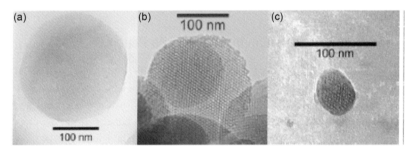

Figure 5.11 How characteristics of mesoporous silica nanoparticles, such as size, porosity, shape, and surface functionalization, can have an influence on the nanoparticle's behavior under biological conditions and also on their therapeutic applicability. (a), (b) and (c) are three different views of these silica nanoparticles at three different magnifications.
Source: Pednekar et al. (2017)

5.7.1   Silica Nanoparticles

Silica nanoparticles can be easily made in various sizes (Wang et al., 2006). During synthesis they can be doped with a wide variety of fluorescent dyes of different colors that, due to their partial protection within the silica matrices, are resistant to photobleaching. Silica nanoparticles tend to be biologically inert and nontoxic, making them good nanomaterial for use in humans (Figure 5.12).

It is also possible to embed fluorescent molecules in these structures. A rapid fluorescence-based immunoassay was developed using 60 nm dye-doped and bioconjugated silica nanoparticles for the detection of single *E. coli* O157:H7 bacterial cells (Wang et al., 2006). This particular strain has caused many foodborne illnesses from contaminated lettuce and other vegetables in the United States. As shown in the SEM images, the nanoparticle–antibody conjugates specifically associate with *E. coli* O157:H7 cells (Figure 5.12a) but not with *E. coli* DH5α cells, which lack the surface O157 antigens (Figure 5.12b). In addition to the fact that each nanoparticle provides a greatly amplified and photostable signal, many surface antigens on a given bacterial cell are available for specific recognition by antibody-conjugated nanoparticles. It is thus feasible to have thousands of nanoparticles bind to each bacterial cell and generate an extremely strong fluorescence signal suitable for rapid detection technologies (Figure 5.13).

5.7.2   Advantages and Disadvantages

Silica nanoparticle cores have many advantages, including inexpensive materials, biocompatibility, and the ability to be fabricated in a wide range of sizes and shapes, as well as mesoporous structure that can secrete drugs over time while protecting those drugs from degradation and elimination by the immune system. They can also be made semi-porous (i.e., mesoporous), allowing drugs to be eluted over time to nearby cells.

Silica Nanoparticles Can Be Easily Made in Different Sizes

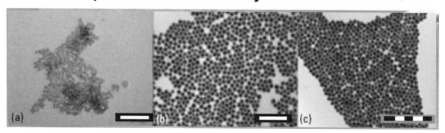

Figure 5.12 Transmission electron micrographs of different sizes of silica nanoparticles prepared in various microemulsion systems: (a) 15 nm nanoparticles, (b) 40 nm nanoparticles, and (c) 120 nm nanoparticles, where the scale bars are 200 nm, 200 nm, and 1 μm, respectively. Source: Wang et al. (2006)

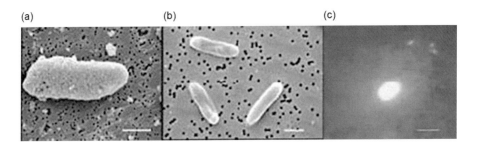

Figure 5.13 (a) SEM image of an *E. coli* O157:H7 cell incubated with nanoparticle–antibody conjugates, showing the nanoparticle binding to the target bacterium. Scale bar is 2.73 μm. (b) SEM image of an *E. coli* DH5 cell (negative control) incubated with nanoparticle–antibody conjugates, showing no nanoparticle binding. Scale bar is 1.5 μm. The small black dots in (a) and (b) are the pores on the surface of the filter membrane, and the white spots are unbound nanoparticles. (c) Fluorescence image of an *E. coli* O157:H7 cell after incubation with nanoparticle–antibody conjugates. Scale bar is 4 μm. The fluorescence intensity is strong, enabling identification of a single bacterial cell in aqueous solution.
Source: Wang et al. (2006)

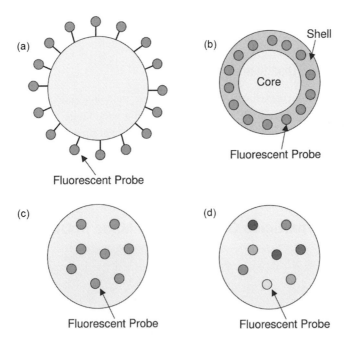

Figure 5.14 Hybrid architectures containing fluorescent nanoprobes include (a) dyes bound to a nanoparticle surface, (b) dyes incorporated into the silica shell of a nanoparticle, (c) fluorescent probes such as quantum dots or dyes incorporated into the core of microparticles, and (d) a multiplexing approach by embedding probes of one or more different colors.
Source: Murcia and Naumann (2005)

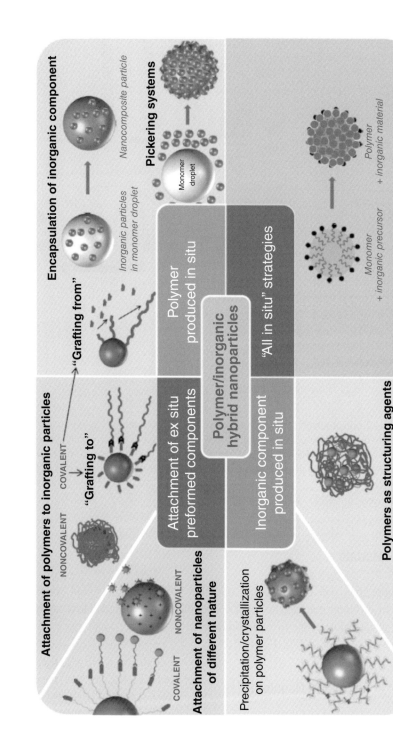

Figure 5.15 Different synthetic strategies in the formation of polymer/inorganic hybrid particles.
Source: Hood et al. (2014)

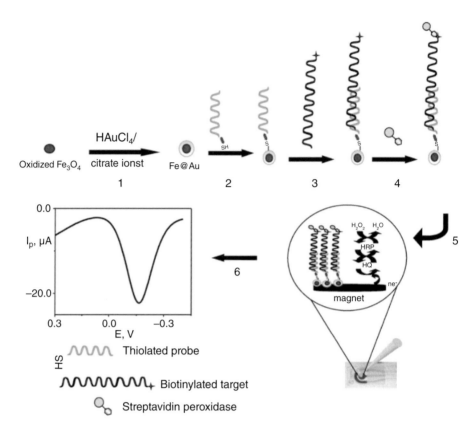

Figure 5.16 Synthesis protocol: (1) formation of the FeAu nanoparticles, (2) modification of the FeAu with the thiolated probe, (3) hybridization with the biotinylated target, (4) enzymatic labeling with streptavidin-HRP (horseradish peroxidase), (5) hybrid-modified magnetic beads deposition on the SPEs, and (6) voltametric detection of the mediated reduction of H_2O_2 with HQ.

Source: Loaiza et al. (2011)

5.8   Hybrid Materials

Use of more than one core material, so-called hybrid materials, can combine the attributes of different materials to yield a superior overall design. There are multiple strategies to incorporate one or more fluorescent dyes within nanoparticle structures (Murcia and Naumann, 2005) (Figure 5.14).

5.8.1   Multifunctional Hybrid Nanoparticles

Multifunctional hybrid nanoparticles combine physical and chemical characteristics of two or more classes of materials – such as polymers, liposomes, gold and silver, quantum dots, and mesoporous silica – to create a new class of nanoparticles that leverage the strength of each type of material. An excellent review of hybrid nano-particles describes the many different types of nanoparticle that can be synthesized (Hood, Mari, and Munoz-Espi, 2014; Morales et al., 2012), as summarized in Figure 5.15.

5.8.2 Gold–Ferric Oxide Nanoparticles and Nanorods

Gold can be added to ferric oxide material to produce gold–ferric oxide nanoparticles (Loaiza et al., 2011). The simple conjugation properties of gold can then be combined with the superparamagnetic advantages of ferric oxide both in manufacturing and in the possibility of additional detection technologies using the unique properties of each material. Having a ferric oxide core that can be superparamagnetic has all of the advantages of that nanomaterial in terms of X-ray and MRI contrast agents. Having a gold outer coating makes surface chemistry easier and permits the use of surface plasmon resonance detection methods (Figure 5.16).

Chapter 5 Study Questions

5.1 How does the composition of the core material affect its mode of detection? Give at least two examples.

5.2 The magnetism of ferric oxide nanoparticles is of two types, paramagnetic and superparamagnetic. What is the difference between the two types? In general, we would try to develop superparamagnetic, rather than paramagnetic cores. Why? What problems could happen with paramagnetic cores in vivo?

5.3 How does a C60 carbon nanoparticle differ from a carbon nanorod? What are the advantages and disadvantages of each as a core material for a nanomedical device?

5.4 For centuries, gold has been used as a material for human implants. What advantages does a gold nanoparticle have for nanomedicine? How can it be detected in vitro and in vivo?

5.5 Why are gold nanoparticles popular in terms of cell surface nanochemistry?

5.6 What advantages do gold nanorods offer over spherical gold nanoparticles in terms of cell uptake and detection?

5.7 What are the advantages and disadvantages of silica as a core material? What is "mesoporous" as a core property that helps it serve as a way of delivering drugs?

5.8 Quantum dots are core materials that have an unusual property in terms of their color. What is the basis of Qdot color?

5.9 Why have Qdots not been used in vivo in humans?

5.10 Can individual Qdots be seen by normal microscopy? Why or why not?

5.11 What are "second-generation Qdots," and why are they potentially important?

5.12 Hybrid core materials have some important advantages. Give some examples and their advantages over single material cores.

5.13 Describe some of the advantages of polymer coatings over various nano core materials.

5.14 Most nanomaterials are hydrophobic. Why is that important to keep in mind when designing nanoparticles?

5.15 Coatings can mask nanotoxicity. Why is that important?

6 Attaching Biomolecules to Nanoparticles

6.1 Attaching Nanomedical Structures to the Core

While nanomedical system design begins with the choice of core materials, the ultimate usefulness of these nanomedical systems is driven by the subsequent attachment of biomolecules in a series of layers. The initial attachment process (i.e., bioconjugation) requires biomolecules to be attached to a core material that is typically nonorganic and may consist of materials quite foreign to these biomolecules. Although there is a tendency to simplify this process by choosing gold cores so that proteins can simply be attached by their cysteine residues, this limits these nanomedical systems' capacity for other interesting properties such as MRI contrast agents. Gold is very electron dense, so it does serve as a good contrast agent for X-ray imaging techniques.

6.1.1 Attachment Strategies Typically Depend on Core Composition

Strategies for attaching biomolecules to the core depend on the composition of core materials. These attachment strategies can have profound effects on the biodistribution of nanoparticles in vivo. For example, macrocyclic DOTA (1,4,7,10-tetraazacyclododecane-1,4,7,10-tetraacetic acid, also known as tetraxetan) is a tetra acid that has the ability to chelate gadolinium (a T1 MRI contrast agent). It also can bind a monoclonal antibody either for targeting or for uptake by a cell (Sun et al., 2005).

6.1.2 Attachment Strategy Should Not Drive the Core Choice

Attachment of biomolecules by cysteine residues is a popular technique (Krpetic et al., 2011). However, the attachment strategy should *not* drive the choice of core materials. Many people choose simple attachment strategies like having cysteine residues bind to gold and then are stuck with using gold, which precludes more interesting choices. InnovaCoat® GOLD nanoparticles have a proprietary surface coating that covalently binds antibodies, peptides, and other molecules to form highly stable conjugates (Di Nardo et al., 2019). These are commercially available in kits from multiple vendors.

6.1.3   Choice of Core Depends on the Desired Overall "Multifunctional" Nanomedical Device

The core should be chosen with the overall design of a multifunctional nanomedical device. Many nanomedical systems are much more powerful if they are theragnostic, meaning they are capable of both diagnostics and therapeutics. As part of the diagnostic arm of the design, location is important information. By analogy to Western blots, which allow us to dismiss false background because the protein has the wrong molecular weight, location allows us to dismiss binding of nanomedical results if it is just background due to mistargeting. By choosing core materials that can provide location information, we can use noninvasive imaging to safely view where the nanomedical systems targeted.

For example, the choice of iron oxide (particularly in its superparamagnetic form) as a core material allows us to carry out noninvasive imaging by any of three noninvasive imaging modalities (or a combination thereof). Iron oxide is electron dense, so it can be used with 2D and 3D X-ray imaging. It is also a T2 contrast agent for MRI imaging. While it is a little more challenging in terms of its surface chemistry, bioconjugation of biomolecules is not that much more difficult. The advantages of iron oxide over many other core materials make the extra effort worth it. One way of beginning the process of attaching biomolecules to iron oxide nanoparticles is to first make them water soluble by using oleic acid as a surfactant to keep the particles from aggregating (Murcia and Naumann, 2005) (Figure 6.1).

Monodisperse iron oxide nanoparticles (MION) are easily synthesized in organic solvents for industrial applications. However, biological applications require that the particles by readily dispersed in aqueous solutions. To improve their dispersion in aqueous solution, MION particles can be conjugated to water-soluble polymers. These water-soluble iron oxide particles (Cooper, 2008; Yu et al., 2006) can then be used for nanomedicine.

After making iron oxide nanoparticles monodisperse in aqueous environments, the next task is to make them suitable for subsequent bioconjugation of biomolecules. There are many ways to accomplish this, but one example of how to also attach subsequent biomolecules is shown (Murcia and Naumann, 2005) in Figure 6.2.

6.2   "Surface Chemistry" Strategies for Attachment of Biomolecules to the Core

Surface chemistry is an important part of the total design of nanomedical systems. It is the critical step of adding either ions or molecules to make hydrophobic nanomaterials (most nanomaterials are hydrophobic) coexist in a hydrophilic environment (e.g., in blood). Critically, it also allows the zeta potential of the nanoparticles, discussed in more detail in Chapter 7, to be adjusted so they do not aggregate inside the human body and cause potentially fatal embolisms. In addition to making inherently hydrophobic nanomaterials hydrophilic, we also need to make them biocompatible with living cells through various biocoating strategies so that their interactions with living cells do not cause problems in those cells.

Example: More Complicated Strategies: Coupling PEG and Folate to an Iron Oxide Nanoparticle

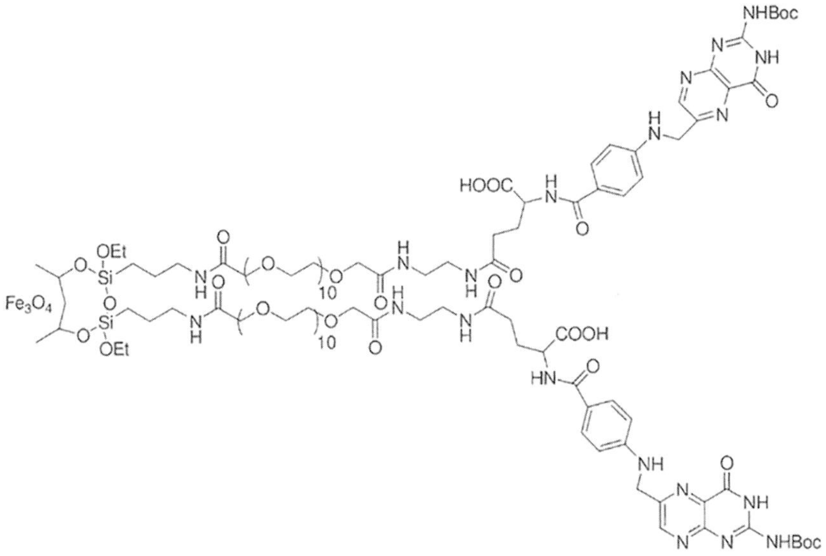

Figure 6.1 Reaction scheme for MION synthesis using iron oxide hydroxide as the iron source and oleic acid as an organic surfactant. In addition, targeting molecules such as folate can be added to the overall structure (Kumar, 2005).
Source: Murcia and Naumann (2005)

6.2.1   Hydrophobic versus Hydrophilic Core Materials

A major factor in the choice of core materials is whether the core is hydrophobic or hydrophilic since most nanomedical systems need to operate in a hydrophilic environment such as blood or interstitial fluids, both of which are hydrophilic microenvironments. If the core is hydrophobic, it typically must be coated with hydrophilic molecules to work in blood. Some strategies for creating hydrophilic cores(Murcia and Naumann, 2005) are shown in Figure 6.3.

6.2.2   Addition of Biomolecules for Biocompatibility

In addition to having core coatings to deal with hydrophilic or hydrophobic environments, these biomolecules need to be biocompatible, meaning they must not, in and of themselves, cause injury to living cells. Some of this biocompatibility involves preventing nanoparticles from aggregating, which can lead to embolisms and other potentially serious consequences. This is accomplished by constructing nanoparticle structures with a better zeta potential, as discussed in great detail in Chapter 7. Addition of appropriate biomolecules can also reduce toxicity.

Stabilizing, Biocompatible Coatings of Core Materials

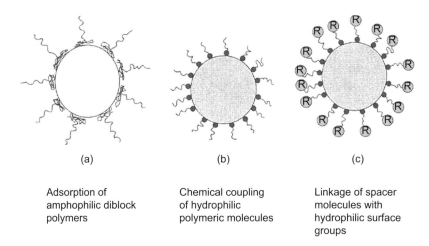

(a)

(b)

(c)

Adsorption of
amphophilic diblock
polymers

Chemical coupling
of hydrophilic
polymeric molecules

Linkage of spacer
molecules with
hydrophilic surface
groups

Figure 6.2 There are at least three ways to make nanoparticles biocompatible as well as stable in aqueous solutions: (a) simple adsorption of amphiphilic polymers, (b) chemical coupling of hydrophilic molecules to the nanoparticle surface, and (c) linkage of spacer molecules containing hydrophilic surface chemical groups (Kumar, 2005).
Source: Murcia and Naumann (2005)

6.2.3   Monofunctional versus Bifunctional Surface Chemistry Strategies

While many surface chemistry strategies are monofunctional, bifunctional strategies can lead to more interesting and adventitious possibilities. Monofunctional strategies are simpler but can result in dead-end design problems whereby there is not an easy strategy for the attachment of additional layers of biomolecules.

In contrast, bifunctional surface chemistry strategies allow for some interesting and creative strategies for adding subsequent layers of biomolecules, including targeting and biosensing molecules.

6.2.4   Pay Attention to Overall Zeta Potential during the Surface Chemistry Process

Throughout the process of adding biomolecules and other molecules to the surface of nanoparticles, the zeta potential needs to be measured after each attachment layer. As will be discussed in much greater detail in Chapter 7, zeta potential is one of the most important factors to be considered in the overall design of nanomedical systems. It is a

Four Common Approaches to Hydrophilic Surface Modification of TOPO-Stabilized Quantum Dots

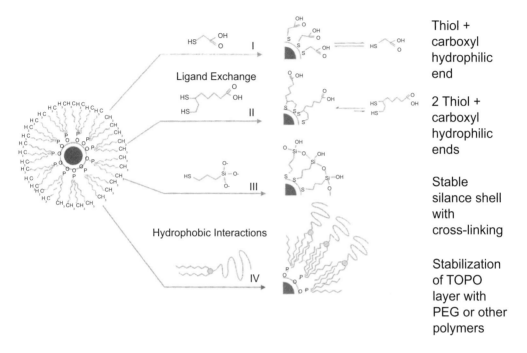

Figure 6.3 At least four different strategies can produce nanoparticles, normally hydrophobic, with hydrophilic surfaces. Schematic representation of four common approaches to hydrophilic surface modification of trioctylphosphine oxide (TOPO)–stabilized quantum dots. Approach I: TOPO replacement with a heterobifunctional linker consisting of a thiol end group, a spacer, and a hydrophilic end group such as carboxylic acid. Approach II: TOPO replacement with a linker consisting of two thiol groups on one end and a hydrophilic end group on the other end. Approach III: TOPO replacement with a silane forming a stable shell via cross-linking. Approach IV: Stabilization of TOPO layer using amphiphilic molecules such as PEG lipopolymers or amphiphilic diblock copolymers that are held on the surface by hydrophobic interaction with the octyl chains of TOPO (Kumar, 2005).
Source: Murcia and Naumann (2005)

critical step to allow for proper targeting as well as uptake of nanomedical systems into living cells in the human body.

6.3   Two Main Attachment Strategies

There are two main attachment strategies for bioconjugating biomolecules in the construction of a nanomedical system. Both of these major strategies have advantages and disadvantages. As with all designs, simplicity is desirable but can result in unacceptable limitations. On the other hand, more sophisticated strategies can result in subsequent complexities that make the additional efforts not worth the additional time and cost.

6.3.1   Noncovalent (Primarily Electrostatic) Bonding Strategies

Noncovalent bonding strategies usually involve charge-based electrostatic formations that allow for systems to just self-assemble. The overall strategy is to perform a layer-by-layer self-assembly of a multilayered structure, as has been performed by Lvov, a pioneer in layer-by-layer assembly of nanomedical devices, in a series of papers (Lvov et al., 1995; Lvov, Decher, and Moehwald, 1993). A strategy of simply adding biomolecule layers through noncovalent, primarily electrostatic, bonding may be initially simpler. A step-by-step synthesis addition of biomolecule layers by electrostatic charge makes the total system spontaneously self-assemble without much experimental effort. It also precludes the need for the use of potentially cytotoxic chemicals necessary for the formation of stable chemical bonds. This allows for the use of gentle chemistries for biocompatibility. The overall chemistry of layer-by-layer assembly can be very simple, including rapid self-assembly.

The ultimate design of multilayered structures should also be considered with the subsequent desired disassembly process in mind. For example, a layer may come off when encountering a more acidic microenvironment – something useful for targeting these nanomedical systems to tumors, which tend to produce acidic microenvironments. Simplicity of design disadvantages can also lead to some problems. Easy assembly and disassembly can lead to instances where the structures fall apart due to unforeseen factors and situations. Instability of these layer-by-layer assembled structures can occur due to changes in pH and ionic strength, which will inevitably be encountered as nanomedical systems travel within the human body.

The zeta potential can suddenly change as layers spontaneously strip off. Layer-by-layer assembly by alternating electrostatically charged materials can lead to potential problems if the disassembly of one layer with a given charge to a subsequent layer with a different charge causes issues. A simple example of such a problem is the disassembly of a negatively charged layer to reveal a positively charged layer, which can lead to the remaining structure simply sticking to all cells since those cells usually have an outer layer that is negatively charged. A more subtle problem is the loss of a charged layer that makes the overall zeta potential of the remaining structure unfavorable for the next step in a multistep targeting process such as binding to a cell surface receptor for subsequent receptor-mediated uptake through the cell membrane.

6.3.2   Covalent Bonding Strategies

In contrast to the preceding strategy, covalent binding strategies can allow for multilayered structures requiring specific disassembly chemistries that can be tightly controlled and insusceptible to unforeseen circumstances. There are a number of advantages to covalent bonding strategies, including stability and more precise control of disassembly. Covalent bonding leads to much more stable structures that are less likely to fail due to changes in pH or ionic strength in new microenvironments. This also allows more sophisticated control over multistep targeting and biosensing processes as the planned disassembly of the structure so that drugs can be delivered to their final destination.

The greatest strengths can become the greatest weaknesses if overdone or misapplied. The very advantages listed previously can also lead to disadvantages. Stability is your design friend until it is not! The creation of some covalent bonds sometimes requires use of potentially cytotoxic organic solvents. Once put into a synthesis process, they can be very difficult to remove. Cells can be sensitive to a few parts per million of these molecules. Covalent structures can be so stable that they can be quite difficult to disassemble, and planned disassembly needs to be a part of the overall design of any multilayered nanomedical system. Use of strong organic solvents can be cytotoxic even at trace concentrations. Many chemical synthesis reactions to form stable covalent bonds between biomolecules require organic solvents, which can be toxic to cells. Even a few parts per million of common organic solvents can be cytotoxic to cells. Once these solvents are included in one or more synthesis steps, they can be difficult, or impossible, to remove. A general design strategy should be to never put in a chemical that you must be able to take out later in the overall synthesis strategy. These green chemistry strategies, which call for avoiding use of these toxic solvents, are discussed in more detail in a subsequent section on safe synthesis and disposal issues in Chapter 14.

6.4 Attaching Targeting and Therapeutic Molecules

To create a useful nanomedical system requires the attachment of specific biomolecules to the core nanoparticle. A good nanomedical system should include an appropriate combination of targeting, sensing and therapeutic molecules.

6.4.1 Preparing the Nanoparticle for Addition of Targeting, Biosensing, and Therapeutic Molecules

One of the most important purposes of bioconjugation is to attach targeting, biosensing, and sometimes even therapeutic molecules to the overall nanomedical system. The bioconjugation process must be done carefully, particularly in the case of targeting molecules. If the bioconjugation process is done incorrectly, the targeting molecule may have its targeting ability compromised, resulting in mistargeting, which can have serious consequences. Likewise, small alterations in biosensing molecules during the bioconjugation process can lead to failure to properly biosense.

6.4.2 What Are the Special Requirements, If Any, for These Molecules?

Obviously, targeting molecules need not be modified adversely during the bioconjugation process in any ways that compromise their targeting specificity and sensitivity.

In addition, the specific targeting area of the targeting molecules must be available for binding to its target molecule. Biosensing molecules must be available and active at the time when they are needed to control the function of the nanomedical system. Therapeutic molecules must not be altered in ways that affect their therapeutic value.

6.4.3   Testing for Targeting at the Single-Cell Level after Bioconjugation

Testing for correct targeting is very important after bioconjugation. Targeting molecules may work fine when they are not attached to other molecules and structures, but this targeting specificity and sensitivity can change dramatically after improper bioconjugation. Target testing procedures and technologies are discussed extensively in Chapter 3.

6.4.4   Testing for Therapeutic Efficacy at the Single-Cell Level after Bioconjugation

It is important to try to compare the therapeutic efficacy of a drug in a nanomedical system to its counterpart in free solution. Sometimes this is not possible because the drug cannot survive in solution or blood without some protective strategy. Drugs are frequently changed by their microenvironments. They can also be changed if they are attached improperly to other molecules.

6.5   Attaching Different Types of Targeting Molecules

There are a wide variety of targeting molecules, as discussed extensively in Chapter 3. Each of these targeting molecules requires different bioconjugation strategies, as described below.

6.5.1   Antibodies

As discussed in Chapter 3, antibodies have antigen recognition substructures that should be avoided during the bioconjugation process to preserve their specificity and sensitivity. Typically, you want to stay as far away from these sites as possible. A logical and preferred bioconjugation attachment point is the Fc portion of the antibody molecule. As discussed in Chapter 3, sometimes only fragments not containing an Fc subelement are used to avoid nonspecific Fc binding. These $F(ab)_2$ fragments and F(ab) fragments are more challenging to bioconjugate because by definition you are making alterations in these molecules closer to the antigen recognition sites.

Because biotin is a small molecule and relatively easy to attach to antibodies, it is used to allow subsequent detection with avidins of various fluorescent colors. Binding biotins to an antibody also tends to not cause losses in binding specificity. Biotinylated antibodies are commercially available for many antibodies, and their additional cost is marginal. For this reason, they are an excellent way to rapidly prototype an overall nanomedical system. Later on, you might want to rid yourself of the large fluorescent avidins that can cause steric hindrance problems.

6.5.2   Peptides

Peptides as targeting molecules, described in greater detail in Chapter 3, are (usually) a relatively small number of amino acids in a specific sequence that have shape/ antigen recognition capabilities. Their ability to recognize antigen shapes is dependent on their protein-folding characteristics, which are affected by a number of variables including the pH and ionic strength of their microenvironments. Care must be taken to attach them in a way that does not change their specificity and sensitivity or cause partial blocking by steric hindrance problems.

6.5.3   Aptamers

Aptamers are DNA or RNA sequences that have, among other properties, shape recognition capability to act as targeting molecules. This shape recognition depends highly not only on the DNA or RNA sequences but also on possible alterations in shape recognition caused by changes in pH, ionic strength, and other microenvironment variables.

There are at least three important advantages of aptamers. First, DNA aptamers are highly stable and DNA in peripheral blood does not tend to engender immune responses, so they can circulate for a long time. Second, aptamers can be generated against epitopes that cannot generate monoclonal antibodies. Aptamers can be generated against almost anything, and the successful candidates can be identified by high-throughput screening and isolation of sequences from very large combinatorial libraries, as described in Chapter 3. Third, aptamers can be PCR amplified. This means that these targeting probes can be used to identify diseased cells over large areas of tissues using in situ PCR techniques described in Chapter 3.

6.5.4   Small-Molecule Ligands

Other small molecules, such as ligands, can serve as targeting molecules. Typically, these molecules bind in a ligand–receptor way. A particularly important example is the use of folate ligands to bind to folate receptors, which are elevated on many types of cancer cells. This is discussed in Chapter 3 in terms of its use for real-time fluorescence-guided surgery (Frigerio et al., 2019).

6.6   Testing the Nanoparticle-Targeting Complex

It is important to be able to test the entire nanoparticle-targeting complex, which may behave differently than the sum of its components. For example, an antibody attached as a targeting molecule to a nanoparticle may behave differently due to steric hindrance problems caused by the larger size of the nanoparticle–antibody complex. For this reason, it is necessary to develop strategies for testing the entire nanoparticle-targeting complex.

6.6.1 Ways of Detecting This Complex

Most conventional assays of targeting molecules do so by directly attaching a fluorescence molecule to the targeting molecule. Care must be taken to not compromise the targeting specificity and sensitivity of the targeting molecule when attaching the fluorescent label. But the nanoparticle offers new and interesting ways to monitor the presence of the nanomedical complex in cells. The nanoparticle itself can be fluorescent as well as an X-ray contrast agent or an MRI contrast agent (as in the case of the nanoparticle having an iron oxide core that is a T2 MRI contrast agent). This permits detection of nanoparticle-targeting complexes deep within the body by a variety of noninvasive imaging techniques.

6.6.2 Ways of Assessing Targeting/Mistargeting Efficiency and Costs of Mistargeting

The same techniques discussed in Chapter 3 should be applied to targeted nanoparticle complexes, with the additional caveat that nanoparticles themselves can be pinocytosed into cells even in the absence of targeting molecules. This is particularly problematic if the nanoparticle complex has the wrong zeta potential. A rule of thumb should be that the nanoparticle complex never has a positive zeta potential. Sometimes, the targeting and other molecules carry enough positive charge that they can give the overall nanoparticle complex a net positive charge, which will cause them to stick nonspecifically to cells that are almost always negatively charged. These complexes should also never have a zeta potential very close to zero because this indicates that they may be unstable and fall apart. For more on zeta potential, refer to Chapter 7, where these issues are discussed in more detail.

6.6.3 Is the Nanoparticle Still Attached to the Targeting Molecule?

An important question that should always be asked, and tested, is whether the targeting molecules are still attached to the nanoparticle complex. This is a rarely discussed problem. To test thoroughly requires more sophisticated testing whereby both the nanoparticle and the targeting molecules are fluorescently labeled in different colors. Then one tests for the simultaneous localization of both fluorescent probes. Even that is a relatively crude, but still useful, measure since the resolution of fluorescence microscopy is limited to the diffraction limit. To really see it at the necessary resolution to prove that the nanoparticle-targeting complex is still intact requires super-resolution microscopy.

6.7   Attaching/Tethering Different Types of Therapeutic Molecules

Conventional attachment techniques, particularly those involving covalent chemical bonds, usually hinder free 3D rotation, which is sometimes necessary to prevent steric hindrance problems. One way to prevent steric hindrance problems is to project targeting or sensing molecules out on "molecular spacer arms" that are simple hydrocarbon chains, typically avoiding double-bonded structures. These molecular spacer arms give the ability of the whole structure to rotate, helping it to more easily bind to targeted molecules by allowing the two molecules to approach each other in a more favorable orientation. For full access, a targeting molecule should be "tethered" so that it has considerable freedom of movement as well as the ability to perform its total functions (Prow, Grebe, et al., 2006).

6.7.1   Antibodies as Therapeutic Molecules

It is particularly important to pay attention to the attachment process if the antibody or other targeting molecule must freely interact with antibodies or proteins in the immune system. Antibody-dependent cell-mediated cytotoxicity (ADCC) attaches an antibody to a cancer cell by the F(ab) portion of the antibody. This leaves the Fc portion of that antibody free to interact with a natural killer (NK) cell, which then initiates the ADCC process, resulting in death to the cancer cell. The first monoclonal anti-cancer antibody approved by the FDA in 1997 was rituximab, which sought out the CD20 receptors on large B-cell lymphoma (non-Hodgkin's lymphoma). There are now dozens of anti-cancer monoclonals that target different types of cancers (Figure 6.4).

6.7.2   Peptides as Therapeutic Molecules

Therapeutic peptides must, if attached, be able to freely interact with other molecules. A particularly important example is the case where the peptide on a nanoparticle complex must interact with other molecules within a cell to initiate apoptosis. This is a powerful way to treat disease. It removes the requirement of getting the drug dose correct. Once the peptide initiates apoptosis, the cell is triggered into a programmed cell death process and the cell knows how to complete all of the necessary steps to safely destroy or repackage itself into reusable subcomponents, without adversely affecting neighboring cells, by apoptosis (Barras and Widmann, 2011) (Figure 6.5).

6.7.3   Therapeutic Aptamers

Aptamers carry the promise of being useful not only for targeting, but also for acting as therapeutic molecules themselves (Keefe, Pai, and Ellington, 2010). Aptamers have shape recognition capabilities similar to monoclonal antibodies or targeting peptides, with the additional advantage that they are comparatively easy to synthesize. Unlike

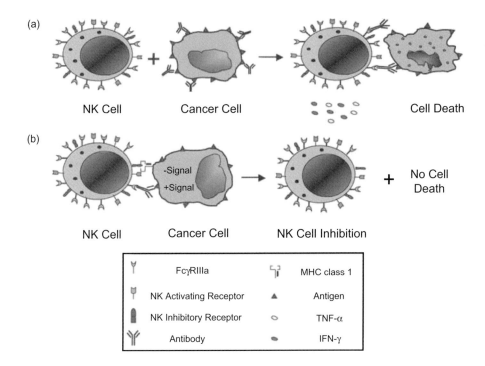

Figure 6.4 "NK (Natural Killer) cell-mediated ADCC against cancer cells. (a) The cancer antigen-specific antibody binds the cancer cell. The Fc region of the antibody binds FcR (Fcγ RIIIa or CD16) on NK cells. The cross-linking of the FcR on NK cells triggers NK cell cytotoxicity and release of soluble mediators. (b) Simultaneous co-ligation of an inhibitory receptor on the surface of the effector cell (e.g., an inhibitory KIR by its MHC class I ligand on the cancer cell) sends negative signal and inhibits NK cell activation and killing of the cancer cell via ADCC."
Source: Iannello and Ahmad (2005, 489)

antibodies, they do not require immunogenicity and can be created by combinatorial chemistry libraries as described in Chapter 3. Examples of aptamers designed against three diseases are shown in Figure 6.6.

6.7.4   Transcribable Sequences

Some sequences, such as mRNA, can be transcribable to act as gene product–manufacturing complexes to literally manufacture a therapeutic molecule within a living cell. Care must be taken to ensure that these transcribable sequences remain transcribable and are not blocked by steric hindrance issues. This has been used to create mRNA vaccine technologies (Schlake et al., 2012) and is the basic idea behind how an mRNA vaccine against COVID-19 was rapidly developed during the 2020 pandemic. An example of mRNA technology is shown in (Pardi et al., 2018) Figure 6.7.

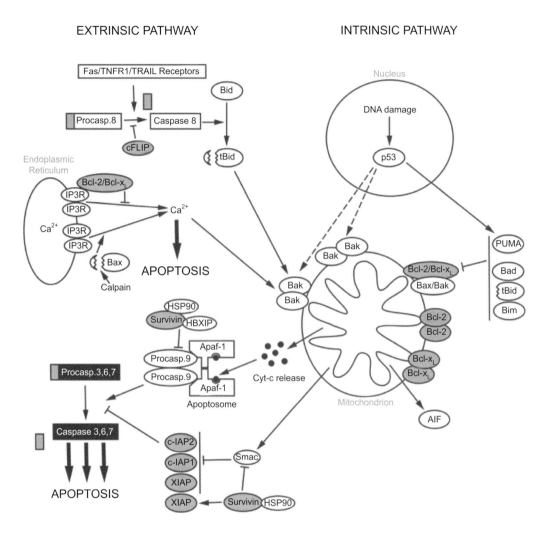

Figure 6.5 Apoptotic signaling pathways. There are both intrinsic and extrinsic pathways. The main molecular players involved in apoptotic signaling are indicated. The anti-apoptotic proteins are grey-colored, while the pro-apoptotic ones are not.

Source: Barras and Widmann (2011)

6.7.5   Small Drugs

Usually, we desire small drugs to be released when they are at the targeted cell. Because they are small, they can be easily encapsulated into nanodrug delivery systems, even small ones like carbon nanotubes or C60 nanodelivery systems. They are particularly useful for nanodrug delivery systems designed to get near diseased cells (e.g., cancer cells) by enhanced permeability and retention (EPR) mechanisms or by acid pH microenvironments near tumors whereby the acidic pH conditions can release drugs from the nanodrug carrier.

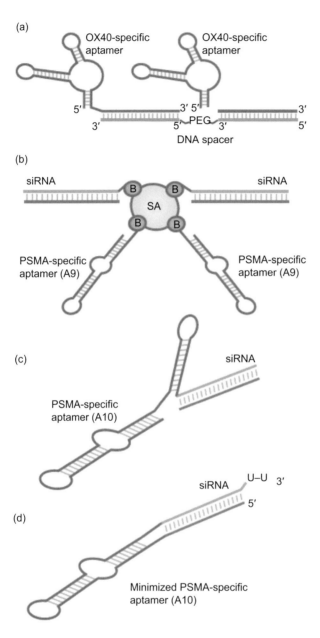

Figure 6.6 "Aptamer architectures for therapy. (a) OX40-specific aptamers were hybridized to an 'organizer' consisting of two antisense oligonucleotides separated by a polyethylene glycol (PEG) spacer. (b) Prostate-specific membrane antigen (PSMA)-specific aptamer A9 and 27-mer small interfering RNAs (siRNAs) were biotinylated and conjugated to streptavidin (SA)124. (c) PSMA-specific aptamer A10 was extended with a sequence that promoted hybridization to the guide strand of a siRNA125. The processed extension of the aptamer can then participate in gene silencing. (d) PSMA-specific aptamer A10 was extended with a short hairpin RNA-like sequence. Processing again leads to gene silencing."

Source: Keefe et al. (2010, 546)

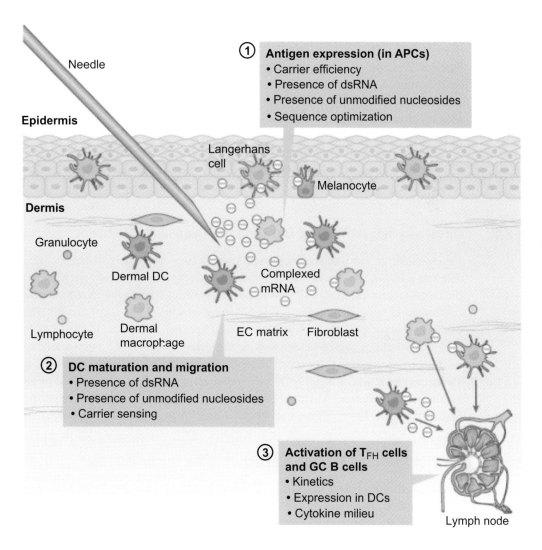

Figure 6.7 "Some considerations for effectiveness of a directly injected mRNA vaccine. For an injected mRNA vaccine, major considerations for effectiveness include the following: the level of antigen expression in professional antigen-presenting cells (APCs), which is influenced by the efficiency of the carrier, by the presence of pathogen-associated molecular patterns (PAMPs) in the form of double-stranded RNA (dsRNA) or unmodified nucleosides and by the level of optimization of the RNA sequence (codon usage, G:C content, 5′ and 3′ untranslated regions (UTRs) and so on); dendritic cell (DC) maturation and migration to secondary lymphoid tissue, which is increased by PAMPs; and the ability of the vaccine to activate robust T follicular helper (TFH) cell and germinal center (GC) B cell responses – an area that remains poorly understood. An intradermal injection is shown as an example. EC=extracellular."

Source: Pardi et al. (2018, 273)

6.8   Testing the Nanoparticle–Therapeutic Molecule Complex for Efficacy

Just delivering the nanoparticle–therapeutic complex to the correct cell does not guarantee usefulness unless the delivered complex has therapeutic efficacy. The therapeutic drug must act on the cell to either eliminate it or change it from diseased to normal, or at least more normal. Tests for efficacy at the single-cell level are still in their infancy. Ultimately, it comes down to changing the gene expression profile of the diseased cell back to a more normal phenotype that is stable. This may not be as complex as it sounds. It may be as simple as changing the output of a few genes so that the cell wants to walk its way back to a more normal phenotype using the rest of its expressed genes.

6.8.1   Direct and Indirect Ways of Detecting the Therapeutic Molecules

First, one should test whether the therapeutic molecules not only reached the correct diseased cells but also that they have the desired effect on the diseased cell. First, in a direct test, one would want to test to see whether the drug is inside the correct cell, and in some cases targeted to the correct part (e.g., nucleus or cytoplasm) of the cell. This can be done most easily by finding a way to attach a small fluorescent molecule to the drug so its location can be determined by fluorescence confocal or super-resolution microscopy. Second, indirect tests can monitor other molecules that are induced to change by the action of the drug. Finally, in the most general indirect method of all, without any specific molecular tests, one can see whether the cell is behaving more normally and has lost or diminished whatever trait is causing it to be "diseased."

6.8.2   Ways of Assessing Therapeutic Efficacy at the Single-Cell Level

As discussed in more detail in Chapter 10, therapeutic efficacy needs to be assessed at the single-cell level. For example, one might detect a phosphorylated protein product that signifies that the protein is functional. If that is not possible, then looking at the gene expression profile (GEP) of the treated cell resembles cells with more normal phenotypes.

6.8.3   Is the Nanoparticle Still Attached to the Therapeutic Molecule? Is That Important?

The therapeutic molecule may or may not still be attached to the nanoparticle. While a nice thing to know, as long as the desired therapeutic effect is reached it is probably not important.

6.9   Nanomedical Pharmacodynamics: The Great Unknown

Virtually all of the field of pharmacodynamics have been studies for molecules, not nanoparticles. Nanoparticles are inherently different due to their much larger size and other factors. There have been relatively few studies of the pharmacodynamics of nanoparticles because they are technically challenging. Scientists tend to shy away from, or do sometimes convoluted work-arounds for, difficult research problems unless they see that the effort will be worth the expense of time and money. My students used to complain that the problems I gave them were problems everyone else wanted to see solved, but by someone else!

6.9.1   Little Is Known about Complex Nanoparticle Pharmacodynamics

At this time, very little is known about nanoparticle pharmacodynamics and the distributions of nanoparticles within tissues in vivo (Meng et al., 2018). Much more work needs to be done in this important area. Unfortunately, the experimental challenges tend to drive many researchers away from this topic.

6.9.2   Obtaining Quantitative Biodistribution Data In Vivo Is Extremely Difficult!

Still relatively few in number are studies comparing the biodistribution of free drugs and drugs encapsulated in nanoparticles in terms of nanoparticle size and shape and their effect on enhanced permeability and retention localization to tumors with leaky vasculature, as well as circulation time, nanoparticle charge properties, the role of stealth molecules (e.g., polyethylene glycol) coating these nanoparticles, and the administration route by which the nanoparticles were administered (Kang et al., 2015; Li and Huang, 2008; Sridhar et al., 2018). Much more work needs to be done in this important area!

6.10   The Importance of "Green Chemistry"

Throughout the design process, it would be wise to employ "green chemistry" (Anastas and Warner, 2000) whenever possible. Not only is that approach better for a total life-cycle analysis of bionanomanufacturing, but also in the absence of those concerns it is still a wise strategy for avoiding the addition of toxic chemicals that are difficult or impossible to remove in later stages of the overall process. Throughout this book, I have tried to emphasize green chemistry syntheses because I think it is critically important to avoid the addition of toxic materials both to nanomedicine for human use and also for the environment. Early in my career, I thought that many scientists indiscriminately overused radioactive probes rather than using fluorescent or chemiluminescent detection technology alternatives. Their use of radioactive probes created huge amounts of low-level radioactive medical waste. At the time, we had no space left for low-level radioactive waste in western New York State. All of our low-level medical radioactive waste was shipped to be buried in west Texas. This is like throwing your garbage in someone else's yard – not a very neighborly thing to do! A little effort with fluorescent and chemiluminescent probe technologies not only avoided these unnecessary and harmful environmental effects of our research efforts, but also showed they were ultimately just as sensitive and much cheaper! Early on in my very long research career, I converted and committed my entire lab and its many diverse research efforts to this nonradioactive approach.

This topic will be discussed again in Chapter 11 with regard to PET probe toxicity in people and Chapter 14 with respect to their disposal in the environment in the sections on total life-cycle analyses, whereby we try to do responsible chemistry that is not going to result in the disposal of toxic materials or by-products.

6.10.1   What Exactly Is "Green Chemistry"?

Green chemistry is the commonsense design of chemical products and processes that try to reduce or eliminate the use and generation of hazardous substances. It makes the product environmentally inherently safer as well as generating less waste in its synthesis. But, in addition, it makes the products safer if administered to humans in terms of reducing or eliminating things like organic solvents that could prove cytotoxic even at a few parts per million. Green chemistry particularly for nanoparticle synthesis has been described in an excellent review and tutorial (Duan, Wang, and Li, 2015). A major difference is that green chemistry finds ways (e.g., soft aqueous chemistry) to perform appropriate nanomedical synthesis chemistry in water rather than in organic solvents. This means that the nanomedical systems need to have hydrophilic components throughout the synthesis process so that the overall device can be water soluble.

6.10.2   Examples of Green Chemistry in the Synthesis of Nanoparticles and Bioconjugations

Sometimes the nanoparticle itself can have therapeutic value. While cisplatin has been a treatment for ovarian cancer for many years, researchers have shown that palladium nanoparticles themselves can have similar anti-ovarian cancer efficacy without affecting normal cells. They used green chemistry to produce palladium nanoparticles without toxic synthesis intermediates and tested these on human ovarian cancer cell line A2780 (Gurunathan et al., 2015).

Lastly, another example involves the use of green chemistry techniques to produce nanoparticles with attached antimicrobial drugs (Malik et al., 2014). They discuss the use of nanoparticles of a number of different compositions for antimicrobial applications.

Chapter 6 Study Questions

6.1   Why are nanomaterial cores frequently first treated with hydrophilic surface coatings?

6.2   Why is it important for these surface coatings to be biocompatible? What factors are important for biocompatibility?

6.3   Two attachment strategies are covalent and noncovalent binding. What are the advantages and disadvantages of each strategy?

6.4   What is soft aqueous chemistry? Why is it generally better to form chemical bonds using soft aqueous chemistry rather than by using organic solvents?

6.5   What is steric hindrance, and why is it important to always keep in mind?

6.6   What are molecular spacer arms, and why are they sometimes important to use?

6.7   Why are gold nanoparticles considered among the simplest to use in terms of surface chemistry?

6.8   Why are bifunctional surface chemistries useful?

6.9   Why is it important to pay attention to the zeta potential for each surface layer added?

6.10   Why is biotin frequently used in surface chemistry designs? What are some of the problems with using biotin?

6.11   What are the advantages and disadvantages of attaching these targeting molecules: (a) antibodies, b) peptides, (c) aptamers, and (d) small-molecule ligands (e.g., folate)?

6.12   What is green chemistry, and why is it important for nanomedicine?

7   Characterizing Nanoparticles

This is the only chapter of the book devoted to the characterizations and measurements of the nanoparticles themselves, although it does include material related to the interaction forces between nanoparticles and cells. Zeta potential is arguably the *most* important factor in determining whether nanoparticles will agglomerate in clusters. If you observe agglomeration happening to your nanoparticles (a *very* common problem), it is likely to be a problem with their zeta potential. The focus of this chapter is on the importance of the zeta potential, which governs the fundamental electrostatic interactions of nanoparticles with each other as well as nanoparticle interactions with cells in an aqueous environment. Zeta potential is perhaps the single most important design consideration of nanomedical systems. It the zeta potential is wrong, the entire nanomedical system is unlikely to work properly!

7.1   Size Characterizations of Nanoparticles

While the size of nanoparticles may seem like an obvious thing, that is not the case. There are many different ways to measure the size of nanoparticles, and they can give different values depending on the measuring technology used.

7.1.1   "Size" Depends on How You Measure It!

Size matters in the performance of nanomedical systems, but size depends on how you measure it. Some measurements (e.g., electron microscopy) are sensitive only to electron density, so the biomolecules in layers on top of an electron-dense core may be invisible to this form of measurement. Measurements of electrical charge (e.g., by zeta potential) measure not only the nanomedical system but also the ordered solution ("counter-ions"), which can vary depending on the fluid in which the nanomedical system resides. An excellent overview and comparison of nanoparticle sizing methods was recently published (Eaton et al., 2017).

7.1.2   Shape May Also Be Important

Shape can also be important to the performance of a nanomedical system. Shape can affect interactions of nanomedical systems with cells as well as cellular uptake. Shape can impact electrostatic interactions between nanoparticles or interactions of nanoparticles with respect to cells. The exact mechanisms for differences in electrostatic interactions between nanoparticles or between nanoparticles and cells are not well understood, but they may involve the effects of radius of curvature at the moment of interaction. An excellent review of a wide variety of techniques for measuring size and shape of nanoparticles is available (Mourdikoudis, Pallares, and Thanh, 2018).

7.1.3   Measures of Nanoparticle Size

There are a number of different technologies used to "size" nanoparticles, and those different technologies will give different measures of size because they are sensitive to different things present, or absent, on the nanoparticles. Near the top of the list for rapidly sizing nanoparticles in suspension is dynamic light scattering (DLS).

7.1.3.1   DLS versus Individual Nanoparticle Measurements

Dynamic light scattering is a measurement dependent on the aggregate of many nanoparticles. There are a number of good recent reviews of DLS techniques (Bhattacharjee, 2016; Leong et al., 2018; Stetefeld, McKenna, and Patel, 2016). DLS is theoretically complicated and is a form of photon correlation spectroscopy (PCS). As such, the DLS size distributions represent average measurements rather than the true individual nanoparticle size distributions as measured by nanoparticle tracking analysis (NTA), which measures the distributions of individual nanoparticles. For this reason, DLS measurements should always be thought of as an inexact measurement performed on an ensemble of many nanoparticles in solution. On the contrary, individual nanoparticle size measurements represent the true distributions of sizes based on individual nanoparticle measurements (Figure 7.1).

Whereas there are many types of DLS instruments, in a typical dynamic light scattering instrumentation there is a single detector fixed at a single angle. The measured data in a DLS experiment is the intensity autocorrelation curve. Embodied within the autocorrelation curve is all of the information regarding the size distribution of the collection of particles in the solution, as opposed to single particle measurements.

Deconvolution of the intensity autocorrelation curve to an intensity distribution is an ill-defined problem. As such, there is an inherent degree of uncertainty in DLS-derived intensity size distributions. This inherent uncertainty translates into peak broadening in the width of the intensity distribution.

DLS measurements compute the hydrodynamic diameter from the Stokes–Einstein equation. The hydrodynamic radius is a measure of the volume associated with a nanoparticle in solution that is not only the nanoparticle but also the fluid that moves with that nanoparticle. The hydrodynamic radius will change in different media according to temperature, viscosity, and the diffusion coefficient (Figure 7.2).

Many different types of data can be obtained from DLS measurements. The distributions by number, relative volume, and DLS intensity can be very different (Figure 7.3).

Actual Nanoparticle Size Distributions Are Usually Better Than Their Measured DLS Distributions

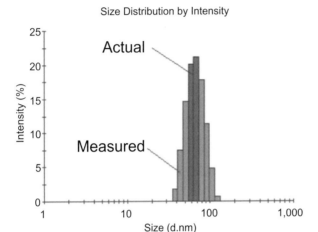

Size Distribution by Intensity

Figure 7.1 Direct comparisons of single-angle ensemble DLS measurements to particle counting NTA measurements which reflects the much higher resolution of individual NTA nanoparticle measurements compared to nanoparticle ensemble measurements by DLS as can be seen from the width of the two distributions overlaid.
Source: Image courtesy of Malvern Panalytical

STOKES-EINSTEIN EQUATION

$$d(H) = \frac{kT}{3\pi\eta D}$$

d(H) = hydrodynamic diameter
k   = Boltzmann's constant
T   = absolute temperature
η = viscosity
D   = diffusion coefficient

Figure 7.2 The hydrodynamic diameter of a nanoparticle can be calculated using the Stokes–Einstein equation.
Source: Image courtesy of Malvern Panalytical

For typical laboratory samples, this increase can amount to as much as 10–15 percent when the Mie theory (or any other volume vs. size algorithm) is applied to the DLS intensity distribution. The algorithm cannot distinguish this apparent increase in width from the true width. Actual nanoparticle size distributions based on individual nanoparticle measurements are usually much better than their measured DLS distributions, as shown in Figure 7.1.

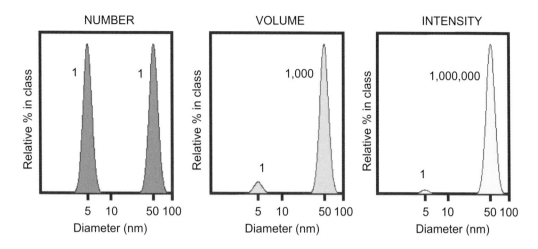

Figure 7.3 How one displays the "size" of nanoparticles can present a very different view of the actual situation.
Source: www.malvern.com

7.1.3.2   Fluorescence Microscopy

Fluorescence microscopy is particularly useful and important when measuring the cell surface, intracellular targeting, and interactions between fluorescent nanomedical systems and their targeted cells. Multicolor fluorescence techniques can help distinguish nanoparticles and cells and are an essential part of studying the multistep targeting process. That said, it should be kept in mind that diffraction limits preclude the resolving of individual fluorescent nanoparticles. Most of what you will see by fluorescence microscopy are aggregates of fluorescent nanoparticles rather than individual ones. A general schematic of a fluorescence microscope is shown in Figure 7.4.

7.1.3.3   TEM and SEM

Transmission electron microscopy (TEM) is a measurement of the transmission of much shorter than optical wavelengths of light passing through the object. Scanning emission microscopy (SEM) measurements are made by reflectance off the electron-dense portions of these nanostructures. Most nanoparticles themselves are electron dense, but if the nanoparticles are not inherently electron dense (e.g., lipid or organic nanoparticles), then electron-dense materials must be added to permit TEM or SEM visualizations (Figure 7.5).

7.1.3.4   AFM

Atomic force microscopy (AFM) uses micro-cantilever technology to "feel" the nanoparticle through electrostatics. The nanoparticle or cell need not be electron dense, but if it is not, then care must be taken not to damage the nanoparticle during the AFM measurement process (Figure 7.6).

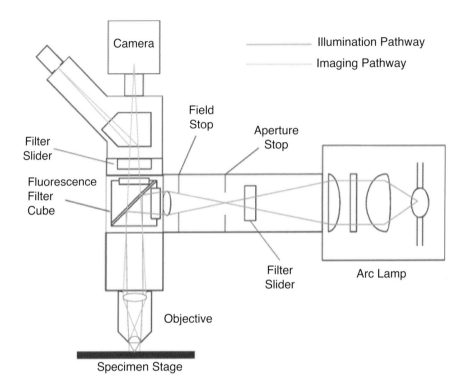

Figure 7.4 Today's most common fluorescence setup is the epifluorescence microscope, named because of the way it uses epi-illumination light to excite the sample. With this configuration, fluorescence filters need only filter out excitation light scattering back from the sample or reflecting off glass surfaces, as opposed to direct excitation light. The use of high-quality oil-immersion objectives (made with materials that have minimal autofluorescence and using low fluorescence oil) eliminates some of these surface reflections, which can reduce the level of backscattered light to as little as 1/100 of the incident light. In addition, the dichroic beam-splitter, which reflects the excitation light into the objective, filters out the backscattered excitation light by another factor of 10–500. An epifluorescence microscope using oil immersion, but without any filters other than a good dichroic beam-splitter, can reduce the amount of observable excitation light relative to observed fluorescence to levels ranging from 1 (for very bright fluorescence) to 10^5 or 10^6 (for very weak fluorescence). If one wants to achieve a background of, say, one-tenth of the fluorescence image, then additional filters in the system are needed to reduce the observed excitation light by 10^6 or 10^7 (for weakly fluorescing specimens) and still transmit almost all of the available fluorescence signal. Fortunately, there are filter technologies available that are able to meet these stringent requirements.
Source: Image courtesy of Chroma Technology Corp

7.1.3.5   Hyperspectral Reflectance Imaging

Hyperspectral reflectance imaging allows a full spectral reflectance emission pattern to be recorded from the nanoparticle. As such, it is sensitive to the many subcomponents composing the nanoparticle or cell and increases contrast between nanoparticles of different compositions and cells (Lu and Fei, 2014) (Figure 7.7).

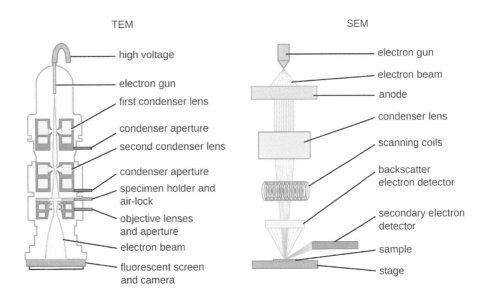

TEM   SEM

- high voltage   - electron gun
- electron gun   - electron beam
- first condenser lens   - anode
- condenser aperture   - condenser lens
- second condenser lens   - scanning coils
- condenser aperture   - backscatter electron detector
- specimen holder and air-lock   - secondary electron detector
- objective lenses and aperture   - sample
- electron beam   - stage
- fluorescent screen and camera

Figure 7.5 In TEM, the electrons pass through the object, whereas in SEM, the electrons scatter off electron-dense surfaces. For biological samples the cells must be precoated with electron-dense contrast agents, either inside or on the surface of the cell. Otherwise, they are effectively invisible on an electron microscope.
Source: https://i.pinimg.com/736x/4a/a5/0a/4aa50a35bcd58df4e14e03aa571f85aa.jpg

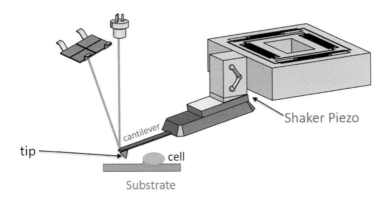

Shaker Piezo

cantilever

tip

cell

Substrate

Figure 7.6 The specimen is "tapped" to sense its presence. Since the movements are on a tiny scale, they are amplified by bouncing a laser beam off a cantilever into a position-sensitive detector.
Source: www.nanosurf.com/images/support/afm-theory-modes-operating-principle-atomic-force-microcope.png

Once the images captured at different wavelengths are acquired, a hypercube of stacked 2D images is constructed (Lu and Fei, 2014), as shown in Figure 7.8.

By taking images at different wavelengths, some of these image wavelengths will better contrast the objects in the overall scene. Hyperspectral imaging has been used for a diverse range of applications, from analysis of Landsat photos of crops on Earth

Hyperspectral Imaging

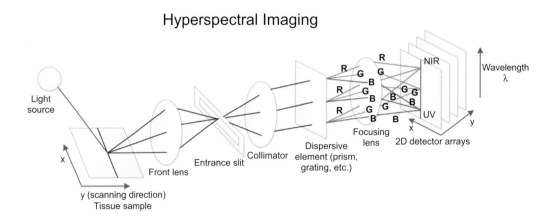

Figure 7.7 Schematic diagram of a "pushbroom" (line-scanning) hyperspectral imaging system. R, G, and B in this grayscale image correspond to the colors red, green, and blue.
Source: Lu and Fei (2014)

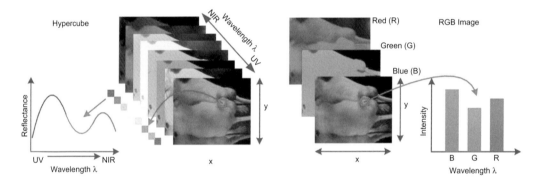

Figure 7.8 Comparison between hypercube and RGB image. The hypercube is a three-dimensional data set of a two-dimensional image on each wavelength. The lower left is the reflectance curve (spectral signature) of a pixel in the image. The RGB color image only has three image bands on red, green, and blue wavelengths, respectively. The lower right is the intensity curve of a pixel in the RGB image.
Source: Lu and Fei (2014)

from space to analysis of tissues for biomedical applications, and even to the ability to see almost invisible writing on the Dead Sea Scrolls by one of the original inventors (Dr. Greg Bearman) of hyperspectral imaging!

7.1.3.6   Super-Resolution Confocal Microscopy

While the diffraction limits of physics cannot be defied, they can be tricked using super-resolution microscopy techniques whereby the Brownian motion of the nano-particle in solution is imaged and a composite view is constructed that allows resolution below the normal optical limits. The reason super-resolution microscopy works is because Brownian motion is a function of the object size. Then all we need to

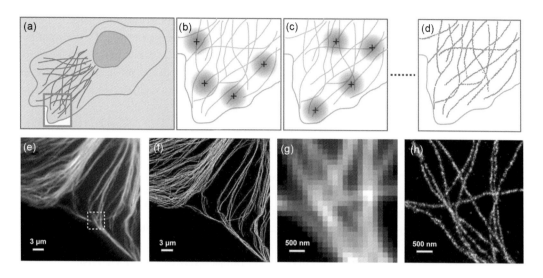

Figure 7.9 "Super-resolution imaging by high-precision localization of photo-switchable fluorophores. (a–d) The imaging concept. (a) Schematic of a cell in which the structure[s] of interest (grey filaments in this case) are labeled with photo-switchable fluorophores (not shown). All fluorophores are initially in the non-fluorescent state. The box in E indicates the area shown in panels b–d. (b) An activation cycle: a sparse set of fluorophores are activated to the fluorescent state, such that their images (large gray circles in b and c) do not overlap. The image of each fluorophore appears as a diffraction-broadened spot, and the position of each activated fluorophore is determined by fitting to find the centroid of the spot (black crosses). (c) A subsequent activation cycle: a different set of fluorophores are activated and their positions are determined as before. (d) After a sufficient number of fluorophores have been localized, a high-resolution image is constructed by plotting the measured positions of the fluorophores. The resolution of this image is not limited by diffraction, but by the accuracy of each fluorophore localization and by the number of fluorophore positions obtained. (e–h) Comparison of conventional immunofluorescence images of microtubules in a BSC1 cell (e and g) and STORM images (f and h) of the same areas. In the STORM images, each localization is rendered as a Gaussian peak whose width corresponds to the theoretical localization accuracy. The areas shown in (g) and (h) are expanded views of the region defined by the dashed box in (e). The microtubules are immuno-labeled with photo-switchable Cy3 – Alexa 647 dye pairs."
Source: Bates et al. (2008, 12)

do is to deconvolve the size information hidden in the Brownian motion of those objects (Bates, Huang, and Zhuang, 2008) (Figure 7.9).

Unlike most other types of images, the super-resolution image is mathematically calculated, which is how it "escapes" the diffraction limit. It is still just science – not magic!

7.2   The Importance of the Zeta Potential

Zeta potential is one of the most important measures of a nanomedical system because it describes the fundamental interactions of those nanoparticles with each other and with the cells to which they are targeted. Since these interactions take place in fluids, the zeta potential describes not only the charge of the nanoparticle itself but also of the counter-ions that remain associated with the nanoparticle as it moves through the fluid. There have been several excellent recent reviews of both the theory and practice of zeta potential measurements of nanoparticles (Bhattacharjee, 2016; Honary and Zahir, 2013; Liao, Wu, and Liao, 2009).

7.2.1   Nanoparticle–Cell Interactions

The fundamental interactions between nanoparticles and cells are governed by their mutual zeta potentials. If the zeta potential is unfavorable, and typically of the opposite charge, then the nanoparticles will nonspecifically stick to cells. Cells are usually negatively charged due to the presence of sialic residues and other negatively charged molecules on the cell surface. In most cases the nanomedical systems should have a negative charge and zeta potential. But as layers of a multilayered nanomedical system are stripped off during multistep targeting and the drug delivery process, the zeta potential of the remaining parts of the nanomedical system will change. This fact means that you must try to anticipate the microenvironments that your nanomedical system will encounter and which layers are on or off at the time. While that sounds like a daunting design task, you do not need to get everything perfect. It just has to be good enough to avoid particularly bad regimes of zeta potential. If the zeta potential charge difference presents too much of a barrier, then nanoparticles may fail to come close enough to the cells to bind to receptors (Campbell and Reece, 2002; Prow et al., 2005) (Figure 7.10).

7.2.2   Zeta Potential Is Part of the Initial Nanomedical System–Cell Targeting Process

Before any targeted receptor binding can occur, the nanoparticle and the cell must overcome electrical repulsion potentials due to their zeta potentials. It is usually wise to have a very slight repulsion so that the nanoparticles will not nonspecifically bind to the cells but will be pulled across that potential barrier by the attractive forces of the receptor ligand binding process. On the other hand, highly negatively charged nanomedical systems may never be able to traverse the potential barrier to the receptor. In general, anything less than about 20 mV zeta potential will tend to bind to its cell receptor. If the nanoparticle is binding too easily, and nonspecifically, you should raise the negative charge slightly to reduce nonspecific binding.

7.2.3   Nanoparticle–Nanoparticle Interactions

Nanoparticle–nanoparticle interactions are governed by their zeta potential as well as any magnetic interactions. It is especially important to characterize these forces in order to prevent agglomeration of nanoparticles in solution. If such agglomeration occurs within the human body, it can cause embolisms, which can be a very serious

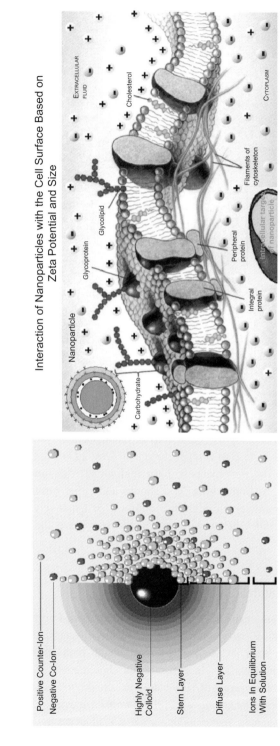

Interaction of Nanoparticles with the Cell Surface Based on Zeta Potential and Size

Figure 7.10 The interaction of charged bodies is a function of their zeta potentials. In an aqueous environment, cells and nanoparticles not only have their own intrinsic charge, but also organize the ions in the surrounding fluid. This zeta potential is a function of many variables, including pH and ionic strength. The zeta potential is important for the stability of the nanoparticles in the various solutions encountered in vivo, and it is also important for determining how the nanoparticle will react electrically with the surface of the cell.

Source: Prow et al. (2005). Adapted from Campbell and Reece (2002)

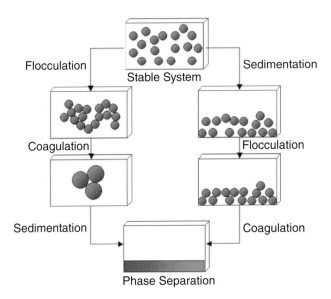

Figure 7.11 Nanoparticles (and cells) can agglomerate in a variety of ways.
Source: Image courtesy of Malvern Panalytical

health hazard. Agglomeration of nonmagnetic nanoparticles is usually due to zeta potentials very close to zero. For most nanoparticle–cell interactions, the zeta potentials should be –10 to –15 mV. The terms *flocculation* and *coagulation* have often been used interchangeably. However, coagulation is irreversible, whereas flocculation can be reversed by the process of deflocculation, as shown in Figure 7.11.

Agglomeration between any objects – cells or nanoparticles – occurs when their zeta potential comes too close to zero. If there is at least 12–15 mV zeta potential, either positive or negative, the objects will not agglomerate due to zeta potential. If your nanoparticles are agglomerating, you should add either positive or negative charges to the nanoparticles through the addition of positively or negatively charged ions or molecules. That will, in most cases, solve the agglomeration problems.

7.2.4   Agglomeration of Magnetic Nanoparticles

If you are using magnetic nanoparticles, agglomeration can and probably will occur if those nanoparticles have cores containing permanently magnetized (i.e., paramagnetic) materials, rather than just zeta potential, although both phenomena may be responsible for the agglomeration. In general, you should *never* use paramagnetic nanomaterials due to this problem. Instead, you should use superparamagnetic nanoparticles that are only magnetic while they are in a magnetic field. However, even some superparamagnetic materials retain some magnetism in the absence of a magnetic field. If this is happening, you should change to a different superparamagnetic material that sheds its magnetism more quickly before agglomeration can occur.

7.2.5 Low Zeta Potential Leads to Low Serum Protein Binding and Longer Circulation

In order to prevent opsonification (i.e., binding of serum proteins to the nanoparticles, leading to their elimination by the immune system), it is wise to adjust the zeta potential to be in the zone where it avoids much uptake of blood protein. If the nanoparticles do not bind serum proteins, then they will circulate much longer in the bloodstream, leading to increased probability of finding the targeted cells. Longer circulation time, a major design goal, allows for much lower drug doses to be given to the patient, thereby also lowering side effects.

7.3 Zeta Potential Basics

The zeta potential is the electric potential at the shear plane, which is the boundary between the Stern layer and the diffuse layer. This electrical property determines whether a colloidal solution will agglomerate or remain a stable suspension. Zeta potential must be determined and controlled at each step in the construction of nanoparticles. The zeta potential of nanoparticles is important for controlling in vivo interactions between nanoparticles and cells.

7.3.1 What Is the Zeta Potential?

Zeta potential describes the electrostatic interactions of cells and particles in a fluid environment. The liquid layer surrounding the particle exists as two parts: an inner region (Stern layer) where the ions are strongly bound and an outer (diffuse) region where they are less firmly associated. Within the diffuse layer, there is a natural boundary inside which the ions and particles form a stable entity. When a particle moves (e.g., due to gravity or an electric field), charged ions within the boundary fluid (i.e., counter-ions) move with it. Those ions beyond the boundary stay with the bulk dispersant. The overall potential (importantly including all internal layers!) at this boundary (i.e., the surface of the hydrodynamic shear) is the zeta potential (Figure 7.12).

Important to the field of nanomedicine, zeta potential represents the potential barrier to cell–nanoparticle interactions. The relationship between zeta potential and surface potential depends on the level of ions in the solution. The net interaction curve between nanoparticles and cells is formed by subtracting the attraction curve from the repulsion curve, as shown in Figure 7.13.

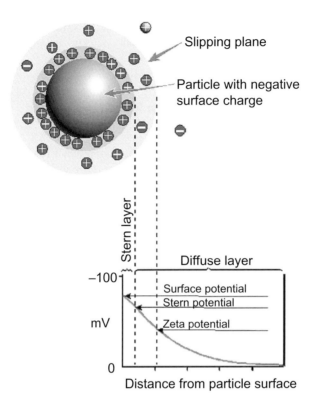

Figure 7.12 The liquid layer surrounding the particle exists as two parts: an inner region (Stern layer) where the ions are strongly bound and an outer (diffuse) region where they are less firmly associated. Within the diffuse layer, there is a notional boundary inside which the ions and particles form a stable entity. When a particle moves (e.g., due to gravity), ions within the boundary move it. Those ions beyond the boundary stay with the bulk dispersant. The potential at this boundary (surface of the hydrodynamic shear) is the zeta potential.
Source: Image courtesy of Malvern Panalytical

7.3.2   How Is Zeta Potential Measured?

While zeta potential can be measured in different ways, the most common way is measuring by electrophoresis. If an electric field is applied across a sample containing charged cells and/or particles, those cells and particles are attracted toward the electrode of opposite charge, as shown in Figure 7.14. In addition to measuring the zeta potential of your nanomedical system, you should also try to determine the approximate zeta potential of your target cells. Different cell types have different zeta potentials. Cells and particles will move in this electric field with a velocity dependent on electric field strength, dielectric constant of the medium, viscosity of the medium, and zeta potential (Figure 7.14).

By measuring the velocity of a nanoparticle in an electric field, its zeta potential can be calculated. This is how most zetasizers work. The velocity of a particle in a unit electric field is referred to as its electrophoretic mobility. Zeta potential is related to the electrophoretic mobility by the Henry equation (Figure 7.15).

Zeta Potential Represents the Potential Barrier to Cell–Nanoparticle Interactions

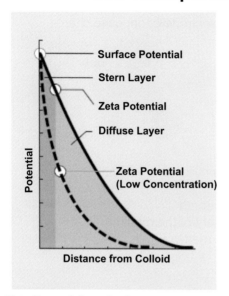

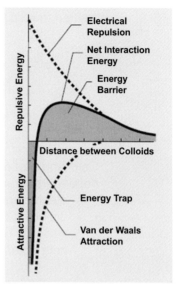

Zeta Potential vs. Surface Potential: The relationship between zeta potential and surface potential depends on the level of ions in the solution.

Interaction: The net interaction curve is formed by subtracting the attraction curve from the repulsion curve.

http://www.malvern.com

Figure 7.13 Schematic showing the basic parameters of the zeta potential. Source: Image courtesy of Malvern Panalytical

We make a number of assumptions about slip layer diameter when calculating Henry's function for the zeta potential. Figure 7.16 illustrates the Huckel approximation and the Smoluchowski approximation used for the conversion of electrophoretic mobility into zeta potential.

7.4 Zetasizers

Many commercial zetasizers use all of the preceding variables to measure both size and zeta potential using electrophoretic light scattering, a combination of electrophoresis and DLS. One such zetasizer is shown in Figure 7.17.

7.5 Some Factors Affecting the Zeta Potential

Zeta potential is the electrical potential at the hydrodynamic plane of shear. Zeta potential depends not only on the particle's surface properties (and including all inner

Measuring Zeta Potential by Electrophoresis

If an electric field is applied across a sample containing charged cells and/or particles, those cells and particles are attracted toward the electrode of opposite charge.

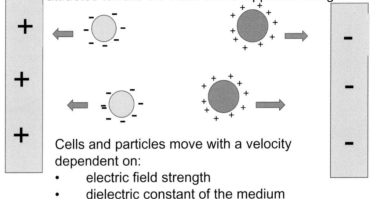

Cells and particles move with a velocity dependent on:

- electric field strength
- dielectric constant of the medium
- viscosity of the medium
- zeta potential

Figure 7.14 Zeta potential can be measured by measuring the velocities of nanoparticles (or cells) in an electric field in a process known as electrophoresis.
Source: Leary teaching

By measuring the velocity of a nanoparticle in an electric field, its zeta potential can be calculated

The velocity of a particle in a unit electric field is referred to as its electrophoretic mobility. Zeta potential is related to the electrophoretic mobility by the Henry equation:

$$U_E = \frac{2\,\varepsilon\,z\,f(\kappa a)}{3\eta}$$

where U_E = electrophoretic mobility, z = zeta potential, ε = dielectric constant, η = viscosity, and $f(\kappa a)$ = Henry's function.

Figure 7.15 Most zetasizers measure the zeta potential of nanoparticles or cells by measuring their velocities in an electric field by electrophoresis.
Source: Image courtesy of Malvern Panalytical

layers!) but also the nature of the solution (e.g., ionic strength, pH). The solution surrounding the nanoparticles becomes part of the zeta potential, so the zeta potential of your nanoparticles will depend on the medium in which they are suspended.

For this reason, zeta potential may be quite different from the particle's surface potential. Small changes in ionic strength and pH of the surrounding fluid can lead to

Assumptions about Slip Layer Diameter When Calculating Henry's Function for the Zeta Potential

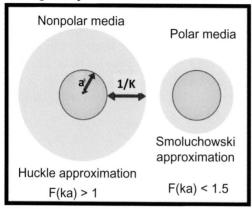

Figure 7.16 The slip layer boundary is where the cell or nanoparticle and the waters of hydration ions that move with it meet.
Source: Image courtesy of Malvern Panalytical

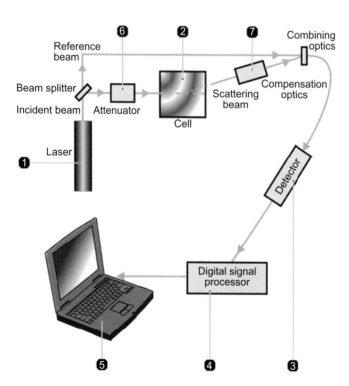

Figure 7.17 Schematic of a zetasizer, the "Zetasizer Nano."
Source: Image courtesy of Malvern Panalytical

large effects in zeta potential. Zeta potential can be used to predict the monodispersity (or agglomeration) of particles. High zeta potential (either positive or negative; >30 mV) can lead to monodispersity. Low zeta potential (<5 mV) can lead to agglomeration. Monodispersity is the quantitative degree to which the objects are singlets as opposed to doublets, triplets, and clumps. The local pH and ionic strength can vary greatly in the different parts of the human body. These factors also change within different regions *inside* human cells. It is a challenge to design nanoparticles that have the optimal zeta potentials throughout the travel time of these nanomedical systems to reach their final destinations.

7.5.1   pH

The relationship between zeta potential and pH is complex, but important. Not only does it govern the interaction between nanoparticles and cells but also the stability of those nanoparticles. There are two stable zones, as shown in Figure 7.18. The zeta potential can be either positive or negative in terms of nanoparticle stability and monodispersity. But at the isoelectric point, where the overall charge is close to zero, the nanoparticle becomes unstable and this can lead to agglomeration of many nanoparticles. This agglomeration is not only undesirable from a design point of view, but also dangerous to the patient if it happens in vivo because it can lead to possibly dangerous embolisms (Figure 7.18).

A phenomenon that most people are unaware of is the effect of simple dilution of the cell/nanoparticle solution with a solution of different ionic strength or even distilled water. Figure 7.19 shows the dramatic effects of both pH and dilution on nanoparticle zeta potential. This is important because once nanoparticles are injected into the human body, they will experience potentially dramatic shifts in zeta potential

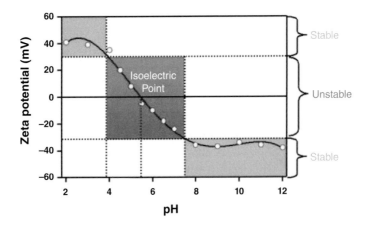

Figure 7.18 Typical plot of zeta potential versus pH showing the position of the isoelectric point and the pH values where the dispersion would be expected to be stable.
Source: Image courtesy of Malvern Panalytical

Effects of pH and Dilution on Nanoparticle Zeta Potential

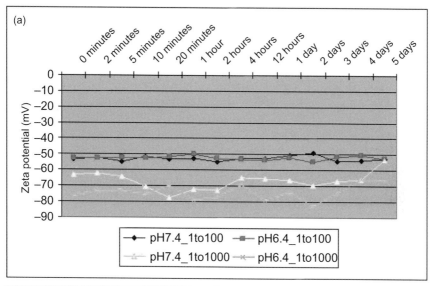

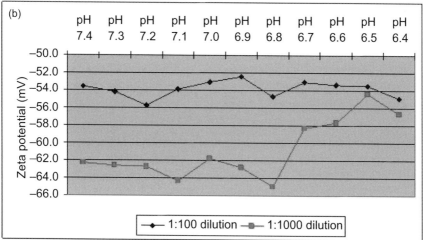

Figure 7.19 Zeta potential testing of SnowTex® (Nissan Chemical Corporation) particles. SnowTex® 40–50 nm silica nanoparticles were diluted to two different concentrations using distilled/deionized water, and utilizing hydrochloric acid (HCl) the solutions were adjusted to two different pH concentrations. Panel a shows that these tested solutions are stable for at least five days when stored at room temperature, which is important for the future storage of the complete nanoparticle solutions. Panel b shows zeta potential testing of SnowTex® particles. SnowTex® 40–50 nm silica nanoparticles were diluted to two different concentrations using distilled/deionized water, and utilizing HCl the solutions were adjusted to different pH concentrations. This figure shows that the pH of the solution and the dilution of the nanoparticles in the solution are important for controlling the zeta potential of the nanoparticle solution.

Source: Prow et al. (2005)

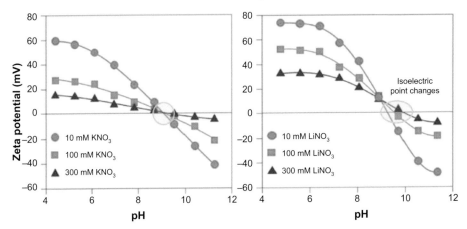

Figure 7.20 Ionic strength of the solution in which the cell or nanoparticle is suspended greatly affects its zeta potential. Nonspecific ion adsorption may or may not have an effect on the isoelectric point. Specific ion adsorption usually leads to a change in the isoelectric point. If the solution containing the cells or nanoparticles is diluted, the zeta potential will change depending on the ionic strength of the diluent.

Source: Image courtesy of Malvern Panalytical

due to the differing pH and ionic strength microenvironments. For this reason, it is important to do in vitro or ex vivo zeta potential measurements to try to simulate the different microenvironments the nanoparticles will experience during a multi step targeting process. The stability of nanoparticles over time may also change, as shown in Figure 7.19.

7.5.2   Ionic Strength

The effect of solution ionic strength and/or conductivity on zeta potential is also important. In general, for in vivo use nanomedical systems should anticipate a major environment of approximately physiological saline, but microenvironments in the human body can vary greatly in ionic strength and pH. Nonspecific ion adsorption may or may not have an effect on the isoelectric point. Specific ion adsorption usually leads to a change in the isoelectric point, as shown in Figure 7.20.

7.5.3   Optimal Zeta Potentials for Nanomedical Systems

What is the best zeta potential to have for nanomedical systems? That is not a simple question, but in general it is good to have a zeta potential of approximately −5 to −15 mV. Since most biological cells have zeta potentials in this range, you want your nanomedical systems to also have slightly negative zeta potentials so that they do not stick nonspecifically to cells but interact through a receptor-mediated interaction that allows binding of nanoparticles only when there is a receptor–ligand bond strong enough to overcome a modest electrical repulsion. But if all you want is to have nanoparticles stick to cells in tissue culture for transfection, the zeta potential can be positive. Just pay attention to the zeta potential of the tissue culture surfaces!

Chapter 7 Study Questions

7.1 What "size" measurement is being made by DLS?

7.2 How is that size determined by DLS different from what is measured by individual nanoparticle sizing?

7.3 What size is measured by TEM?

7.4 Where are the (a) surface, (b) Stern, and (c) zeta potential boundaries? Draw a picture of a nanoparticle and label these boundaries.

7.5 The cell surface charge is usually negative. The zeta potential of a cell in most physiological solvents is usually still negative. Should a nanoparticle be designed with a net negative or positive zeta potential? Why?

7.6 If a nanoparticle surface is highly charged, it will recruit a large number of oppositely charged counter-ions. If a nanoparticle is to avoid sticking non-specifically to a cell, what approximate regime of zeta potential would be best to prevent nonspecific binding and to ultimately be used for a cell-targeted nanoparticle device?

7.7 At what types of zeta potentials will nanoparticle aggregate (i.e., agglomerate) in solution?

7.8 Why would a nanoparticle's zeta potential change with pH and ionic strength? Why would this be important when designing a nanomedical device for tumor targeting?

7.9 What kinds of zeta potential values lead to monodispersity? Why is mono-dispersity important?

7.10 What happens if the zeta potential gets very close to zero (i.e., the isoelectric point)?

7.11 Why are there two different stable charge regions?

7.12 How is zeta potential measured? What are some of the important variables to keep in mind?

7.13 How is the zeta potential related to electrophoretic mobility and viscosity?

7.14 If you construct a multilayered nanomedical device out of charged polymer layers in a layer-by-layer self-assembly, what happens to its zeta potential during this process? Why?

7.15 What happens to the zeta potential of nanoparticles in distilled water if you dilute them with more distilled water? Why?

7.16 Why is DLS called a "computed measurement"? What are the important variables in this computed measurement?

7.17 How can we measure polydispersity or degree of agglomeration with DLS?

8 Nanomedicine Drug Dosing

8.1 Overview of Drug-Dosing Problem

There are many challenges related to proper drug dosing with nanodelivery systems. The first part of this discussion concerns how experts think about drug dosing with conventional drugs. In the second part, we need to consider the differences between nanodevices and traditional drug delivery, along with pharmacokinetics using nano drug delivery.

8.1.1 Are Animals Good Models for Humans in Terms of Nanomedical Applications?

Perhaps the first question we should ask is whether a particular animal is a good model for predicting human responses to drugs. Animals have, in some cases, vastly dissimilar organs and metabolism as well as clear genetic differences from humans. All of these factors affect drug metabolism, efficacy, and side reactions. They also have different body compositions in terms of fat and muscle – an issue critically important to consider even for human children as opposed to adults. Perhaps the best animal for modeling a human is the pig, followed by the dog. In both of those cases, the animals may or may not be eating food similar to that of humans. Dogs who eat cooked meat and other "family" foods are one of the only animals that contract stomach and bladder cancer. Dogs who do not eat cooked meat do not get these cancers. Nutrition and environment are complex and confounding factors in disease.

The extrapolation of animal data to humans is only a rough predictor because we have not only differences in body weights in area but also different organs and metabolism, such that some animals are good models for drug dosing in humans and many are not. This chapter will address genetic responses to drug dosing and how humans obviously are not all genetically or genomically equivalent.

8.1.2 Problems of Scaling Up Doses from Animal Systems

Scaling up doses from animal systems introduces problems because doses are based on body size as measured by a mysterious "area calculation" and sometimes the weight of the recipient. While at best a very rough estimate of dose, there is at least some logic to this idea. Nonetheless, scaling to humans from animal data is, at best, a gross approximation. At worst, it can lead to potentially harmful doses due to unexpected differences in human responses and particularly metabolic reactions to the drugs given to animals.

8.1.3   Basing Dosing on Size, Area, and Weight of Recipient

The mysterious "area calculation" is based on the assumption that since drugs cross two-dimensional tissue areas, rough estimates of a person's "area" provide some prediction about proper dosing. Dosing is also sometimes based on the weight of the recipient. There is some justification for this since weight is a measure of the total tissues and organs. This weight is also a measure of the total volume of blood, lymphatics, and interstitial tissue fluids, which dilute those drugs. One attempt to take size, weight, and body mass into account is the Dedrick plot, which has been performed on many different animals as well as humans to show the decline of concentration of drug as a function of time through metabolism and excretion (Wajima et al., 2004) (Figure 8.1).

Other simpler models showing just the rate of drug clearance with time reveal the general problem of drug residence in both animals and humans. Many drugs have a residence time of only minutes or, at best, a few hours before being cleared by the kidneys or liver, or simply degraded in the blood or tissue. This failure to protect the drug from being rapidly excreted or degraded leads to the patient being exposed to a much higher drug dose (typically 10 times or more) than should be necessary. This is a strong argument for use of "stealth factors" to protect the drug and increase its total circulation time. Polyethylene glycol (PEG) is commonly referred to as a time release and prevents proteins in the blood from sticking to the drug, a process called opsonification. Through opsonification, these proteins, coating anything perceived by the body as foreign, cause the immune system to attack the drug and remove it from blood via one or more mechanisms (Wajima et al., 2004) (Figure 8.2).

8.1.4   Vast Differences between Adults in Terms of Genetics and Metabolism

There are vast differences between adults due to their genetic differences and underlying health issues, in addition to their different body weights and sizes. People of similar size and weight can be vastly dissimilar in terms of their response to drugs. Genetic differences in metabolism, and also excretion rates, can produce large variations in responsiveness to drugs and drug dosages.

8.1.5   Dosing in Children: Children Are *Not* Smaller Adults!

Children are *not* just small adults (Baber and Pritchard, 2003). This fact cannot be overemphasized! In addition to differences in children besides weight and size, there are large adolescent hormonal differences. Interestingly, differences in their actual percentage of water and fat weight, which is much higher in young children than it is

Dedrick Plots to Predict Concentration–Time Profiles of Drugs in Humans from Animal Data

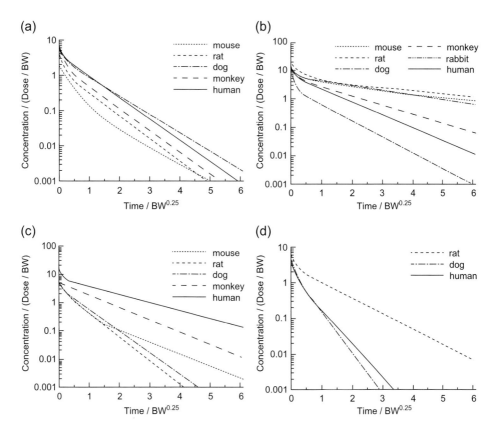

Figure 8.1 Dedrick plots, shown at two different concentration maximum scales (10 and 100), attempt to show the relationship between drug concentration and dose with individual animals or humans as a function of time and body weight.
Source: Wajima et al. (2004)

in adults, can have large effects on the behavior of water-soluble drugs and make a huge difference in terms of the actual delivered drug dose. Also, the ability of people to process drugs is quite different. Metabolism, excretion, and breakdown can be quite different from person to person.

Some facts and data are shown in Figure 8.3 that will help dispel a bit of the dangerous assumption that children are small adults. A premature baby is over 80 percent water by body weight, as opposed to an older child, who is about 54 percent water. That is a huge difference in terms of water-soluble drugs. This is a normal part of human development, but it also means water solubility goes down with time and stays remarkably constant throughout people's lives until late in life, when older citizens have a slight shrinkage of their body volume, largely through fluid loss. There is also a difference in total body protein levels, but again they are fairly constant

Comparing Drug Clearance in Humans and Other Animals

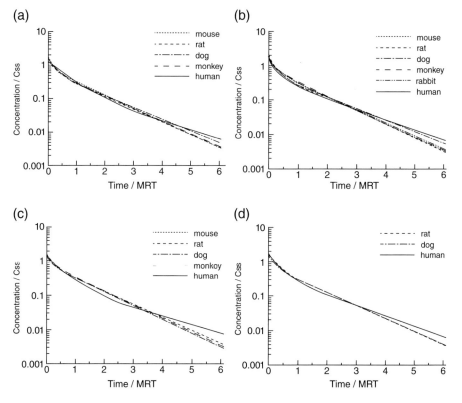

Figure 8.2 Experimental data from various drugs. The Dedrick plots for (a) ceftizoxime, (b) cefodizime, (c) cefotetan, and (d) cefmenoxime show the elimination of each drug from the blood as a function of time. MRT = mean residence time.
Source: Wajima et al. (2004)

until adolescence, when a relatively large increase in protein occurs as hormones kick in. In postpuberty, women have two to three times the body fat percentage of men, so these differences can affect the proper doses for men and women for fat-soluble drugs. These differences as a function of age are shown in Figure 8.3.

8.1.6   Pharmacokinetics: Drug Distribution, Metabolism, Excretion, and Breakdown

One of the new ways that people are actually trying to calculate the pharmacokinetics is by using compartment models. They can include various assumptions about conventional dosing, such as assuming that the drug goes everywhere in a diffusive model, as opposed to targeted delivery, where things are quite different.

(a)

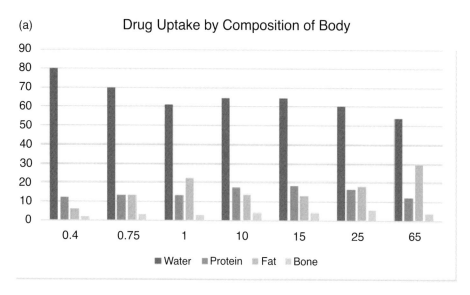

Drug Uptake by Composition of Body

■ Water   ■ Protein   ▨ Fat   ▨ Bone

(b)

BODY FAT CHART FOR MEN (%)

AGE																	
18-20	2.0	3.9	6.2	8.5	10.5	12.5	14.3	16.0	17.5	18.9	20.2	21.3	22.3	23.1	23.8	24.3	24.9
21-25	2.5	4.9	7.3	9.5	11.6	13.6	15.4	17.0	18.6	20.0	21.2	22.3	23.3	24.2	24.9	25.4	25.8
26-30	3.5	6.0	8.4	10.6	12.7	14.6	16.4	18.1	19.6	21.0	22.3	23.4	24.4	25.2	25.9	26.5	26.9
31-35	4.5	7.1	9.4	11.7	13.7	15.7	17.5	19.2	20.7	22.1	23.4	24.5	25.5	26.3	27.0	27.5	28.0
36-40	5.6	8.1	10.5	12.7	14.8	16.8	18.6	20.2	21.8	23.2	24.4	25.6	26.5	27.4	28.1	28.6	29.0
41-45	6.7	9.2	11.5	13.8	15.9	17.8	19.6	21.3	22.8	24.7	25.5	26.6	27.6	28.4	29.1	29.7	30.1
46-50	7.7	10.2	12.6	14.8	16.9	18.9	20.7	22.4	23.9	25.3	26.6	27.7	28.7	29.5	30.2	30.7	31.2
51-55	8.8	11.3	13.7	15.9	18.0	20.0	21.8	23.4	25.0	26.4	27.6	28.7	29.7	30.6	31.2	31.8	32.2
56 & UP	9.9	12.4	14.7	17.0	19.1	21.0	22.8	24.5	26.0	27.4	28.7	29.8	30.8	31.6	32.3	32.9	33.3
	LEAN			IDEAL			AVERAGE				ABOVE AVERAGE						

BODY FAT CHART FOR WOMEN (%)

AGE																	
18-20	11.3	13.5	15.7	17.7	19.7	21.5	23.2	24.8	26.3	27.7	29.0	30.2	31.3	32.3	33.1	33.9	34.6
21-25	11.9	14.2	16.3	18.4	20.3	22.1	23.8	25.5	27.0	29.4	28.6	30.8	31.9	32.9	33.8	34.5	35.2
26-30	12.5	14.8	16.9	19.0	20.9	22.7	24.5	26.1	27.6	29.0	30.3	31.5	32.5	33.5	34.4	35.2	35.5
31-35	13.2	15.4	17.6	19.6	21.5	23.4	25.1	26.7	28.2	28.6	30.9	32.1	33.2	34.1	35.0	35.8	36.4
36-40	13.8	16.0	18.2	20.2	22.2	24.0	25.7	27.3	28.8	30.2	31.5	32.7	33.8	34.8	35.8	36.4	37.0
41-45	14.4	16.7	18.8	20.8	22.8	24.6	26.3	27.9	29.4	30.8	32.1	33.3	34.4	35.4	36.3	37.0	37.7
46-50	15.0	17.3	19.4	21.5	23.4	25.2	26.9	28.6	30.1	31.5	32.8	34.0	35.0	36.0	36.9	37.6	38.3
51-55	15.6	17.9	20.0	22.1	24.0	25.9	27.6	29.2	30.7	32.1	33.4	34.6	35.6	36.6	37.5	38.3	38.9
56 & UP	16.3	18.5	20.7	22.7	24.6	26.5	28.2	29.8	31.3	32.7	34.0	35.2	36.3	37.2	38.1	38.9	39.5
	LEAN			IDEAL			AVERAGE				ABOVE AVERAGE						

Figure 8.3 (a) Since many drugs are either water-soluble or fat-soluble, differences in these percentages in humans of varying ages can make a big impact on how the drugs are sequestered in tissues and organs. Redrawn from www.merck.com/mmpe/sec19/ch270/ch270b.html. (b) Differences in men and women of different body fat compositions. Courtesy Accufit.com

8.1.7   Conventional Dosing Assumes Drug Goes Everywhere in the Body

Conventional dosing assumes that the only mechanism is diffusion, which is not rigorously true. More sophisticated drug dosing assumes compartments that model specific human organs. Then the total human response to drug doses is a systems analysis of linked compartments.

8.1.8   Targeted Therapies: A Model for Future Nanomedical Systems?

Targeted therapies provide another alternative. The targeting process changes the dynamics of drug delivery from diffusion. It does encompass some of the features of the compartment model in that targeting changes the relative amount of drug and success in reaching a particular organ.

8.2   From Animal Dosing to Human Clinical Trials

Animal dosing can provide some useful rough estimates of proper dosing, particularly allowing human drug trial dosing to stay away from dangerous regions of drug toxicity. But attempts to simply scale up drug dosing from animals to humans based on surface area and weight have had many problems. Animals, particularly mouse models, suffer from unpredictable differences due to differences in genetics and metabolism.

8.2.1   Importance of Picking an Appropriate Animal Model System

This leads us to important factors in picking an appropriate animal model system. Pigs, followed by dogs, are usually more appropriate animal models for drug dosing. Pig organs and metabolism are similar, in many respects, to humans. Dogs can provide a useful model of human bladder cancer and stomach cancer disease because they are one of the only animals that have these human cancers, presumably due to their living in close proximity to humans and eating similar foods, particularly cooked meat.

8.2.2   Does Drug Dosing Really Scale?

It is legitimate to ask whether drug dosing really scales up. The many differences in animal genetics and metabolism call into question the very idea of scaling. Even within humans, even of the same gender, but particularly in humans of different genders, there are vast differences between humans in terms of metabolism and genetic differences. Children are another source of scaling issues because their body fat distributions are quite different from adults, causing many drugs to sequester in fat-rich regions, changing their bioavailability. Beyond that, children are rapidly changing in terms of weight and size and are susceptible to rapid changes in hormones during adolescence. One of the big problems is that children are frequently not included in clinical trials due to obvious ethical issues in terms of human consent. Can parents really provide consent for children? This question presupposes that parents are responsible and caring. Frequently, parents do not want to give consent for their children to be exposed to a new drug. That means that we have only a small or nonexistent amount of data about children's responses to many drugs and drug dosing. For many years, there was a paucity of drug-dosing data on females. Many clinical trials avoided using females in their studies due to hormonal variations and concerns about whether the females could possibly be pregnant and not aware of that pregnancy. Now National Institutes of Health (NIH) clinical trials require a significant presence of females. Still, pregnant females remain underrepresented due to concerns about both the mother and the fetus.

8.2.3   The "Human Guinea Pig" in Clinical Trials and Beyond

The most important result from all of the previous discussion is that drug testing in humans, no matter how many prior animal studies, means that people in these initial tests are human guinea pigs since human responses to new drugs at any given dose are largely unknown and unpredictable. Hopefully, we can reduce risk in the future by first conducting drug tests on human "organs-on-a-chip" technology that gives us a partial alternative to animal testing, as discussed later in this book.

8.3   Some Drug-Dosing Methods

Most of medicine is a statistical prediction based on the human population. A doctor may tell you that the probability of a given drug helping you is 80 percent and that the probability of it not helping you is 20 percent. But the details of that 20 percent figure can be critical. It may be that 19 percent of the time it may not be effective in helping you, but 1 percent of the time it may seriously injure you or even kill you! As in many things in life, the devil is in the details! You may have some very bad side reactions, including death, so what matters to you is not what the statistics are, but what it is for *you* personally. Nothing else really counts!

8.3.1   Attempts to Scale Up Based on Area

Does drug dosing really scale, and what do we actually mean by "scale"? We have an intuitive sense that a larger person probably needs more drug than a smaller person. What we really mean is that a larger person has a greater weight and volume. Larger volume tends to scale according to the weight. But when you think about the actual drug delivery, the drugs go across membrane surfaces. That is the reason why people try developing models for drug dosing based on surface area. We try to predict the surface area of a person. You cannot take a person apart and measure the surface area of all the pieces, so you try to come up with some simple models that predict surface area from other measures such as height and weight, which are used to estimate body fat percentage, another important factor in estimating proper drug dosage for an individual. Body fat can also be quickly estimated by weight scales that include an electrical impedance measurement. That is why in your doctor's office they measure your height and your weight – those two variables allow the doctor to calculate an estimate of surface area (actually usually read from a chart) for you. This calculation can help predict the proper dose for you based on your height and weight, which will provide a rough approximation of what membrane surface area is inside you and your body organs. Obviously, it is a *very* crude estimate!

8.3.2 Attempts to Scale Up Based on Weight/Volume

Scaling up on the basis of weight/volume makes the most sense when comparing two adults because a reasonable expectation is that the drug will be diluted according to the size/volume of the individual. But that type of comparison does *not* make sense when comparing an adult and a child (Baber and Pritchard, 2003).

8.3.3 Attempts to Use Control Engineering Principles

Engineering principles are sometimes used in attempts to predict drug-dosing issues. Most engineering models use systems analysis of interacting tissue/organ compartments (Bailey and Haddad, 2005) (Figure 8.4).

8.4 Genetic Responses to Drug Dosing

Genetic differences from human to human can cause vast differences in drug uptake, targeting effectiveness, and drug metabolism, including excretion rates. While most researchers are concerned about genetic differences causing extreme side effects,

Multicompartment Model of Drug Delivery and Dosing

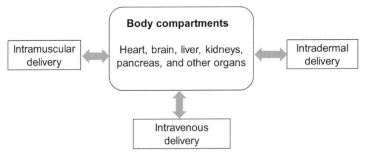

Figure 8.4 If the body can be thought of as a series of compartments, each with different uptake and retention characteristics, then the entire body can be modeled by control theory.
Source: Leary teaching

normal differences in general metabolism and excretion rates can also be important factors.

8.4.1 All Humans Are Not Genomically Equivalent!

Genomic differences between people can be quite large. Hence, we realize the importance of "personalized medicine," which attempts to predict the response of individual humans to a drug based on each individual's genome rather than basing the prediction on total human population statistics.

8.4.2 Predicting on the Basis of Family Tree Responses

In the absence of individual genomic knowledge, some measure of an individual's response to a drug can be based on the responses of other family members to a particular drug or history of disease. Although sometimes useful, the genomic differences even within a family are enough to sometimes produce surprising differences in drug response.

8.4.3 SNPs, Chips, and Beyond: Predicting Individual Drug Response

Single-nucleotide polymorphism (SNP) chips were the first use of individual human genomic differences as they looked at particular regions of the total human genome thought to be of importance in predicting an individual's response to drugs. As we gather more and more data on individual differences in response to drugs based on differences in gene sequences, there will be more SNP chips because they are relatively inexpensive to produce.

8.4.4 Cost/Benefit of Individual Genome Sequencing

The initial costs of sequencing an individual human being's genome, accomplished several years after the "general human genome," based on pieces of human DNA from various sources, was more than $100,000 per genome – too costly to be practical. Now the genome sequencing costs are approximately $1,000 per genome; that is a reasonable cost to justify its use in personalized medicine. But the other important factor is that only recently, after tens of thousands of humans have been sequenced, have we been able to possibly make good use of individual gene sequencing by knowing which differences are, or are not, important and relevant to human disease. Those individual genome sequences will become more and more valuable as we learn the importance, or not, of genome differences from individual to individual in terms of human disease and clinical outcomes.

Earlier statements about the human genome being sequenced should have generated the question "whose DNA was it?" The answer was it was no individual's DNA, but rather a collection of DNA sequences from various human cell lines spliced together computationally at Los Alamos National Laboratory. The first complete gene sequences of individual humans were later performed on two individuals, James Watson and Craig Venter, at a rough cost of more than $100,000 each. Since that time, the cost to sequence individuals has dropped to the order of a few thousand US dollars. While that may seem to be expensive, we are approaching the time when that will be seen as a reasonable expense, in terms of both reducing risks of bad drug interactions and also predicting which drugs will be effective for that individual. We are fast approaching a time when it makes good sense for every human to have their whole genome sequenced. The factors driving this decision will be the cost/benefit of beneficial and adverse drug reactions as well as predictions of drug efficacy. Failure to choose the most effective drug can be costly, not only in terms of delays in effective treatment but also because the initial incorrect drug may change the effectiveness of subsequent drugs, something not often considered by many doctors.

8.5   Dosing in the Era of Directed Therapies: A Future Model for Nanomedical Systems?

As discussed earlier, directed therapies involving targeted drugs can completely change the drug-dosing paradigm by bringing a higher fraction of the total drug dose to the site of the diseased cells. This may allow us to reduce the total drug administered to a patient by 10–100 times. When chemical drugs were cheaply produced, this was not a relevant economic factor. But for many new special drugs, the cost of the drug itself starts to be an economic factor.

8.5.1   How Directed Therapies Change the Dosing

A combination of more effective targeting and also stealth factors for preventing rapid elimination of the drug can result in much lower total drug exposure in individuals. This should result in reduction or even elimination of many side effects. Side effects are the reason why many people decide to stop taking a drug, even if that drug is effective against the disease. It is one of the major reasons why drugs fail to effectively treat human disease and is also a major source of frustration to doctors. But rather than blame patients for not wanting to take their medications because those medications produce unpleasant side effects, we should strive to reduce or eliminate those side effects. This is one area where nanomedical drug delivery through directed therapies can make a huge difference, even with current drugs.

8.5.2   Current Generation of Directed Antibody Therapies Dosing

The current generation of directed antibody therapies greatly reduces total drug exposure and reduces side effects. It also helps reduce the use of chemotherapies, which are, at best, blunt-edged instruments for dealing with cancer. Chemotherapy is warranted if there are no directed therapies available. The first such therapy, approved by the FDA in 1997, was a humanized monoclonal antibody Rituximab against non-Hodgkin's lymphoma.

8.5.3   Some Typical Side Effects of Directed Therapies

Simple use of directed therapies, usually injected intravenously, still have some side effects due to the fairly vast number of antibodies being injected. The reason for this large injection bolus is that the antibodies themselves are rapidly eliminated from the body either directly or through neutrophil uptake, which can also result in a potentially dangerous risk of infection due to neutropenia. Care to provide stealth packaging could reduce the total amount of antibodies administered by maybe 90 percent or more. These antibodies are very expensive to produce, so investment in directed therapy research could provide a handsome return on investment for drug companies. This exposure to large amounts of antibodies can also cause risk of anaphylactic shock in some patients. This is also a problem in some patients receiving vaccinations for various infectious diseases.

8.5.4   Nanomedical Systems Are the Next Generation of Directed Therapies

Nanomedical systems address many of these defects in delivery of conventional drugs and directed antibody therapies in a variety of ways. First, nanomedical systems can protect against rapid elimination of these drugs using stealth factors to protect the nanomedical system from either degradation in the blood and tissue or elimination of the drug by the liver and kidneys. Second, nanomedical systems can provide for single- or even multistep targeting to deliver the drugs to diseased cells, reducing exposure of other normal cells. Third, such multistep targeting can even help target the drug to the proper intracellular region of the cell for the most effective treatment at the lowest possible dose.

8.6   Most Directed Therapies Are Nonlinear Processes

One factor not discussed previously is that most directed therapies are nonlinear processes. The importance of this cannot be overemphasized. Most humans have difficulty visualizing nonlinear processes, and some nonlinear processes can lead to large differences in outcome based on small differences in inputs. In general, processes that are nonlinear are notoriously hard to scale properly. Some of these nonlinear processes are "chaotic" in nature, meaning that they are governed by chaos theory. This means there are "tipping points" whereby not much happens until the process reaches a certain dosage level. Thereafter, the effects of small changes in drug dosage can be very large!

8.6.1   Current and Pending FDA-Approved Monoclonal Antibody Therapies

The first FDA-approved directed antibody therapy was for rituximab in 1997. It is a humanized monoclonal antibody that recognizes CD20 on the surface of non-Hodgkin's lymphoma cells. It has proved to be a very effective treatment for non-Hodgkin's lymphoma for more than 20 years (Pierpont, Limper, and Richards, 2018).

Names for directed monoclonal antibodies usually end in "mab" – as compared to those ending in "ib," which are intracellular biochemical inhibitors, a totally different type of drug. There is an increasing number of these inhibitors, which are capable of directly or indirectly altering biochemical pathways within individual cells. These methods are comparatively new but will grow in importance. There are now more than 20 different human cancers for which there are FDA-approved monoclonal antibody directed therapies. These targeted therapies provide a spectacularly successful alternative to conventional chemotherapeutic drugs, many of which have serious side effects as well as more serious negative impacts on the immune system.

8.6.2 WHO Attempts to Standardize Therapeutic Antibody Names Worldwide

The World Health Organization (WHO), as part of its International Nonproprietary Names (INN) Programme, has attempted to adopt an international monoclonal naming convention based on the target organ and the type of antibody (WHO, 2019). It is important that we standardize our naming protocols so we know whether we are talking about the same thing (Table 8.1).

Table 8.1. WHO Has Adopted Naming Conventions for Therapeutic Monoclonal Antibodies

Target Site of Antibody		Source of Antibody	
o(s)	bone	u	human
vi(r)	viral	o	mouse
ba(c)	bacterial	a	rat
li(m)	immune	e	hamster
le(s)	infectious lesions		
ci(r)	cardiovascular	i	primate
mu(l)	musculoskeletal	xi	chimeric
ki(n)	interleukin as target	zu	humanized
co(l)	colonic tumor	axo	rat/murine hybrid
me(l)	melanoma		
ma(r)	mammary tumor		
go(t)	testicular tumor		
go(v)	ovarian tumor		
pr(o)	prostate tumor		
tu(m)	miscellaneous tumor		
neu(r)	nervous system		
tox(a)	toxin as target		
fu(ng)	fungal		

Source: Adapted from WHO website

8.7   Other Ways of Controlling Dose Locally

Thus far, we have described how directed therapies deliver a drug dose near, or actually to, diseased cells. Such systems are fully autonomous, or at least not requiring further external intervention. But it is possible to administer nanomedical systems and then use external means to manipulate their behavior inside the human body, as described below. These methods provide some clear advantages, as well as a few disadvantages.

8.7.1   Magnetic Field Release of Drugs

One of the most attractive externally triggered possibilities is the use of superparamagnetic nanomedical systems that are responsive to externally applied magnetic fields. Such magnetic fields can help concentrate superparamagnetic nanomedical systems at a particular internal site of the diseased cells. Alternatively, an oscillating magnetic field can vibrate the magnetic nanomedical systems, producing heat that can be localized down to the individual diseased cell level and thereby producing single-cell hyperthermia (Thomas et al., 2010) (Figure 8.5).

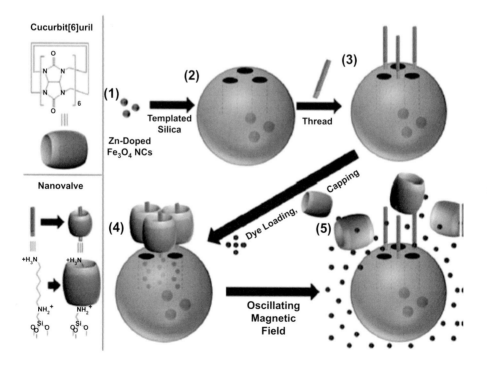

Figure 8.5 Magnetic nanomedical systems can be used to either release a drug or produce localized hyperthermia if exposed to oscillating magnetic fields.
Source: Thomas et al. (2010)

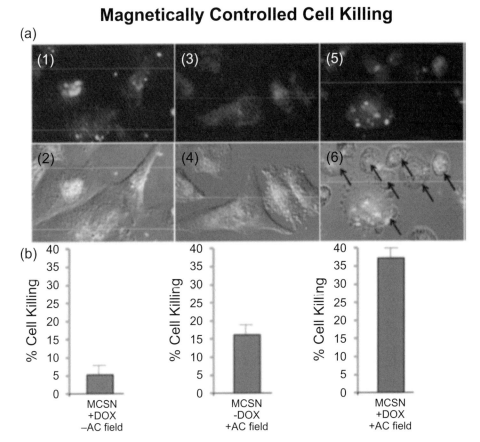

Figure 8.6 Magnetic fields can be used to control selective cell killing.
Source: Thomas et al. (2010)

It is also possible to create systems that are capable of magnetically controlled cell killing (Thomas et al., 2010), as shown from data in Figure 8.6.

8.7.2   Light-Triggered Release of Drugs

Light can be used to externally trigger the release of drugs (McCoy et al., 2010; McCoy et al., 2007). Though this is mainly useful for activating drugs on the skin, it can also be used to trigger drug release to more internal body parts through endoscopic techniques. Additionally, it can be used during more invasive surgery whereby internal organs are made temporarily accessible to these light therapies (Figure 8.7).

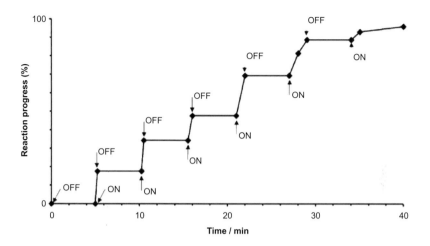

Figure 8.7 Light can be used to closely control the rate and amount of drug release. It can even be done successfully, creating mini-doses to reduce overall toxicity.
Source: McCoy et al. (2007)

Chapter 8 Study Questions

8.1   What are some of the problems that occur when we attempt to scale up drug doses from animals based on simple differences in weight?

8.2   What are some of the reasons why proper doses to different humans may not be predictable on the basis of weight alone?

8.3   Why are children not just "smaller adults"? What are the sources of these nonscalable differences?

8.4   What are four fundamental components of drug pharmacokinetics in vivo? Which one of these components is least likely to be comparable between two different humans?

8.5   Why is it so important to choose an appropriate animal model?

8.6   What are some of the traditional methods used to scale up drug doses?

8.7   Why is individual genomics so important to the future of drug dosing?

8.8   Directed therapies provide targeting of therapies. What are some of the remaining problems in these current "pre-nanomedicine" therapies?

8.9   How can in vivo drug dosing be controlled externally? Give two examples.

8.10   What are Dedrick plots and why are they used?

8.11 How is control theory used to predict drug dosing? Why is it theoretically better to use than simple weight or area estimations? Why is it more difficult to use in practice?

8.12 According to current naming conventions, what is the difference in the type of drug between one ending in "mab" versus "ib"?

8.13 What are some of the common side effects of targeted therapies?

8.14 How can nanomedicine design principles be used to minimize these side effects?

8.15 What are some other ways of controlling drug release locally?

9   Nanodelivery of Therapeutic Genes

In this chapter, I will introduce some concepts that you may initially think are in the realm of science fiction. But I assure you that all of this is real science! These fascinating recent developments represent paradigm shifts in the basic concept of nanodrug delivery systems. I hope that this chapter will particularly inspire the next generation of scientists working on new nanodrug delivery systems!

9.1   Introduction and Overview

Typically, there is an optimal dose for treating a diseased cell. But most conventional drug delivery systems have either exponential or bi-exponential decay characteristics. So-called timed-release drugs create a sawtooth drug delivery pattern, which is a big improvement, but still less than optimal. What we really want is to be able to hold the drug delivery to as close to the optimal dose as possible over time, as shown in Figure 9.1.

As we can see in Figure 9.1, there are three main paradigms of drug delivery. The problem with drugs obeying the conventional exponential drug decay characteristics is that they start out much higher than the optimal dose. When very high, they can lead to bystander cell injury at the cellular level and undesirable side effects at the patient level. When in the regime below the optimal level, there is little or no therapeutic effect. The sawtooth pattern of timed-release drug delivery stays closer to the optimal dose and represents a major advance in the history of drug delivery. Many conventional medicines now have timed-release versions to prevent too high doses from causing injury or side effects and too low doses from having no positive therapeutic effect. But what we really want is to be able to set and maintain a stable drug release dose as near as possible to the optimal drug dose. To do so usually requires the use of feedback loop control over drug delivery. The closest paradigm we currently have is an insulin pump, which samples the patient's insulin levels and then injects insulin accordingly. It is a tremendous advance. But we need to miniaturize, actually to the nanolevel, this type of feedback control into individual nanomedical devices for drug delivery. This chapter will show not only that such an approach is possible but also that it has actually been experimentally demonstrated.

We can try inserting a therapeutic gene sequence directly into the genome of the cell, but that might have some pretty drastic consequences. However, with recent

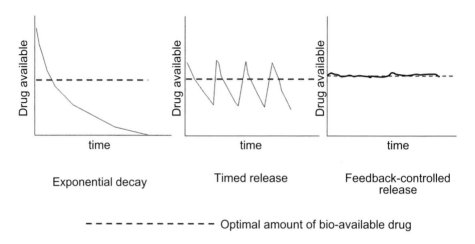

Figure 9.1 Three basic paradigms of drug delivery. Conventional drug delivery has an initial bolus of drug followed by an exponential or bi-exponential decay phase. Newer timed-release drug delivery has a sawtooth pattern of drug available over time that keeps the overall drug dose closer to the optimal dose. Feedback-controlled drug release of the future allows drug availability to be kept near the optimal level.
Source: Leary teaching

advances in CRISPR gene-editing technology, this approach has been shown to be effective in individual patients. This is a topic we will discuss in more detail later in this chapter.

An alternative approach to gene editing is to simply have that gene expressed with a plasmid-type delivery in the cytoplasm where the drug delivery system is just another drug. The drug then is transient and you can turn it on and off, as needed. This approach will be demonstrated later in this chapter.

9.1.1   A Multistep Gene Delivery Process

Delivering a therapeutic gene construct is a multistep process because it also involves a multistep targeting process to get the package to the proper place inside the cell where it can work in consort with the host cell genes (Figure 9.2).

9.1.2   Some of the Advantages of Therapeutic Genes

A therapeutic gene is basically a gene construct that is capable of being manufactured under host cell control. There may or may not be feedback control elements built into the therapeutic gene construct. In other words, it has a start switch and a stop switch

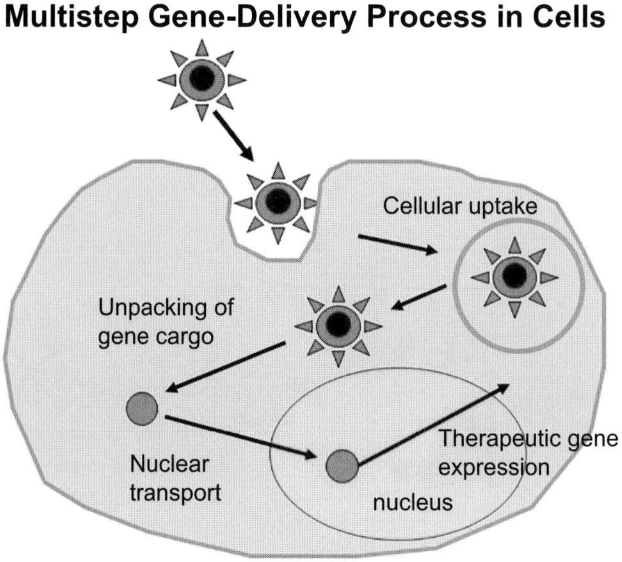

Multistep Gene-Delivery Process in Cells

Figure 9.2 Gene delivery can be either stably integrated into the host genome or expressed in the cytoplasm, without integration, using a plasmid. If not integrated and expressed only in the cytoplasm, such transient gene therapy should be thought of as another kind of drug therapy. Source: Leary teaching

that allow you to modify the expression levels, so it gives you a lot of opportunities for control. You can also have on-off switches, which can be responsive to small drugs that can be added from external sources, such as the one described later in this chapter that responds to tetracycline.

9.1.3 Concept of "Nanofactories" Manufacturing Therapeutic Products Inside a Living Cell

One of the things we can do is manufacture a drug dose inside a cell. A therapeutic gene construct is capable of being replicated, so it knows how to use the host cell machinery to make copies of itself. Almost 20 years ago, my laboratory demonstrated that this is possible in some model systems. This is now large-scale, demonstrated science. It is the basis of the latest vaccines against COVID-19 introduced by scientists in 2020 to fight the deadly worldwide pandemic. An mRNA template is delivered via liposomal nanoparticles to a patient's cells, which then manufacture copies of the coronavirus spike protein that is responsible for developing an immune response to COVID-19 infection.

Gene constructs can be completely self-packaged, particularly when we take plasmid approaches where the entire construct – including all the control machinery and on-off switches – is put into a neat package that can be manufactured inside a cell. We can

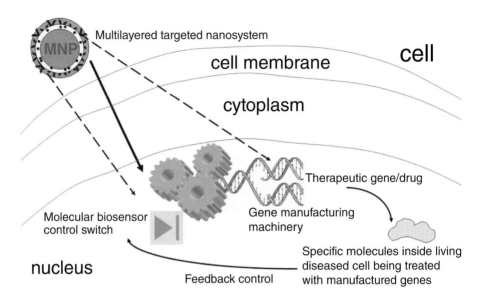

Figure 9.3 The nanoparticle delivery system delivers the therapeutic gene template, which uses the host cell machinery and local materials to manufacture therapeutic gene sequences that are expressed under biosensor-controlled delivery.
Source: Prow et al. (2005)

include a set of control switches inside these plasmid constructs that allow you to modify levels and to turn things on or off with an external trigger. In this chapter, I will demonstrate how it is possible to set up "nanofactories" inside living cells to manufacture therapeutic genes. The concept is shown (Prow et al., 2005) in Figure 9.3.

At the end of this chapter, in Section 9.5.3, an experimental demonstration of this concept is presented.

9.1.4   Some of the Advantages of Molecular Biosensor Feedback Control Systems

Biosensor feedback control systems can go beyond simply killing cells. By including a negative feedback loop system, the nanomedical system can use nanosensor molecules to sense the production of disease molecules and then basically knock down those disease molecules to keep them under control within the cell and allow us to achieve homeostasis in the body. This approach will be important for the development of new nanomedical strategies to treat chronic diseases.

9.1.5   Why a Nanodelivery Approach to Gene Therapy Is Appropriate

Gene therapy began with the use of adenoviruses (Benihoud, Yeh, and Perricaudet, 1999; Luisoni and Greber, 2016) and continues to be used. But the fundamental problem is that adenoviruses have their own agendas, which are often not conducive to the overall health of the patient. It is difficult to completely eliminate these unfavorable aspects of adenovirus as a gene delivery method.

Nanomedical delivery approaches offer a promising alternative to adenoviruses and other virus-based systems. These systems build the favorable aspects of gene delivery of a virus while using biomimicry to provide a totally synthetic virus-like alternative. These synthetic gene delivery systems have only the characteristics built into them. They also have no capacity for self-replication or undergoing mutations, removing that risk completely.

9.2   The Therapeutic Gene Approach

The therapeutic gene approach involves inserting a gene to produce a specific product that is therapeutic in terms of treating the diseased cell. We really have to start thinking about disease in very different terms. We need to start thinking about disease as an abnormal expression of genes within a single cell. In medical school or veterinary school, you are taught to think of diseases essentially as disease at one or more organ levels. To think of disease as something occurring at the single-cell level is a very foreign concept to the field of medicine. However, if we are to make full use of the promise of nanomedicine to medicine, we must make this major conceptual paradigm transition.

If you start looking deeper, there has to be a cellular-level expression of disease-related molecules for most diseases. Some diseases may be caused by over- or underexpression of specific molecules. Instead of saying that a disease is a misfunctioning at an entire organ level, we should be looking at the causes of this organ misfunction at the level of individual cells in that organ. Indeed, we know that many, if not most, of the cells in an organ are still normal. That is exactly why surgeons frequently remove only a subregion of the total organ in order to restore proper organ function. The problem is that by the time a surgeon tries to treat the disease, frequently diagnosed through patient symptoms, the disease or injury to the organ is late-stage such that the organ is irreversibly damaged. What we really need to do is to identify the actual underlying molecular events that have caused that organ to move from a normal state to a diseased state.

9.2.1   What Constitutes a "Therapeutic Gene" at the Single-Cell Level?

The general concept is to use therapeutic genes, in the broadest sense of the word, to try to move the cell from a diseased state to either a normal state or a "more normal state" that is less consequential. What constitutes a therapeutic gene? I think we have to be pretty liberal with our definitions here. A therapeutic gene should be seen as something that produces one or more molecules that either modify the expression of existing genes in the cell or produce a gene product that positively affects a diseased cell in the direction toward becoming a normal, healthy cell. Examples of such therapeutic genes will be discussed in subsequent sections. A particularly clear case of a therapeutic gene is the mutated gene responsible for sickle cell disease and beta thalassemia. Replacing that specific gene mutation via CRISPR gene-editing technology moves these patients from a diseased state to a normal, or at least close to normal, patient, as has recently been accomplished in the clinic (Frangoul et al., 2020) and is described in more detail in Section 9.7.

9.2.2 Transient versus Stable Expression Modes

One choice is whether the therapeutic gene needs to be transiently expressed or expressed at a constant level. That decision turns out to be consequential and drives the design of the therapeutic gene and its delivery method.

A therapeutic gene that is expressed in the cytoplasm without integrating into the host cell genome is usually at the level of transient expression. In that case, it just becomes a drug-manufacturing vehicle. It knows how to replicate what it needs, has everything that it needs in it, and simply then produces that product. Since it is not integrated into the host cell genome, it is less potentially dangerous and consequential than something fundamentally altering the genome with potentially unforeseen future consequences. This is the basis of the new mRNA COVID-19 vaccine, which represents a revolutionary advance in the development of new vaccines.

But if we are going to stably integrate these gene sequences into the host genome, then we potentially have another whole complex series of steps with hard-to-see outcomes. We definitely do not want to randomly insert those genes into the host cell genome because we might inadvertently interfere with the functioning of other normal genes in the cell. This has led to the development of fairly sophisticated methods for highly targeted gene delivery to specific insertion points in the DNA host genome, such that you do not disturb the functioning of other genes. In subsequent sections, we will discuss a number of gene insertion methods, such as zinc finger constructs, TALENS, and CRISPR technologies (Gaj, Gersbach, and Barbas, 2013), to insert genes at specific locations within the host cell genome. Section 9.7 provides a more detailed discussion of CRISPR technology. Such methods are in the early stages of being used on human patients to treat systemic-level diseases such as muscular dystrophy, which is obviously a body-wide, systemic disease rather than a disease specific to a particular organ.

9.3 Molecular Feedback Control Systems

Drug delivery systems have traditionally not used feedback control for drug delivery. Whereas most engineers would immediately think in terms of using feedback control systems for drug delivery, they have traditionally not been part of the drug delivery design process. This is unfortunate because it has hindered the development of more sophisticated drug delivery systems. Feedback control systems can offer a much safer avenue for drug delivery. Since many drug delivery systems tend to be nonlinear, it is dangerous to use them without some restraining feedback control system to prevent harm to the patient. Also, the same amount of drug in two different patients can lead to major differences in patient response. The most extreme examples have some patients benefiting from a drug and others becoming seriously injured or even killed by a similar dose.

When you think about designing a drug delivery system using a feedback control system, it is not a simple thing in most cases. There usually needs to be a way of biosensing the presence and number of specific molecules, molecules that must be either activated or deactivated. Direct or indirect interaction of naturally occurring, disease-related molecules within or in the vicinity of the diseased cell must typically occur with the biosensor feedback control drug delivery system to make the total system function properly.

Timed-release drug delivery systems usually have some layering design that allows drug release each time a layer comes off, essentially giving you a fresh release of drug. An example is the timed-release form of the drug melatonin to assist sleep. Such systems tend to oscillate around a particular drug concentration level, avoiding the very high extremes as well as the very low drug concentration extremes. It certainly represents a big improvement over straight exponential-decay drug delivery systems. Nanotechnology, even within simple liposome encapsulation systems, has made some big contributions to the development of timed-release systems. One way to accomplish this is through the use of structures (Narayan et al., 2018; Slowing et al., 2007; White-Schenk et al., 2015) built with mesoporous silica nanoparticles. As another example, glaucoma treatments for intraocular pressure control to prevent optic nerve damage have reached a "zeroth-order drug delivery," essentially being a straight-line drug delivery system at a near-optimal dose level for at least 30 days (Yadav, Rajpurohit, and Sharma, 2019).

One can target a specific disease molecule inside a cell, which can then bind to a sensor switch that is not normally off. This molecular biosensor controller is typically a negative feedback switch that sits upstream from a gene promoter, which is itself upstream from a therapeutic gene that can be produced (or not) under control of that overall feedback control system. You can also add a reporter gene (e.g., green fluorescent protein) to the overall complex to quantitatively measure the entire process. The required, or perhaps most important, components of an integrated molecular biosensor/gene delivery system are shown in Figure 9.4.

Reporter genes are typically fluorescent molecules, usually with a very small number of important transcribable sequences, which can make nice fluorescent products of different colors. The fluorescent reporter gene you may be most familiar with is a jellyfish protein called green fluorescent protein (GFP). There have been many mutations introduced to that protein to produce almost any color you desire. The Nobel Prize in Chemistry 2008 was awarded jointly to Osamu Shimomura, Martin Chalfie, and Roger Y. Tsien "for their discovery and development of green fluorescent protein (GFP) technology. This technology has represented a major new tool for cell and molecular biology" (Tsien, 1998). My laboratory has used reporter genes to monitor experimental feedback control drug delivery systems, as will be shown later in this chapter.

Basic Components of a Molecular Biosensor

Targeting to initial region of interest	Cleavage domain sensitive to target molecule	Transactivator or reporter molecule (e.g., GFP or luciferase)

For example, this biosensor can be designed to sense a target in the cell and then target to the nucleus to allow for transcription of a gene for therapy.

TMD   CD   tTA

TMD = transmembrane domain, CD = cleavage domain to allow transport to nucleus if target is present, tTA = transactivator (with tetracycline molecular on-off switch) with gene to be transcribed

Figure 9.4 Many molecular biosensors used for diagnostics have targeting regions, a cleavage domain that is sensitive to enzymes or proteases, and a fluorescent reporter molecule. Source: Prow (2004)

Green fluorescent protein has a small enough gene sequence to produce a bright green fluorescent product that can easily be incorporated within a feedback control drug delivery system. Green fluorescent protein comes from jellyfish. It is relatively nontoxic unless you produce large amounts, so it can usually be left to accumulate within a cell and without worry of it becoming very toxic to that cell.

If you have a system with no feedback control, thus producing more and more green fluorescent protein, at some point it can produce enough to be toxic to a cell. While we tend to think of GFP as completely nontoxic, that is not true. Another thing to keep in mind is that you can have more than one sensor, each using a mutant variation of green fluorescent protein of a different color, so you can look at the interaction of two or more genes in a cell in more sophisticated studies of gene interactions, as will be shown later in this chapter.

A therapeutic gene product is designed to either inactivate or eliminate a disease molecule. As you eliminate the disease molecule, it gets down to a low enough concentration inside the cell to turn the switch off and stop producing more therapeutic molecules, which may be harmful if overproduced. Theoretically, you could make something that would survive in the individual cells for a long time and treat chronic diseases. Although that is still a challenging task, think about the benefits of being able to do it for a matter of days or even a few weeks. It is exactly the kind of thing that you want for most chronic diseases where you do not want to kill the cells.

Feedback Control of Therapeutic Genes in Nanomedical Systems

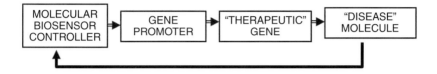

Figure 9.5 Placing a molecular biosensor upstream from a gene promoter allows that promoter to be turned on or off according to the presence and/or quantity of a sensed molecule. Source: Prow (2004)

9.3.1 Drug Delivery Has Traditionally Not Used Feedback Controls

Although drug delivery has not traditionally used feedback control systems, it is a shortcoming that needs to be corrected if nanodrug delivery systems are to become commonplace. It is also a much more powerful use of nanomedicine for drug delivery beyond comparatively simple timed-release approaches to drug delivery. The basic components of a typical feedback control system for nanomedical systems to treat disease (Prow, 2004) are shown in Figure 9.5.

9.3.2 Why Feedback Control Might Be a Very Good Idea!

Feedback control turns out to be a good idea because it can prevent the delivery of too much drug, which could be toxic or otherwise injure a patient. It can also prevent the drug release rate from falling below the level of any positive therapeutic effect. Most engineers would not even consider building nonlinear systems without some form of feedback control to prevent them from going out of control. Many drug delivery systems are inherently nonlinear and need some form of feedback control system to operate properly and safely.

9.3.3 Positive or Negative Feedback?

Feedback systems can be positive or negative. A positive feedback system tends to grow on itself. A negative control system means that the product is turned off due to the excess of the product. The advantage in having some kind of a negative feedback loop system is that it can sense the production of disease molecules and then basically knock down those disease molecules to keep them under control within the cell and allow us to achieve a homeostasis in the body. Feedback control systems, and typically negative feedback systems, are inherently safer!

9.4   Molecular Biosensors as a Component of a Nanomedicine Feedback Control System

Molecular biosensors are an important part of any advanced gene delivery system. If you are going to manufacture a therapeutic drug or gene within a cell, there must be a way for it to turn its production on and off. Molecular biosensors can be simple and direct, sensing a specific molecule related to the disease at a cellular level. Alternatively, it can sense other molecules that are involved in a more complicated biochemical pathway to influence that pathway overall.

9.4.1   What Is a Molecular Biosensor?

A molecular biosensor is a molecule that has sensing capabilities. What is "sensing"? Sensing merely means that something is sensitive to the presence of something else. At the molecular "biosensor" level, it means that molecule(s) respond in some fundamental way to the proximity of other molecule(s). Such sensing can involve things like a conformational change in either the sensing or sensed protein molecule. Simpler sensing systems for diagnostics can involve enzymatic cleavage of a molecule from being nonfluorescent to being fluorescent.

9.4.2   How a Molecular Biosensor Functions as a Therapeutic Gene Switch

If a molecular biosensor is included as part of a therapeutic gene construct, when the therapeutic gene is activated, the biosensor can function as a "therapeutic gene switch" to report that the therapeutic gene has been turned on. Then one can know it is time to see whether the therapeutic gene is indeed providing a therapeutic response. While not necessary at the clinical stage, such molecular biosensors with reporter characteristics are extremely useful at the research level in prototyping new therapeutic gene complexes.

9.5   Building Integrated Molecular Biosensor/Gene Delivery Systems

Lest the reader think that the preceding is just an abstraction, let me illustrate several examples of working systems for very different purposes. There are many more examples, but I have chosen these three because they help illustrate the tremendous range of possibilities. The first example shows how these approaches can be useful for antiviral strategies in the treatment of hepatitis C infection to destroy the virus without destroying the infected cell. It is an excellent example of a Trojan horse approach to nanomedicine whereby a nanomedical system looks like a naturally occurring, friendly messenger and then delivers a therapeutic package decidedly unfriendly to its recipient – in this case, a virus! The second example shows how these approaches can be used to find and treat radiation-damaged cells, including making decisions about whether to treat or kill the irradiated cells depending on their likelihood of having become cancerous. The third example shows how these approaches can be used to treat glaucoma and other eye diseases.

9.5.1   Example of a Ribozyme/Antivirus System

Hepatitis C infection uses different molecules of the virus to accomplish specific steps in the infection process. Figure 9.6 shows some of the major proteins responsible for these steps in the infection process (Prow, 2004).

The following is an example of a particular nanomedical approach to combatting hepatitis infection. The trick is to use biomimicry to have the artificial nanomedical construct perceived by the body as an actual hepatitis virus that has the components necessary to take over host cell machinery. But this time, instead of delivering a virus, it delivers a therapeutic package to fight against the virus. The nanomedical system shown in Figure 9.7 masquerades as a hepatitis C virus in a Trojan horse approach, masquerading as an actual hepatitis C virus, so that it goes through all the multiple normal steps of hepatitis C virus infection of human cells (Prow, 2004).

We started out by constructing a nanoparticle roughly the size of a hepatitis C viral particle and then coated its surface with E2 protein that the actual virus uses to home

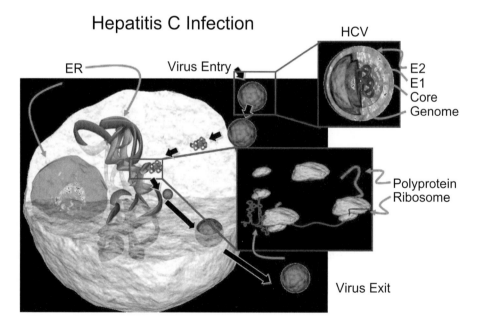

Figure 9.6 The molecular components of a hepatitis C virus responsible for infecting human cells. A nanomedical system needs to appear as if it were actual hepatitis viral particles with similar molecules to go where the hepatitis virus goes in the body in terms of targeting specific organs. It should also contain some of the hepatitis proteins that allow it entry to those specific cell types.
Source: Prow (2004)

Nanoparticle That Mimics Hepatitis C Virus to Find and Destroy That Virus inside Living Cells

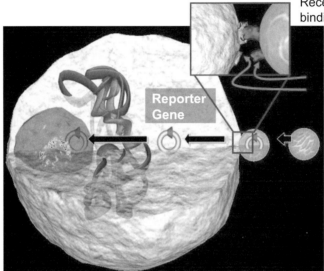

Figure 9.7 Using the principle of biomimicry, we can mimic nature by making a nanoparticle look, at least at first encounter, like a virus. But the nanoparticle can contain antiviral molecules inside to combat viral infection at the single-cell level. A nanoparticle that mimics the structure and function of a hepatitis C virus can be used as a Trojan horse to deliver antiviral therapies to cells infected with hepatitis C.
Source: Prow (2004)

to various organs, particularly the liver, but also to other organs just like a real hepatitis virus does. Next, we used a targeted ribozyme, which acts as a pair of "ribozyme molecular scissors" to search for a very particular hepatitis viral sequence, called the IRES sequence in this case, inside the cell. The process involves binding this ribozyme to the IRES region so that the ribozyme's molecular scissors can cut it out and thereby inactivate the virus (Prow et al., 2005). All of this is accomplished without killing the cell. The virus is now no longer able to translate sequences required to replicate itself (Figure 9.8)!

Usually once you are infected with a virus, you are infected with that virus for life because it hides from the immune system inside cells and lies dormant, sometimes for years, to be reactivated by a variety of environmental factors. A more familiar example of this is herpes viruses. Once you have a mouth sore caused by herpes type I virus, you will always have that herpes type I residing in your body. It may not flare up often while it hides in your nerve endings, but when it does flare up it really hurts because it sits right on your nerve endings!

If you merely try to kill infected cells, you might be just releasing those viruses out to neighboring cells and actually help spread the original infection. However, you

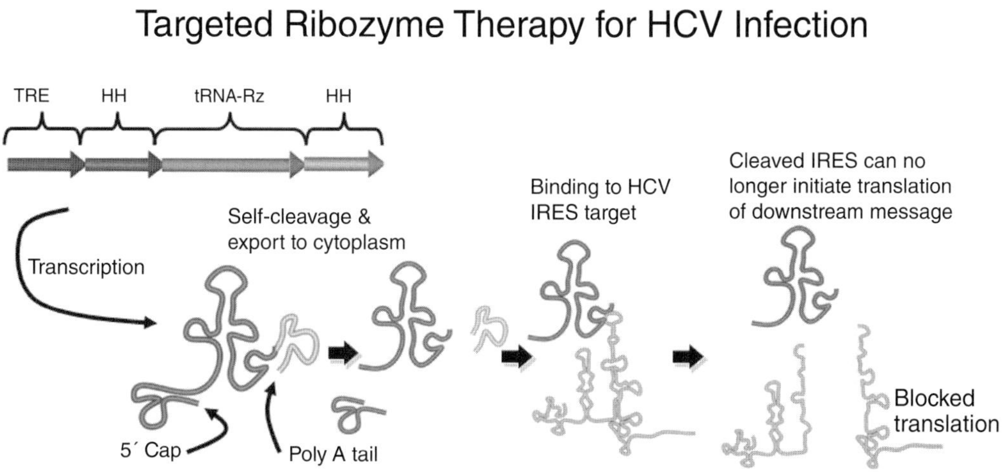

Targeted Ribozyme Therapy for HCV Infection

Figure 9.8 Ribozymes can provide nanomedical therapy at the single-cell level. These molecules function as "molecular scissors" that can cut recognized viral sequences to inactivate viruses inside single cells.

Source: Prow et al., (2005)

could take a different approach, with an antiviral therapy inside a single cell that can actually inactivate the virus without killing the cell. This is one way you might actually eliminate the virus without killing the host cell.

We used biomimicry so we could go inside a cell and use reporter genes to watch what was happening inside single cells. In the hepatitis infection, the tetracycline switch works in two different ways. In one case, we used it as an off switch. In another case, we used it to turn *on* our construct. Sometimes it is useful to have an "emergency *off* switch" to shut everything down in case the system is running out of control. An example of this is shown in Figure 9.9 (Prow, 2004), whereby a sensor switch responds to treatment with tetracycline.

As a simulation of actual gene therapy involving binary gene delivery, we constructed nanocapsules containing two jellyfish reporter genes of different colors (red and green) inserted into plasmids, and we then transfected these plasmids into cells (Prow, 2004). Binary gene therapy can allow for overall treatments that go beyond what can be achieved with a single therapy (Figure 9.10).

9.5.2 Example of an ARE Biosensor/DNA Repair System for Radiation-Damaged Cells

The following is an example of an assisted DNA repair nanomedical system with an antioxidant response element (ARE) molecular biosensor switch to control the process of DNA repair in radiation-damaged skin, which can lead to skin cancer in humans (Prow et al., 2004) (Figure 9.11).

Biosensors for Hepatitis C Treatment

(a) Tet-Off System: tetracycline transactivator protein-based sensor mechanism with "emergency off switch" using tetracycline

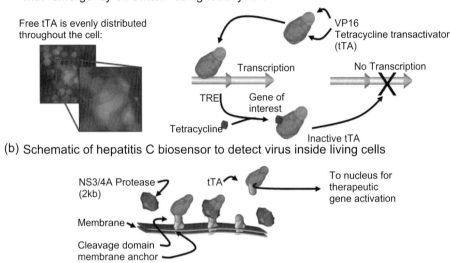

Free tTA is evenly distributed throughout the cell:

VP16
Tetracycline transactivator (tTA)

Transcription   No Transcription

TRE   Gene of interest

Tetracycline

Inactive tTA

(b) Schematic of hepatitis C biosensor to detect virus inside living cells

NS3/4A Protease (2kb)

tTA

To nucleus for therapeutic gene activation

Membrane

Cleavage domain membrane anchor

Figure 9.9 Design of an antiviral nanomedical system for hepatitis C infection. (a) The Tet-Off system serves as an "emergency switch" controlled by injections of tetracycline. (b) Schematic of a protease-activated biosensor for hepatitis C virus. The biosensor protein is a triple-fusion protein containing a tetracycline-inactivated transactivator (tTA), a protease-specific cleavage domain, and a localization domain. The activated protease cleaves the cleavage domain, releasing the tTA. The tTA then localizes to the nucleus and activates transcription of a predetermined gene.
Source: Prow (2004)

It was originally developed in my laboratory under a NASA-sponsored grant to try to find a way to provide continuous repair of cells in astronauts traveling to Mars, where they would be exposed to significant space radiation once they left the protective magnetosphere of Earth. This is the nanomedicine application that got me started in the field of nanomedicine way back in the year 2000. However, like many "space applications," it also has an important terrestrial application, namely, the treatment of radiation-damaged skin and potential production of malignant melanoma skin cancer. The actual design is complicated by the need to distinguish between repairable and nonrepairable radiation-damaged cells. If the radiation-damaged cells become cancerous, then it is probably desirable to just eliminate them rather than trying to repair the cells. This leads to a nanomedical system that must make more sophisticated decisions (i.e., fix or eliminate), an example of a binary decision process (Figure 9.12).

Nanocapsule Delivery of Two Reporter Gene Plasmids for Transient Gene Therapy

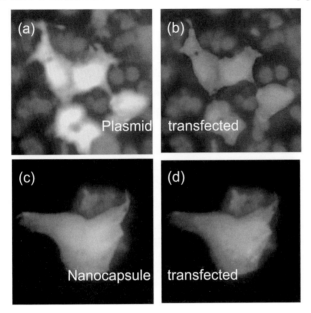

Figure 9.10 Example of nanocapsule-mediated gene delivery for hepatitis infection of cells. Source: Prow (2004)

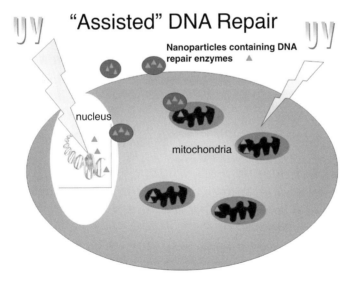

Figure 9.11 DNA repair of cells may or may not involve treating both the cellular DNA and the mitochondrial DNA. Source: Prow et al. (2004)

Nanomedicine for DNA Repair in Response to Radiation Damage

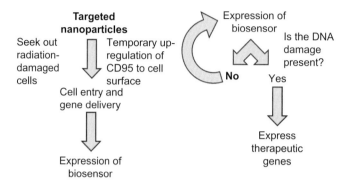

Figure 9.12 This nanomedical system is designed to first recognize cells exposed to high levels of radiation, as measured by transient expression of CD95 on the cell's surface. Then, depending on the level of expression, the nanomedical system makes the decision to either repair the cells, if repairable, or to kill the cells if nonrepairable (i.e., if they are transformed into cancer cells that are nonrepairable and should be killed).
Source: Prow et al. (2004)

Ionizing radiation is equivalent to a potent carcinogen due to oxidative stress–induced DNA damage. We used the antioxidant response element to initiate expression of foreign DNA repair and then, when activated, the biosensor drove the synthesis of our repair enzyme. Since radiation is a bit difficult to control in terms of flux and dose, we used the chemical tBHQ to produce reactive oxygen species inside the cell. We also were unable to use a "space radiation" source because we were doing this research before the construction of a "space radiation" simulated light source at Brookhaven National Laboratory. Instead, we used near-UV radiation sources that also simulated a portion of natural sunlight, which was perhaps a more immediately relevant radiation source for a terrestrial spin-off application of this research, since we performed the research in laboratories on Galveston Island, the home of the University of Texas Medical Branch, where I was working at the time. Galveston is a college student spring break destination where the students spend much time sunbathing on the beaches, exposing themselves to considerable UVA and UVB rays in natural sunlight, as part of their process of encouraging future skin cancer and melanoma (Prow et al., 2004) (Figure 9.13)!

This situation also provided us with an opportunity to study assisted DNA repair by collaboration with a colleague who was studying DNA repair enzymes in bacteria. The interesting thing about DNA repair is that you have to worry about two totally different types of DNA in a human cell. We always think about the normal genomic DNA. But we also have completely separate mitochondrial DNA in our cells, inherited entirely from our mothers. Mitochondrial DNA may also need to be repaired

ROS (Reactive Oxygen Species) Activated Biosensor in T24 Cells

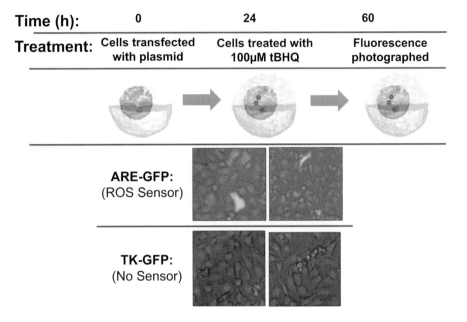

Time (h):	0	24	60
Treatment:	Cells transfected with plasmid	Cells treated with 100μM tBHQ	Fluorescence photographed

Figure 9.13 During oxidative stress, a cell will produce molecules that bind to the antioxidant response element (ARE). The binding of these molecules can be used as a molecular biosensor to drive the process of DNA repair enzymes downstream in the gene sequence construct. Source: Prow et al. (2004)

independently of the repair process of the cellular DNA in the cell nucleus. Every time a cell divides, it not only divides but produces twice as many copies of its nuclear DNA. But it also has to produce twice as much mitochondrial DNA. Cell reproduction also results in mitochondrial reproduction if we are to have two daughter cells with their normal complements of mitochondria, the so-called batteries of our cells. In fact, there are mitochondrial-specific diseases that are basically caused by DNA damage to or other problems with mitochondria. So when we think about cellular DNA repair, we also need to consider including the mitochondrial DNA. Since radiation is going to damage potentially both of them, we would want to not only target these DNA repair enzymes to the nucleus but also target the mitochondria using "localization sequences" that will take them to the mitochondria. Many localization sequences (e.g., nuclear localization and mitochondrial localization sequences) are known to guide things to particular subregions of a cell.

UV from the UVA and UVB parts of the spectrum in daylight on Earth induces damage to DNA, primarily to exposed skin, in humans. It turns out that humans, unlike many other animals, have only one intact, functioning DNA repair pathway. Humans actually have a complete secondary DNA repair pathway, but it is missing

Ionizing Radiation Activates Biosensor-Mediated DNA Repair Enzyme Expression

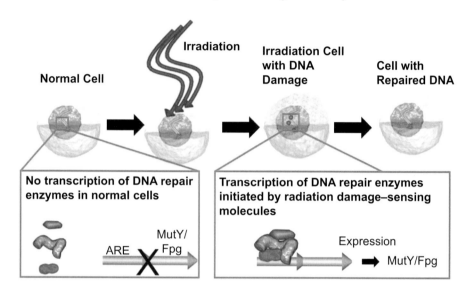

Figure 9.14 Cells were transiently transfected at 60 percent confluence with either ARE-GFP or TK-GFP; 24 hours later, the cells were treated with tert-butylhydroquinone (tBHQ), an ROS-inducing agent, to simulate the ROS effects of radiation doses. The cells were examined every 12 hours posttreatment. Weak fluorescence (ARE-GFP), shown in white, was present at hour 48 and photographs were taken at hour 60.
Source: Prow et al. (2004)

one critical enzyme to make it all work. Perhaps one of our female ancestors stepped out of the cave at the wrong time, one of her eggs was hit by a cosmic ray, and her subsequent baby lacked this particular enzyme. In all seriousness, how humans lost this enzyme is unknown and really does not matter – except when we really need that secondary repair pathway! Humans are not very robust in radiation environments. On Earth we are protected by the magnetosphere, plus a protective layer of ozone. But if we are to someday become a space-faring species, dealing with sensitivity to DNA damage caused by space radiation is going to be an important problem (Prow et al., 2004) (Figure 9.14)!

However, this gave us the opportunity to think about designing nanomedical systems to activate this secondary pathway by supplying the missing enzyme using a strategy (Prow et al., 2004) shown in Figure 9.15.

Our normal intact UV-DNA repair pathway takes 24–36 hours to repair cells. But when we add the one missing enzyme to complete the secondary repair pathway, the DNA repair can be done in about six hours. In our experiments, we demonstrated highly accelerated DNA repair by activating the second DNA repair pathway, as can be seen by the comet assays (Prow et al., 2004) in Figure 9.16.

A Strategy for Improving DNA Repair in Human Cells

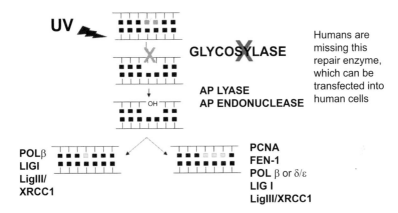

GLYCOSYLASE

AP LYASE
AP ENDONUCLEASE

Humans are missing this repair enzyme, which can be transfected into human cells

POLβ
LIGI
LigIII/
XRCC1

PCNA
FEN-1
POL β or δ/ε
LIG I
LigIII/XRCC1

Figure 9.15 In humans, there is only one mechanism to repair UV-induced damage to DNA due to loss of a glycosylase enzyme needed to make this repair pathway operational.
Source: Prow et al. (2004)

DNA Comet Assays Show Evidence of DNA Repair in Transfected Human Cells

Hxpa DNA repair-deficient human cells   NLSI DNA repairable human cells
No T4 treatment   T4 treated   No T4 treatment   T4 treated

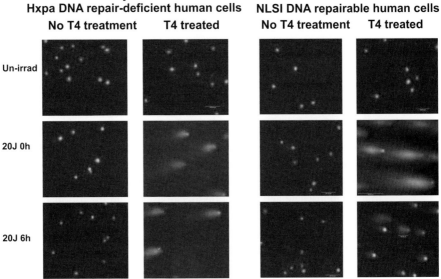

Un-irrad

20J 0h

20J 6h

Figure 9.16 A comet assay shows that DNA damage is successfully repaired at an accelerated rate due to successful activation of a second repair pathway in a transient gene therapy approach to repairing DNA damage as it occurs.
Source: Prow et al. (2004)

Each of these dots in the comet assay is a single cell. The fuzzy regions coming away from those dots, which look like a comet tail (hence the name *comet assay*), are pieces of DNA that have been broken into smaller pieces by near-UV radiation. Some cells are repair deficient, but other cells are repairable. We put these constructs into the cells and then six hours later found repaired cells on these gels. Since the normally intact, native DNA repair system takes 36 hours, we can see that the secondary repair pathway must have been active since it took only the six hours required by this secondary, now activated, pathway.

9.5.3   Nanomedicine Approaches to Eye Diseases

Many of you may think of glaucoma as a problem of intraocular pressure inside the eye. Actually, this is too simplistic to describe what is really going on. Glaucoma is really a problem of optic nerve damage. Approximately 30 percent of people with glaucoma have no elevated intraocular pressure. Interestingly, many of these people originally examined in New York City were super-fit women marathon runners. It turns out that their intraocular pressures were low such that they tested negative by the normal air-puff test. But their intracranial pressure was even lower, so even normal intraocular pressures were causing damage to the optic nerve as a result of this intraocular-intracranial pressure difference. Experts figured this out during an annual Think Tank meeting of the Glaucoma Foundation, convened by founder Dr. Robert Ritch, and one of the many in which I participated. Interestingly, ophthalmologists consider the brain an extension of the eye, whereas brain scientists consider the eye an extension of the brain! It took both groups brainstorming the vexing problem to figure out the problem.

In addition to glaucoma causing problems with the optic nerve, it can also cause retinal damage. To study this retinal damage, we made a nanoparticle with a tethered antioxidant response element sensor switch (Prow, Grebe, et al., 2006; Prow, Smith, et al., 2006), as shown in Figure 9.17.

For simplicity, we took our reporter gene sequence and showed that we could tether it to a magnetic nanoparticle to produce this product inside a living cell, as shown by the fluorescent protein reporter gene sequences in the gene product. To show that it is really tethered to the to the nanoparticle, and not released, we analyzed the products of our experiments on gels and showed that the tethered reporter gene remained attached to the nanoparticle and did not dissociate itself from the nanoparticle.

9.6   The Promise of Regenerative Medicine

Regenerative medicine refers to the idea of regenerating, or fixing, a cell. It is a little different from thinking of the cell as diseased. Often as a cell ages and stops conducting its biochemistry efficiently, it needs to be repaired or restored to its pre-aging self. Even though we do not yet tend to think of aging as a disease, it really is – just a different kind of disease programmed into the genes. We think of aging as

Tethered Gene Expression of Magnetic Nanoparticles for Nanomedicine

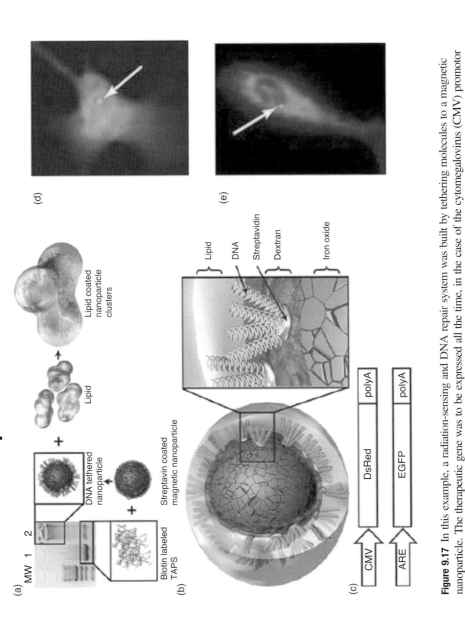

Figure 9.17 In this example, a radiation-sensing and DNA repair system was built by tethering molecules to a magnetic nanoparticle. The therapeutic gene was to be expressed all the time, in the case of the cytomegalovirus (CMV) promotor switch, or expressed only if DNA damage is sensed by an antioxidant response element (ARE) biosensor switch.

(a) Schematic of the construction of the MNP. (b) The layered anatomy of a lipid coated DNA tethered nanoparticle.

(c) Schematics of the two DNA constructs used to assess general transfection with Cytomegalovirus (CMV) and Anti-oxident Response Element (ARE) specific activity.

Source: Prow, Grebe, et al. (2006); Prow, Smith, et al. (2006)

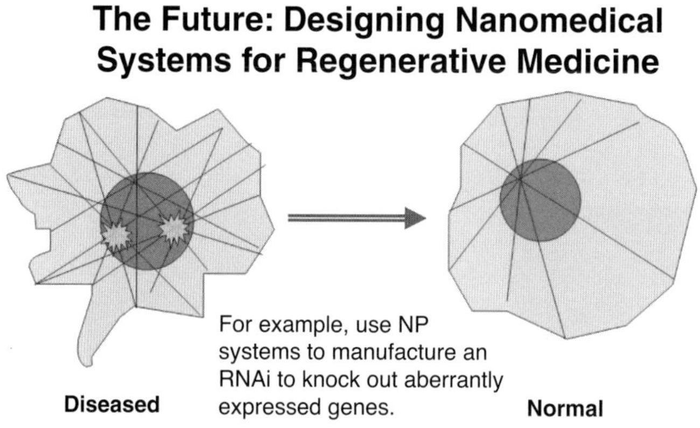

The Future: Designing Nanomedical Systems for Regenerative Medicine

For example, use NP systems to manufacture an RNAi to knock out aberrantly expressed genes.

Diseased   **Normal**

Figure 9.18 Nanomedical systems can, and should be, more than just a way to deliver drugs to kill diseased cells. Whenever possible, we should be trying cure disease by restoring diseased cells to a normal phenotype. It may not be necessary to restore all aberrant gene expressions. It is only necessary to change a relatively small subset of critical ones in a complex gene network to cause a cell to go by itself from a diseased state to a normal state, using a systems biology approach with data from genome-level microarrays to study the full transformation process. Source: Leary teaching

natural, but it is mostly unrepaired cell damage, often to its DNA, or the malfunctioning of its organelles. Is it possible to use nanomedicine techniques to combat the declines of aging in a new form of regenerative medicine whereby genes are reprogrammed at the single-cell level (Figure 9.18)?

9.6.1   Repairing Aberrant Gene Expression through siRNA Techniques

One way to correct aberrant gene expression is to use gene-silencing siRNA constructs to silence the expression of one of more genes. One can envision a strategy of constructing an appropriate cocktail of these gene-silencing probes to drive the cell from an abnormal phenotype to a "more normal" one, which becomes a "new normal" metastable phenotype. While this on the surface seems complex, it may be sufficient to change a few critical expressed genes that will cause the cell to want to return to a normal, or at least quasi-normal, state that is less of a problem in terms of disease outcome.

9.6.2   Using Gene-Editing Techniques for Regenerative Medicine

A recent, exciting advance is the use of gene-editing techniques to perhaps fix gene defects in one or more organs of an adult. Since not all genes are expressed in different types of cells, it is not necessary to change all cells to treat a disease or the effects of aging. Gene-editing techniques have evolved over the past decades to a point where new gene-editing techniques such as CRISPR can edit genes down to the resolution of a single base pair mutation. This can be done at the somatic rather than germ-cell level, which addresses ethical concerns about altering germ cells themselves.

9.7   The Promise of CRISPR Gene Editing

When zinc-finger gene editing was the current state of the art, the overall technique had such a degree of imprecision and experimental challenges that gene-editing solutions in nanomedicine seemed remote. Transcription activator-like effector nucleases (TALENs) was another way to improve on the problems of zinc-finger technology (Gaj et al., 2013). Then a remarkable advance took place with the invention of clustered regularly interspaced short palindromic repeats (CRISPR) gene editing, whereby it appears not only possible, but even practical, to perform gene editing at precise locations anywhere in the genome of somatic cells, with an ability to edit down to a single base pair (Cong et al., 2013; Jinek et al., 2012). The Nobel Prize in Chemistry 2020 was awarded to Emmanuelle Charpentier and Jennifer Doudna for discovering one of gene technology's most powerful tools, "CRISPR/Cas9 genetic scissors." Dr. Doudna has written a fascinating lay-level book (Doudna and Sternberg, 2018) on the history of CRISPR technology as well as addressing ethical concerns about her scientific achievement.

9.7.1   What Is CRISPR Gene Editing, and How Does It Work?

In no way will I even attempt in this book to explain the full complexities and variations of CRISPR technology. That is the task of more comprehensive sources on the subject. One very readable handbook version has been provided to the research community by Genscript (http://genscript.com/CRISPR-handbook.html). A recent publication shows how CRISPR technology may provide a new way to perform gene therapy within the field of nanomedicine, as shown by the use of liposomes containing CRISPR technology to target cells (Rosenblum et al., 2020). But I would be remiss if I failed to at least introduce the technology to the level where readers can appreciate its importance and try to envision how it might work in a nanomedical system of their own design. As seen from the full name behind the CRISPR acronym, the laboratory technique started as yet another case of biomimicry whereby viruses in nature insert their DNA into host cell genomes and the host develops a defense mechanism to remove those virus sequences. In our case, CRISPR techniques can remove bad sequences of its DNA and replace them with normal sequences. Nanomedical drug delivery methods could be used to target CRISPR gene-editing tools to the appropriate cell type or organ as part of an in vivo gene-editing therapy. Aberrant genes in some organs do not matter because those proteins or functions are not needed for those organs to properly function. But in other organs, proper functioning of those genes may be essential. For this reason, a targeting of gene-editing nanomedical systems would seem to be a very good idea!

This is yet another example of biomimicry from nature pointing the way by seeing how bacteria manage to protect themselves from invading bacteriophage viruses. Interestingly, some of the initial ideas for CRISPR technology were first observed in the food industry. Yogurt tends to get easily infected with bacteria, which can change the taste of the yogurt. When scientists studied the process, they discovered that the bacteria had developed a single-cell immune system whereby the cells found a way to recognize bacterial infection and edit it back out.

How does CRISPR/Cas9 work? There are two main components to the CRISPR/Cas9 genome-editing system. The CRISPR arrays allow the bacteria to "remember" the viruses (or closely related ones). If the viruses attack again, the bacteria produce RNA segments from the CRISPR arrays to target the viruses' DNA. The bacteria then use Cas9 or a similar enzyme to cut the DNA apart, which disables the virus. When CRISPR responds to an invading bacteriophage, the bacteria transcribe the spacers and the palindromic DNA into a long RNA molecule that the cell then cuts into short spacer-derived RNAs. Cas9 protein is a nuclease, an enzyme specialized for cutting DNA, with two active cutting sites, one site for each strand of the DNA, without interfering with the ability of the complex to home in on its target DNA. But to control where Cas9 cuts, it is necessary to create a single-guide RNA molecule from early fragments created during the CRISPR process. This control of CRISPR is easier and more precise than the TALENs technique, which itself improved greatly over the precision of zinc-finger technology.

The Cas9 protein initially recognizes the DNA and also acts like a pair of "molecular scissors" (Pennisi, 2013) that precisely cleaves the targeted DNA sequence. The guide RNA guides the Cas9 scissors to the desired target DNA sequence and activates the scissors so they cut. A simplified view of CRISPR/Cas9 technology (Chadwick and Musunuru, 2018) is shown in Figure 9.19.

It is possible to alter virtually any gene with Cas9 by exploiting its DNA-cutting ability to either disable the gene or cut it apart and then allowing any substitute DNA desired to be inserted. This can be done multiple times on the same cells to edit out multiple sequences at different locations in the human genome, so its applications are virtually limitless!

9.7.2   How Will CRISPR Technology Impact the Field of Nanomedicine?

Sickle cell disease affects millions of black people. It is caused by a mutation to the hemoglobin gene, which then causes red blood cells to have a sickle shape that can cause these cells to become stuck in blood vessels, causing great pain and sometimes more serious consequences if blood vessels are completely occluded by the buildup of sickled cells. Sickle cells also have a shorter lifetime, which causes some affected people to have anemia.

Beta-thalassemia, an autosomal–recessive gene mutation, affects approximately 1 in 100,000 people of Mediterranean, African, and South Asian ancestry. The only cure has previously been a bone marrow transplant of a closely matched donor – a sibling or, preferably, an identical twin – obviously not possible for most sufferers!

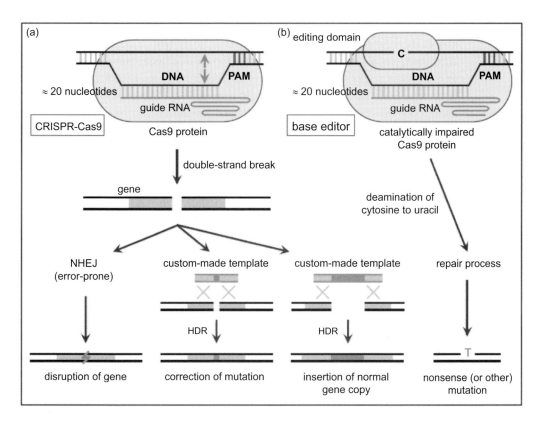

Figure 9.19 Schematic of possible outcomes of genome editing that are relevant to the prevention and treatment of atherosclerosis. (a) Standard CRISPR-Cas9 is directed to a genomic site by the protospacer-adjacent motif (PAM) in the DNA and protospacer sequence in the guide RNA, whereupon it generates a double-strand break (location indicated by arrows pointing to DNA strands). The break is repaired by either nonhomologous end-joining (NHEJ) or homology-directed repair (HDR), which yield different outcomes. (b) The base editor does not generate a double-strand break but rather converts a cytosine base (or bases) into a uracil base, which is ultimately replaced with a thymine base (with a complementary change of guanine to adenine on the opposite strand).
Source: Chadwick and Musunuru (2018)

Any transplant from another individual means that the patient must take immuno-suppressive drugs for the rest of his or her life to prevent organ rejection. Such immunosuppressive treatments over time greatly increase the incidence of lymph-omas in those individuals, as well as increasing susceptibility to infections and infectious diseases.

Recently, CRISPR gene editing has been used on a small number of patients. Stem cells from the patients are taken from blood and then treated with CRISPR technology to knock out the mutated hemoglobin gene (Frangoul et al., 2020). Successful treatment with CRISPR technology precludes the need for immunosuppression, thereby eliminating those increased risks of infection or lymphoma.

While it is still early to fully see exactly how and where CRISPR gene editing will impact the field of nanomedicine, there is little doubt that its overall impact will be significant. An important advance needed for in vivo gene therapy using CRISPR technology is a way to package this technology inside a nanomedical complex and target it to the cells of interest. A targeted nanomedical delivery system would allow intravenous injection of these systems for gene-editing therapy throughout the human body or, alternatively, to specific tissues and organs in a targeted delivery approach.

9.7.3 Ethical Issues about Application of Gene-Editing Technologies to Minorities and Ethnic Groups

Every new technology has the potential for good or bad uses. Already there has been an unethical use of CRISPR by a researcher on germ cells in human babies. This has led to ethicists seeing the need to lay out some ground rules for ethically appropriate uses of CRISPR technology.

First, it is important to separate facts from fiction when discussing ethical issues of technology (Cribbs and Perera, 2017). It should also be seen as a moving ethical target, whereby some ethical issues are no longer relevant due to advances in the field that make the use of less ethically dubious methods no longer necessary. That was the case in the early use of embryonic stem cells – a subject of some major ethical concerns. Scientists eventually learned that use of embryonic stem cells was not as good as finding pluripotent (non-embryonic) stem cells present in most organs of the human body. Interestingly, these pluripotent stem cells in a particular organ already know through their gene expression profile and how to help regenerate that organ. Pluripotent stem cells just need to have the proper signals sent to them by injured organs to start to repair those organs.

Second, one of the challenges of treating patients with these therapies is that certain minorities and ethnic groups treated unfairly or unethically in the past are reluctant to participate as research subjects or as volunteers in clinical trials. They are also underrepresented in genetic databases. This means that special efforts will be needed to address these health system disparities in CRISPR therapies (Hildebrandt and Marron, 2018).

Chapter 9 Study Questions

9.1 What are two different ways that a gene can be delivered to a cell for therapy? What are the advantages and disadvantages to each approach?

9.2 What are two advantages of a therapeutic gene approach to therapy?

9.3 Why might feedback control be a good idea? Should that feedback control be positive or negative, and why?

9.4 What is a molecular biosensor, and why is it important for therapy?

9.5   What is transient versus stable expression? Why do we usually prefer transient expression systems?

9.6   What are some different types of nanoparticles for drug/gene delivery?

9.7   How does a molecular biosensor provide for programmability or decision making?

9.8   Why are fluorescent reporter genes useful in developing new gene therapy systems?

9.9   How might CRISPR gene editing be applied to certain types of diseases?

9.10 Why might using nanomedical system targeting of gene-editing tools make sense in treating certain diseases?

10 Assessing Nanomedical Therapies at the Single-Cell Level

10.1 Introduction and Overview

How do you assess whether your nanomedical system is actually doing what you wanted it to do? More specifically, is it having a positive therapeutic effect on the cells that it is treating? To answer this question, it is necessary to make quantitative single-cell measurements of efficacy. "Efficacy" is broadly defined as anything that is helpful or is providing some kind of positive therapeutic effect, either directly or indirectly, that can occur on a single-cell level. While traditional medical measures of efficacy remain relevant and useful, it is important for a nanomedical approach to remain focused mostly on what is happening at the single-cell level.

This efficacy can be measured either as a structural or functional effect. Usually, we try to make simpler structural measurements, rather than more complicated functional measurements. We have some faith that there is a relationship between structure and function, but this is not always the case and sometimes not possible.

One way we can start assessing function is by using so-called phospho-specific antibodies because when proteins undergo phosphorylation, they go from an inactive state to an active state. By measuring the state of phosphorylation, we can determine whether the protein is switched *on* or *off*. Proteins are the agents of change within a cell. The science of protein action from the gene level outward is known as proteomics.

A typical human cell contains 20,000–30,000 different functioning proteins (they can be different based on cell type) expressed both independently and cooperatively as gene clusters. Different types of cells have different clustered genes, leading to different gene expression profiles (GEP). What we are trying to see, and understand, is the movement of a cell from a diseased state to a normal state. To do this, we must first characterize the normal state of the cell in terms of all these gene products. The totality of a cell's expressed gene products is known as its gene expression profile. First, we must define the normal state of a specific cell type by its GEP. Then we must learn how to define the diseased state in terms of its own GEP.

Most proteomics is currently being done on serum and serum proteins but not at the cellular level or, even more rarely, the single-cell level. Nanomedicine efficacy is perhaps best evaluated using single-cell proteomics techniques, some of which will be discussed in this chapter. The most powerful technology for studying differences in protein expression (single-cell proteomics) is flow cytometry, which is why this chapter will discuss these measures of proteins in their activated, or inactivated, states by flow cytometry.

10.1.1   Nanomedical Treatment at the Single-Cell Level Requires Evaluation at the Single-Cell Level

Evaluation of treatment efficacy requires knowledge of disease at the single-cell level. Such knowledge is still in very short supply. Some effects, like premature truncation of a protein due to gene defects, are obvious. But many diseases involve biochemical perturbations or imbalances that are more difficult to detect and fix.

10.1.2   Does Structure Reveal Function?

A fundamental belief in science is that "structure reveals function." Structure is also much easier to evaluate than function. Proteins change their structure when they are either in an activated or deactivated (i.e., inactive) form. The normal process of these changes in a protein's structure for it to become functional is described as phosphorylation.

10.1.3   The Difficulty of Anything but Simple Functional Assays

It is extremely difficult to quantitate anything other than fairly simple functional assays. But we must be careful that these simple assays truly reflect a change from a diseased state to a normal state. That said, the most important aspect of efficacy is when a cell is functioning properly. Although it has not fully returned to normal, causing a cell to stop carrying out the most destructive functions as a diseased cell should be seen as a partial success. The treated cell may not return completely to its previous normal state, but it can at least be less harmful in terms of disease.

10.1.4   The Need for Assays to Show Correlation to Functional Activity

The quantitative assays must directly, or indirectly, lead back to single-cell evaluations because nanomedicine is single-cell medicine. We need to assess things one cell at a time, correlated to functional activity. Functional assays are, in general, very difficult to do. Many functional assays require large numbers of cells. Another challenge is that living cells may change quite rapidly, negating any concept of a group of cells that are really the same. A general dictum is that measurements must always be much faster than the rate of the change in the cells to be meaningful – a good thing to always keep in mind!

10.2   Quantitative Single-Cell Measurements of One or More Proteins per Cell

One of the most powerful ways of quantitatively measuring the number of specific proteins on or within a single cell is by single-cell flow cytometry. Nanomedicine is

A Flow Cytometer Is a Virtual Fluorescence Microscope

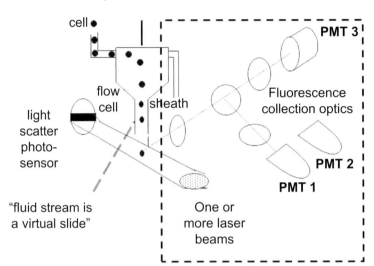

Figure 10.1 A flow cytometer is an instrument capable of making thousands of quantitative measurements per second on single cells as they pass through the light source.
Source: Leary teaching

single-cell medicine. Flow cytometry is a technology for measurements on large numbers of single cells. It is a natural wedding of problem and solution! We can measure thousands of individual cells per second and can distinguish between different types of cells in a cell mixture. For this reason, a great deal of this chapter will discuss the use of flow cytometry to study proteins associated with single cells, including whether those particular proteins are phosphorylated.

A flow cytometer in its simplest interpretation is just another type of fluorescence microscope, except that instead of putting cells on a slide we have them flow one at a time past a light source (e.g., a laser beam) and measure different colors of fluorescence, each marking the presence of a particular molecule on, or within, a single cell (Figure 10.1).

10.2.1   The Power of Multiple Correlated Measurements per Cell

While measurement of a single parameter of a cell is powerful, a much more important capability of a flow cytometer is to make multiple correlated measurements on the same cell. Such multiparameter measurements are capable of seeing specific cell subpopulations that are otherwise impossible to view. Figure 10.2 shows the power in seeing many more cell subpopulations using only two correlated measurements per cell. Imagine the power of making 20 or more simultaneous measurements per cell!

Importance of Quantitative Single-Cell Molecular Measurements

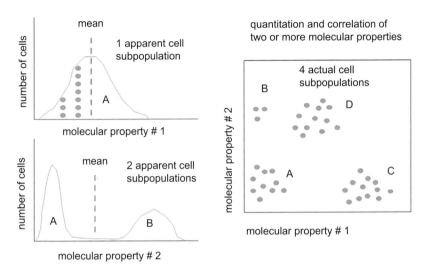

Figure 10.2 What may appear to be just one or two subpopulations may actually be many more if one looks at multiple correlated measurements per cell.

Source: Leary teaching

10.2.2 Why Single-Cell Measurements by Flow Cytometry Are Superior to Bulk Cell Measurements

Many assays require a large number of cells to make the measurement. An inherent problem with such measurements is that heterogeneous cell mixtures can contain cells with vastly different expressions of specific proteins. Hence, a bulk measurement is the sum of the numerically weighted presence of each different protein and can lead to a distorted picture of the protein expression levels of the cells in that mixture. For people accustomed to quantifying cells on Western blots, this is a good illustration of why Western blots can fail to capture the presence of two or more cell subpopulations having very different amounts of a specific protein (Krutzik et al., 2004) (Figure 10.3).

10.2.3 Use of Cell Sorting to Purify Cell Subpopulations

One way that flow cytometry has tried to solve this problem is to sort what appear to be similar cells out from the cell mixture, obtaining "purified" cell subpopulations. The result is a purified cell subpopulation hopefully containing similar cells in

Why Single-Cell Analysis Is Superior to Bulk Cell Analyses

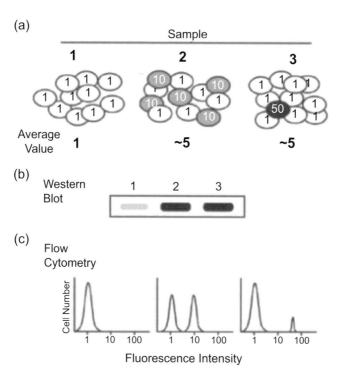

Figure 10.3 This figure illustrates why flow cytometry can measure vastly different protein expression levels on cell subpopulations not discoverable by Western blot analyses. Source: Krutzik et al. (2004)

sufficient number to allow subsequent assays to be performed or for the live sorted cells to be further grown in tissue culture. We can even sort to the single-cell level in plates to obtain different cell clones. Depending on the fraction of cells having desired properties, several thousand cells per second can be sorted into separate containers for subsequent assays. Importantly, this allows us to study drug effects on single cells of a particular type isolated from a heterogeneous mixture of cell types. But we must be careful to also include "accessory cells," which must be present for those cells to function.

The cell sorter is just a flow cytometer making the initial measurements joined to a second-stage cell-sorting module that can subsequently sort cells according to those upstream measurements. Currently, most cell sorters use inkjet technology to charge droplets containing specific cells and then electrostatically deflect droplets containing

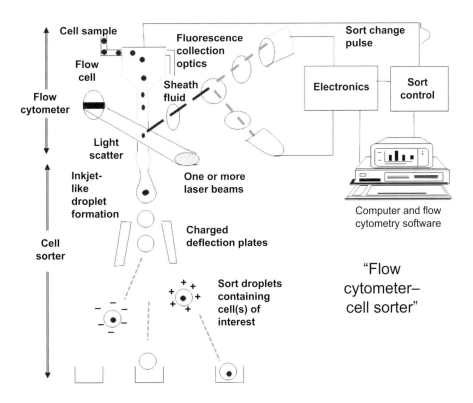

Figure 10.4 A cell sorter is just a flow cytometer to which has been attached some components that allow the separation of specific cell types. Usually, the sorting is performed by inkjet-based electrostatic deflection of droplets containing specific cells, but this can also be done with microfluidic valves.
Source: Leary teaching

similar cells, based on those flow cytometric measurements, into different containers. However, other methods of sorting, such as fluidic switching or laser ablation, can also be used, as shown in Figure 10.4.

10.2.4   Cell Surface or Intracellular Measures of Protein Expression on Live, Single Cells

Molecules on or within a cell can be fluorescently labeled in a variety of different colors. These fluorescently labeled cells can be simultaneously measured for multiple different colors of fluorescence either on the individual cell surface or intracellularly. More than 17 different colors of fluorescence had been simultaneously measured as of 2004 (Perfetto, Chattopadhyay, and Roederer, 2004). Since that time, many more colors per cell have been measured by flow cytometry. With the invention of mass cytometry (Bandura et al., 2009), based on inductively coupled plasma time-of-flight mass spectrometry, the number of potential parameters is virtually limitless. However, mass cytometry is destructive in its procedure, with single cells destroyed during the process, meaning that sorting of live, or even fixed, cells by this technique is not possible.

The main challenge now is how to analyze such high dimensionality dataspaces. Gone are the days when a human data analyst could visualize the data by human pattern recognition! Luckily, there are now a number of data-mining techniques for high-dimensionality flow cytometry data, as was discussed earlier in Chapter 3, Section 3.9. The problems of visualization and replacing human experts using pattern recognition alone drove the field of flow cytometry to finally(!) embrace bioinformatic techniques using R software (Montante and Brinkman, 2019). Repositories for a growing number of data-mining techniques using software written in R are now available (www.bioconductor.org/packages/release/bioc/html/flowCore.html).

There are now many thousands of commercially available antibodies that can identify specific proteins. If those proteins are intracellular, then the cell must first be chemically fixed. Such chemical fixation can, and sometimes does, change the shape of a protein, so this must be taken into account. If the 3D structure of a protein is changed during fixation, an antibody may or may not still recognize that protein. Sometimes there is a different antibody needed to identify a fixed versus natural form of a specific protein. The antibodies that recognize the fixed form of a protein are called "phospho-specific antibodies." Many phospho-specific antibodies are now commercially available, allowing us to perform single-cell proteomics by flow cytometry.

10.2.5   High-Throughput Flow Cytometric Screening of Bioactive Compounds

In addition to the wealth of thousands of antibodies available to identify molecules (mostly proteins), virtually any molecule can be detected using "biomimetic" (meaning analogous to recognition by antibodies but without using antibodies) peptides or aptamers. Some molecules may not be detectable using conventional antibodies. Instead, these cellular proteins may be identified using high-throughput flow cytometry of bioactive compounds (Blind and Blank, 2015; Yang et al., 2002, 2003, 2011) in combinatorial chemistry libraries. There are vast libraries of aptamers (Lee et al., 2004) whose sequences are published and which can be readily synthesized.

10.2.6   Challenges of Measuring Protein Expression Inside Fixed, Single Cells

Chemical fixation used to access intracellular proteins can change the shape of a protein, so care must be taken. Sometimes there is a different antibody needed to identify a fixed versus natural form of a protein. Recently, that has become critical to the development of phospho-flow cytometry, which allows the distinction of phosphorylated versus nonphosphorylated proteins on or within a cell by using antibodies that recognize those different forms of proteins. It is effectively a way of determining not only which genes are active versus non-active, by using antibodies labeled with different fluorescent colors for phosphorylated and nonphosphorylated proteins, but also what fraction of a given protein is phosphorylated in a single cell. This provides a faster, much less expensive way of assessing gene expression profiles as different colors of flow cytometry on single cells as opposed to gene expression profiles, which are normally only done on many cells. Single-cell gene expression profiles are possible, but expensive and hard to do in sufficient quantities to see the variations between cell subpopulations or the variation of GEP within specific cell subpopulations.

10.2.7 High-Resolution 3D Imaging Spatial Location of Proteins Inside Cells If Location Is Important

Two-dimensional and three-dimensional image analysis methods are also important tools for measuring the location of proteins inside a cell. One of the remarkable discoveries of the Human Genome Project was the unexpectedly low total number of human genes. It turns out that the same proteins can have different functions according to where they are located within a cell. This has led to a whole subfield of proteomics known as "locational proteomics" (Murphy, 2005). It requires high-resolution 3D microscopic techniques on single cells to locate the proteins, along with sophisticated bioinformatics techniques to combine this with proteomics and other databases. With locational proteomics, the location of a protein within a cell is important. Due to the surprisingly limited size of the human genome, proteins in different locations within a cell serve different purposes, allowing the same proteins to be used over and over to perform different cellular functions.

10.3 Quantitative Multiparameter Phospho-Specific Flow Cytometry

As mentioned earlier, it is possible to construct monoclonal antibodies that can distinguish between phosphorylated and nonphosphorylated forms of the same protein. These special antibodies are called "phospho-specific antibodies." We can use these two different antibodies, each labeled with a different color of fluorescence, to measure the relative fraction of a particular protein that is phosphorylated.

10.3.1 Attempts to Measure Functional Proteins by Detecting Phosphorylation in Single Cells

By using phospho-specific antibodies (Krutzik et al., 2004; Krutzik and Nolan, 2003), we can attempt to determine whether a protein is "functional." As discussed previously, measuring differences in structure, such as phosphorylated proteins, with phospho-specific flow cytometry is a far easier way to try to assay functional changes in a single cell. Usually, such functional proteins are identified by phospho-specific antibodies, but sometimes the protein is switched on, in terms of function, only if the phosphorylated form is not expressed!

10.3.2   Examples of Phospho-Specific Antibody Staining by Multiparameter Flow Cytometry

An important scientific advance was the use of phosphorylated proteins and their single-cell detection by multiparameter flow cytometry (Krutzik et al., 2004; Krutzik and Nolan, 2003). A general schematic of the idea of phospho-specific antibodies to identify functional proteins is shown in Figure 10.5 (Krutzik et al., 2004).

An example of detection of actual cell subpopulations of actual stimulated or unstimulated cell functional proteins is shown in Figure 10.6 (Krutzik et al., 2004; Krutzik and Nolan, 2003).

10.3.3   Measuring Single-Cell Gene Silencing

One way of approaching disease at the single-cell level is by gene silencing, whereby undesirable properties of a cell can be stopped by interruption or suppression of the expression of a gene at transcriptional or translational levels. A type of gene therapy is "gene silencing" whereby one or more gene expressions are blocked. We can measure this gene silencing by using phospho-specific antibodies to detect the flipping of active (on) genes to an inactive (off) state, or in some cases, flipping of inactive (off) genes to an active (on) state. An example of gene silencing is shown in Figure 10.7 (Chan, Olson, and Utz, 2006). These methods are very useful for applications of drug or gene delivery therapies that silence specific genes.

10.4   Quantitative Measures of Gene Expression: The Promises and the Realities

A way to approach the general problem of using gene therapy to treat a diseased state is to use gene arrays. By examining the gene expression profile of normal and diseased cells, one can begin to understand how to move a cell in a diseased state back into its normal state. This may not be as daunting as it sounds. It probably is not necessary to change all the genes. Pioneering work from the lab of Mina Bissell has shown that altering the microenvironment of a cell can change its degree of malignancy without actually changing the genotype itself (Bissell et al., 2002) – a phenomenon she has referred to as "phenotype is dominant over genotype." This has very important implications for the field of nanomedicine. For example, if an organ has a high number of cancer cells, then even eliminating them and sparing normal cells may not leave that organ in a functioning state. The ability to allow cancer cells to remain, while reducing or eliminating their ability to metastasize, may be the better treatment for some patients.

The cell has evolved to maintain groups of processes in a kind of "single-cell homeostasis" that helps keeps the cell in a normal functioning state. Disease can perturb the cell out of this more natural state to a diseased state. But once enough genes have been altered, the cell will itself try to finish the process and walk its way

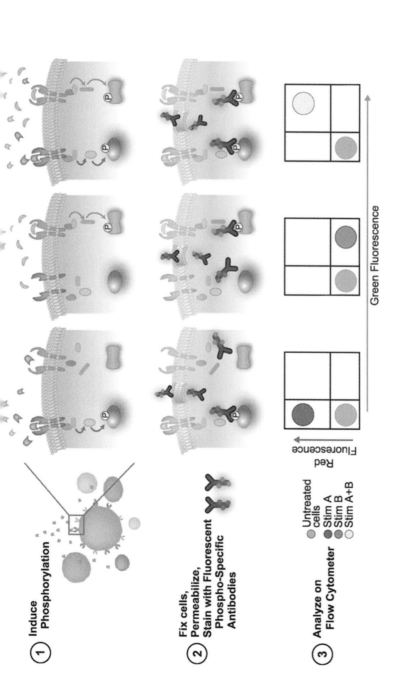

Figure 10.5 "General phospho-protein staining technique for flow cytometry. (Step 1) A heterogeneous sample of cells is treated with two different stimuli, A and B (i.e., cytokines, growth factors, drugs, inhibitors), to induce distinct signaling cascades and phosphorylation of two target proteins. A third sample is treated with both stimuli simultaneously to induce phosphorylation of both proteins of interest. (Step 2) The cells are then fixed, permeabilized, and stained with fluorophore-conjugated phospho-specific antibodies to the phosphorylated (and typically active) forms of the two proteins (surface markers can also be stained during this step with appropriate antibodies and fluorophore combinations). (Step 3) Finally, the cells are analyzed on a flow cytometer with two or more fluorescence channels. Because the antibodies bind only to the phosphorylated form of the proteins, an increase in fluorescence correlates with an increase in phosphorylation. Therefore, stimulus A produces an increase in red fluorescence because the red protein is phosphorylated. The combination of stimuli A and B induces phosphorylation of both proteins making the cells both green and red fluorescent. This technique can also be applied to patient samples to help characterize aberrant signaling events that occur during disease progression or determine the efficacy of signaling pathway-specific drugs in vivo. In this case, samples must be isolated from patients and immediately subjected to fixation and permeabilization conditions that will maintain phospho-epitope integrity."

Source: Krutzik et al. (2004, 212)

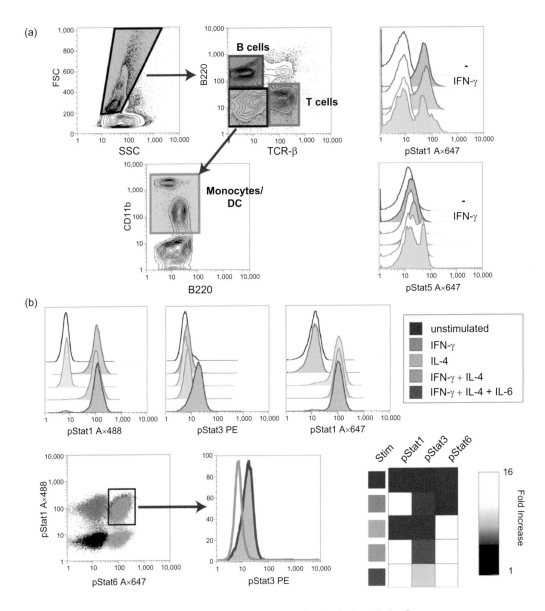

Figure 10.6 Differences in phosphorylated proteins in single cells by flow cytometry. "Multidimensional analyses with phospho-specific flow cytometry. (A) Surface markers with phospho-epitope staining. Murine splenocytes were subjected to IFN-g stimulation (filled histograms) or left unstimulated (open histograms), then stained with B220, TCR-h, and CD11b to distinguish B cells (blue), T cells (red), and monocytes or dendritic cells (green), respectively. The cell types were simultaneously analyzed for induction of Stat1 and Stat5 phosphorylation with phospho-specific antibodies. B cells and T cells showed clear Stat1 responses to IFN-g, but the CD11b-positive population was heterogeneous in its response. Only minor inductions of phospho-Stat5 are seen. (B) Multiple kinases: U937 cells were treated with IFN-g, IL-4, and IL-6 in the combinations shown. The cells were then analyzed for pStat1, pStat3, and pStat6 simultaneously after fixation and permeabilization. The top panel shows histograms of each

back from a diseased state to a normal, or at least more normal, state. For this reason, we will now discuss how gene expression analysis is performed because it is an important way of assessing the general health and function of single cells.

Gene expression array chips have evolved from quirky, somewhat unreliable measures of gene activity to much more precise measures of gene activity.

Detection of siRNA-Mediated Protein Knockdown with Flow Cytometry

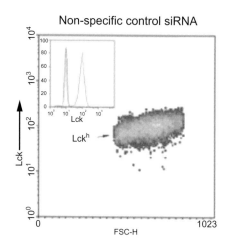
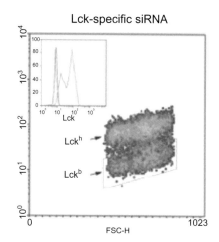

Figure 10.7 "Detection of siRNA-mediated protein knockdown with flow cytometry, Western blotting, and immunofluorescence microscopy. (A) Density plot of Jurkat T cells 48 h post-transfection with either a non-specific control or Lck-specific pool of siRNA. Forward scatter (FSC-H) is on the X-axis and intracellular Lck staining on the Y-axis. A representative gate circumscribing the Lcklo subpopulation is shown. Histograms of the corresponding cell populations for Lck are shown in the insets (green histograms). Red histogram plots represent background staining with purified rabbit gamma globulin. These plots are representative of ten independent experiments.
Source: Chan, Olson, and Utz (2006)

Caption for Figure 10.6 (*cont.*) channel individually and clearly shows the expected induction of Stat1 with IFN-g, Stat3 with IL-6, and Stat6 with IL-4. When plotted in two dimensions (lower left panel), two samples appear coincidentally in the pStat1/pStat6 positive quadrant. However, when one analyzes these samples for pStat3, only the sample treated with IL-6 shows an induction. Therefore, samples that appear homogeneous within two dimensions can be separated clearly with simultaneous staining in three dimensions. Such correlations are not possible with Western blotting. The lower right panel is a representation of the data generated by a FACS analysis tool being developed in our laboratory. Each row represents a different stimulus, and each column represents a phospho-protein. The color of each block is indicative of the fold change in median fluorescence intensity in that channel. The data are easily visualized and compared without needing to plot all 15 samples. Larger screening experiments will require this form of analysis."
Source: Krutzik et al. (2004, 214)

Affymetrix Microarray Images of CEM and A2780 Cells

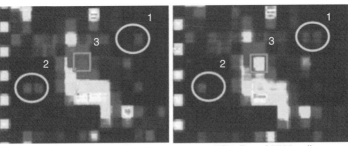

PS1. Pure CEM cells   PS7. Pure A2780 cells

Figure 10.8 Each small square represents the expression level of an individual gene sequence. Examples for gene sequences expressed only in one of the cell types are marked with ovals 1 and 2. The sequence to the left in oval 1 is only expressed in A2780 cells; the sequence to the right in oval 2 is only expressed in CEM cells. A differentially expressed gene sequence with much higher expression level in A2780 cells than in CEM cells is visible in square 3. (To visualize the raw image data of microarray analysis, corresponding segments of Affymetrix images were cropped, magnified, and pseudo-colored using the Affymetrix Microarray Suite 5.0 software).
Source: Szaniszlo (2007)

Figure 10.8 shows a magnified image of the array surface of an Affymetrix microarray we used in early experiments to study the gene differences between two human cell lines. These images show magnified regions of the tiny elements of the arrays, each corresponding to particular genes, with pseudo-color measures of differences in levels of gene expression (Szaniszlo, 2007).

10.4.1   Is Measuring Gene Expression at the Single-Cell Level Really Possible?

When we began our own studies of changes in gene expressions many years ago, it was not possible to look at the gene expression levels of a single cell. For this reason, we had to use cell sorting to attempt to provide pools of purified cell subpopulations so we would have enough DNA to assay with the chips. As shown in Figure 10.9, we had to ask ourselves two basic questions. First, could we identify cell subpopulations containing truly similar cells so that studies of these cell subpopulations were truly meaningful? It became abundantly obvious early on that no two cells were exactly alike. But they were at least similar in meaningful ways. The second question was how purified the cell subpopulations needed to be, and how many cells were required to give reliable results for those sorted cell subpopulations (Szaniszlo et al., 2004).

The answers to the two questions are that it is indeed possible to derive useful information about differences in gene expression between two cell subpopulations, and it only requires a realistic number of cells of a realistic purity. That said, it was also obvious that better measures of single-cell variations in gene expression levels required analyzing single cells using technologies to amplify the sequences of those

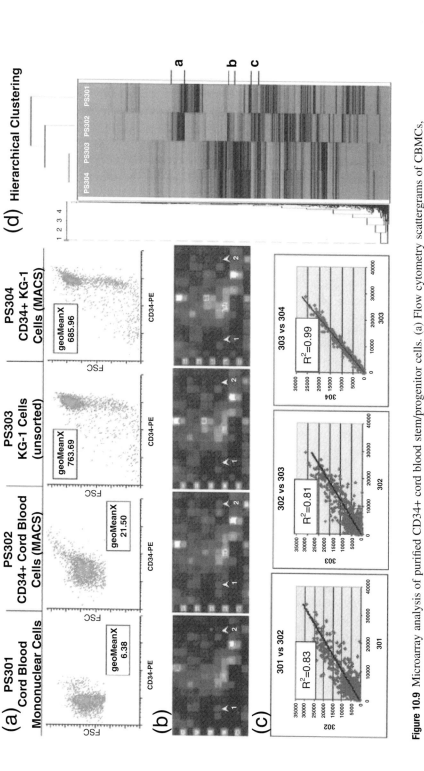

Figure 10.9 Microarray analysis of purified CD34+ cord blood stem/progenitor cells. (a) Flow cytometry scattergrams of CBMCs, MACS-purified CD34+ cord blood stem/progenitor cells, unsorted KG-1a cells, and MACS-sorted KG-1a cells. (b) Affymetrix microarray images of the same four samples. (c) The scatterplots shown confirm that the sort process did not introduce any distortion into the overall GEPs since the sorted and unsorted KG-1a cells were truly identical (R2 = 0.99). With an R2-value of 0.83, the overall GEP of purified CD34+ stem/progenitor cells was very different from the GEP of unsorted CBMCs. (d) Heat map of the same four samples. Samples and genes are ordered by Spotfire hierarchical clustering analysis based on normalized expression levels. Some groups of genes (a, b, and c) are differentially expressed in CD34 cord blood stem/progenitor cells.

Source: Szaniszlo et al. (2004)

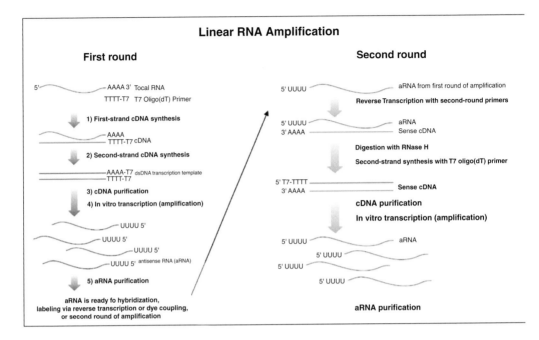

Figure 10.10 Reverse transcription with oligo(dT) primers only converts the poly-A-tailed mRNAs into cDNA. The reverse transcription reaction is also used to label the cDNA with a T7 promoter sequence (T7 oligo[dT]primer). In vitro transcription utilizing T7 RNA polymerase is used to amplify each reverse-transcribed sequence starting transcription at the T7 promoter sequence. The resulting product is single-stranded, antisense RNA (aRNA or cRNA). This product can be used for a second round of amplification if necessary. (Adapted from www .ambion.com.)

Source: Szaniszlo (2007)

single cells to the point where the gene expression profile of single cells could be reproducibly done. This led us to efforts to linearly amplify the RNA of single cells. A general strategy for this amplification process is shown (Szaniszlo, 2007) in Figure 10.10.

After many efforts, we were able to linearly amplify most of the gene expression of single cells, as shown in Figure 10.11 (Szaniszlo, 2007). I include these results from early studies to show the difficulty and the immense efforts from us and many other people over a period of years to accomplish this task. We subsequently were able to get reasonably reliable and reproducible results on single cells (Szaniszlo et al., 2004).

10.4.2   Gene Arrays of Purified Cell Subpopulations

Since cell subpopulations have their own distinct GEPs, it is important to use technologies such as multiparameter flow cytometry with cell sorting to isolate enough cells to perform GEP analyses. In the beginning, these efforts were tedious, particularly for rare cell subpopulations, because it required sorting live cells that were

In General, Linear RNA Amplification Preserves Many of the Differences between Samples

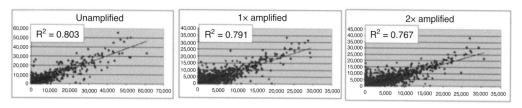

Figure 10.11 The GEPs of CEM and A2780 cells were plotted against each other after each round of T7 amplification. Similar R^2 values indicate similar correlations between the two cell types after each round. The R^2 value did not increase after amplification.
Source: Szaniszlo et al. (2004)

changing and dying as they were being sorted. Later use of reagents such as RNAlater® (Ambion, Inc., Austin, TX) (Sikulu et al., 2011) allowed the chemical fixation of these cells so that they could be sorted for longer periods of time without concerns that they were changing during the time that it took to sort these cells.

10.4.3   Is It Even Possible, or Useful, to Try to Measure a Single Gene's Changes Inside a Single Cell?

While this may appear to be an almost philosophical question, it is important to try to approach this question because nanomedicine is designed to treat single cells. If any two cells are sufficiently different, it may or may not be possible to design nanomedical systems that can treat multiple diseased cells. We do not yet know the answer to this question, but it appears that no two cells are identical in their GEP. However, they are probably similar enough for the cell to walk itself from a diseased to normal state because the cell itself knows how to do this.

10.4.4   RNA Amplification Techniques to Attempt to Perform Single-Cell Gene Arrays

The science of RNA amplification has progressed considerably since our lab began trying in 2004 to perform single-cell gene arrays (Szaniszlo et al., 2004). It was difficult, clumsy, and expensive to try to do single-cell gene arrays using a full Affymetrix gene array. We tried first using linear RNA amplification techniques to obtain enough RNA for a gene array system that was never designed for single-cell analysis. The technology at the time was just not there yet, so we eventually moved on to other research problems. But now the technology has greatly advanced to allow the measurement of single-cell gene profiles in a reasonable way (Efroni et al., 2015; Sachs et al., 2020). Our job was to point the way and then let others, using more advanced technology, make it practical.

Chapter 10 Study Questions

10.1   Why, or why not, do structure and size reveal function?

10.2   Why are structural measurements usually easier than functional measurements?

10.3   How sensitive are in vivo imaging techniques in detecting growth or shrinkage of a tumor during treatment?

10.4   Why is it important to assess drug-induced changes at the single-cell level?

10.5   Why is it a challenge to measure protein expression inside single cells?

10.6   The location of a protein is important in determining its functions. How can we get location information of proteins in single cells?

10.7   Phospho-specific antibodies have been useful in learning about functional, as opposed to latent, proteins in single cells. How do phospho-specific antibodies work?

10.8   How does detection of proteins by Western blotting compare with detection by flow cytometry?

10.9   How can phosphor-specific antibodies be used to detect gene silencing in single cells? How does it compare with detection of gene silencing by Western immunoblots?

10.10   Is gene expression possible at the single-cell level? If so, is it meaningful? Why or why not?

10.11   Why is it important to perform gene expression analysis ("transcriptomics") on purified cell subpopulations rather than on cell mixtures?

11 Nanotoxicity at the Single-Cell Level

11.1 Toxicity of Nanomaterials

While the toxicity of nanomaterials is not yet well established, there are a number of increasingly sophisticated recent publications (Chan, Shiao, and Lu, 2006; Eustaquio and Leary, 2012a, 2012b; Leary, 2014; Monteiro-Riviere and Tran, 2014). The issues are varied but mainly fall into one or more of the categories described in the following sections.

11.1.1 Nanomaterials Are Potentially a Nanotoxicity Problem due to Large Surface-to-Volume Ratios

One potential source of toxicity of nanomaterials is the large surface-to-volume ratios inherent in materials on the nanoscale size range. Many chemical reactions are aided by surfaces that can bring together different molecules for potential inter-actions and aid in their reorientations for potential interactions. In this way, small nanoparticles can act as catalysts and like enzymes to amplify the effects of interactions between molecules. Reactivity of nanomaterials also varies widely with size. For all of these reasons, nanotoxicity should be measured using nanoparticles of specific size ranges. The same nanomaterials can be more or less cytotoxic depending on their size.

11.1.2 Nanomaterial Toxicity Coatings Can Mask True Toxicity

Most nanomaterials are hydrophobic, meaning that they will not readily dissolve in water or other common aqueous solutions unless a hydrophilic coating is added to the nanomaterials. The presence of this hydrophilic layer or biocoating can mask potential toxicity of nanomaterial to a cell if or when the coating is subsequently degraded. That said, one should not depend on a biocoating to always be intact to prevent nanotoxicity!

If the biocoating is stable in its microenvironment, then the nanotoxicity may measure very low, even if it is really not, if the biocoating later is stripped off. For example, cadmium-derived quantum dots usually have biocoatings that allow them to be monodispersed in aqueous solutions. With the biocoatings intact, we are measuring the toxicity of the biocoating, not the nanomaterials underneath. But if the biocoating is subsequently removed, the nanomaterial may become much more toxic, in this particular case if those quantum dots are irradiated for some time by a bright light source that can release elemental cadmium. Different biological cell types may cause disruption of these biocoatings such that the nanoparticles appear to be safe and nontoxic in one cell type but very toxic to another cell type. Hence, nanoparticle toxicity should always be measured in realistic and relevant environments and tested for a variety of cell types likely to be exposed in the eventual use (Haglund et al., 2009).

11.1.3   Size, Shape, and Electrical Charge Matter

As discussed earlier, the size of nanomaterials can affect overall reactivity. But shape also matters. Charge properties and interaction surfaces frequently vary with the local radius of curvature, which is a measure of nanoparticle shape. For example, rod-shaped nanomaterials may have different toxicity than spherically shaped nanoparticles of the same material (Wani et al., 2011). One possible reason is that the shape of the nanoparticle affects the local charge density of the counter-ions in the aqueous solvent.

Nanotoxicity measurements on living cells should always be carried out in these aqueous solutions, which have the proper composition to be suitable for living cells. Although the toxicity of the nanomaterials must be deduced by elimination of the toxicity of the solvent alone as a control sample (e.g., many cells find phosphate-buffered saline at least mildly cytotoxic), it should not be assumed that the two such toxicities are entirely independent. The nanotoxicity of the net system may be either enhanced or decreased by the interactions between nanomaterials and the solutions in which the cells are exposed. This makes determination of nanotoxicity more complicated to avoid nonlinear effects of each of these combined components.

11.1.4   Some Nanomaterials Contain Very Toxic Elemental Components

Nanomaterials may contain atoms that are not very toxic while they are embedded in the total nanostructure. However, changes in environmental conditions (e.g., exposure to light, pH changes) may release those atoms from their initial nanostructures, making the remaining nanostructure more toxic. For example, cadmium semiconductor-derived quantum dots in which cadmium is tightly linked to the nanostructure often have low toxicity to many cell types. But partial breakdown of the nanostructure by exposure to light can cause free elemental cadmium to be released. Released elemental cadmium is much more toxic than the original composite nanostructure (Pons et al., 2010). In an appeal to "green chemistry" (Anastas and Warner, 2000; Duan et al., 2015; Gurunathan et al., 2015; Malik et al., 2014) and "total life cycle" principles, we probably should not be making nanomaterials that may not be immediately cytotoxic but become more so over time and after disposal.

11.1.5   Measuring Biodistribution of Nanomaterials Can Be Challenging

Nanotoxicity can be partially dependent not only on the relative distribution within tissues but even its location within individual cells. This makes for the quite challenging experimental testing situation of trying to determine the number of nanoparticles over vast areas of tissues and organs and determining whether the nanoparticles are inside or outside of individual cells. This would require high-throughput, high-resolution 3D confocal microscopy, which by nature of the conflicting requirements does not yet exist. Nanoparticles are suboptical, meaning they cannot be measured individually by conventional light or fluorescence microscopy. Individual nanoparticles are also very small, making it difficult or impossible to tag them with enough fluorescent molecules per nanoparticle to be individually seen.

Though single-molecule detection technology can now measure individual nanoparticles, the measurement process cannot currently be done at high-enough speeds to realistically permit high-speed image analysis of cellular tissues of any significant size or cross-sectional area. Atomic force microscopy has a similar problem. New super-resolution microscopy may or may not be able to simultaneously measure suboptical objects fast enough to provide realistic measurements across larger tissues. Not enough work with super-resolution microscopy has been performed to determine whether it is possible to assay large areas in a reasonable time frame.

One recent attempt to deal with these issues, using nanobarcoding and in situ PCR amplification (i.e., polymerase chain reaction amplification of nucleotides performed inside a single cell), of oligonucleotides attached to nanoparticles, tried to solve the problem by finding the location of nanoparticles across large areas of tissues using optically sized spots first created by in situ hybridization followed by scanning image cytometry to measure large numbers of individual cells containing nanoparticles. Since the nanoparticles can be labeled with different barcodes, which can be "read" by PCR expansion of individual nanoparticle barcodes, this raises the intriguing possibility of performing multiple experiments simultaneously in single animals, thereby reducing both the number of animals needed and inter-animal measurement variations (Eustaquio and Leary, 2012a). We demonstrated the possibility of this, but it may or may not prove practical for performing multiple experiments within a single animal.

11.1.6   Measuring in Single Cells, Rather Than in Organs, Is Important but Not Sufficient

We tend to measure what we can measure quickly and easily rather than what we should measure, which may be much more difficult but also more important! Many nanotoxicity assays involve measuring two-dimensional monolayers of tissue culture cells of one type rather than the actual and more natural three-dimensional structures of multiple cell types of non-immortalized, primary cells in the bodies of animals or humans. While thin slices of real human or animal tissue can be a better model, they still lack many of the complex functions of thick sections of real tissue. Tissue culture cells are by definition immortalized, and similar to cancer cells, are no longer similar to cells as they exist in the body. Different cell types send signals to one another to maintain the integrity of morphology and function that constitute tissues and sometimes are even necessary to maintain properly differentiated cells in tissue cultures using co-cultures. In the body, different tissues act cooperatively to constitute organs with highly specialized functions. Rapid measurements on 3D tissues and organs are challenging and produce huge amounts of information that is difficult and time-consuming to analyze.

Measurements of cells at the single-cell level are frequently less complex than performing functional tests of a tissue or an organ. Unfortunately, cells in tissues and organs interact with each other, and the overall behavior of the tissue or organ may or may not be predictable based on the collection of single-cell measurements. But high-speed measurements of large numbers of single cells, of multiple cell types distinguished from one another by cell subpopulation-specific biomarkers, are now possible and can provide at least some prediction of overall nanotoxicity. Single-cell measurements can involve many cellular parameters simultaneously and can be quite sophisticated. For example, many measurements per cell can be made simultaneously using multicolor fluorescence techniques measured by either flow cytometry or scanning image cytometry. Since science is an iterative process, it will never be perfect. For this reason, nanotoxicity measurements are never final and can always be improved through progressively better measurements or by using newer technologies that conquer the limitations of older technologies.

11.2   Concept of Single-Cell Nanotoxicity

We are not used to thinking about toxicity at the single-cell level. Mainly it is envisioned at the total organism or at least the specific organ level. Traditional ways of measuring toxicity, and nanotoxicity, are in so-called LD50 experiments whereby we find the toxic dose level where 50 percent of the animals die. Measuring the nanotoxicity of single biological cells can be challenging due to the variations within single-cell types (e.g., specific cell cycle sensitivities), cell–cell interactions between different cell types (necessitating the ability to distinguish between different cell types in heterogeneous, rather than homogeneous, cell types), and the need for high-throughput (and perhaps simultaneous high-content) screening methods. Some of these challenges are outlined in the following subsections. To simulate the whole animal, we must test major cell types in vitro using 3D engineered "organ-on-a-chip" technologies that better mimic organ behavior than 2D monolayer cultures, as discussed later in this chapter.

11.2.1   There Is More Than One Way for a Cell to Die!

From all of the previous discussion, it should be evident that there is more than one way for a cell to die. Dying is not a simple process, and there are multiple different ways a cell can die. How a cell dies matters! It is sometimes difficult to distinguish whether cell death is due to nanomaterial exposure or to many other factors. Aside from using careful controls, it is important to pay attention to the effect of hydrophilic molecules attached to nanomaterials, which are usually themselves hydrophobic in the absence of a biocoating. For example, we have observed differences in the effect on cells of nanoparticles coated with carboxyl groups as opposed to just DNA oligomers.

11.2.2 Necrosis Is Unplanned Cell Injury

Necrosis is cell death due to injury. Often this involves failure of the cell to maintain its integrity due to leaky cell membranes. In necrosis, toxic intracellular molecules (e.g., hydrogen peroxide in intracellular vacuoles) leak out through damaged membranes and can cause injury to surrounding cells. Necrotic cells can also send the immune system into an inflammatory response that can sometimes lead to even greater injury. In general, necrosis is a bad thing that can lead to other bad things happening. For this reason, we should try to avoid causing necrosis with our nanomaterials.

Many of the single-cell assays for necrosis involve use of fluorescent dyes that are normally nonpermeant to live cells but which gain entry to cells through leaky membranes. For this reason, they are called "dye exclusion" viability probes because they are excluded by live, viable cells. To provide brightly fluorescent assays, many of these assays involve DNA-specific fluorescent probes that gain orders of magnitude in quantum efficiency when bound to DNA. For this reason, all cells (live and dead) can be incubated in suitable "live cell–friendly" buffers. Free probe outside cells has very low quantum efficiency and is mostly invisible by fluorescence microscopy or flow cytometry. But when the probe is able to enter dead or damaged cells, it binds tightly (usually irreversibly) to the DNA within these cells, rendering the nuclei brightly fluorescent. The most common fluorescent probe of this type is propidium iodide, an analog of ethidium bromide, a common DNA stain for electrophoresis gels. An analogous brightfield dye exclusion probe is trypan blue, which interestingly is also a deep red fluorescent probe, although few people are aware of that fact, which is why it is only rarely used as a fluorescent probe.

Another type of viability dye used to assay for cell death takes advantage of the fact that virtually all living cells have intracellular esterase enzymes of varying types. Dyes have been constructed with acetate groups that quench the naturally fluorescent other parts of the molecule. An example of this dye is calcein AM, a dye that has been rendered membrane permeant due to the AM part (an acetomethoxy derivate of calcein) of the molecule. When it gains entry to live cells, the nonfluorescent calcein AM is converted to a green fluorescent calcein after acetoxymethyl ester hydrolysis by these intracellular esterases. The dye is now also commercially available in a modified version that gives blue or orange-red fluorescence so that it can be used in better combination with other fluorescent dyes being used in a specific application.

11.2.3   Apoptosis Is Planned Programmed Cell Death

Apoptosis is very different from necrosis. It can best be described as a natural recycling system whereby cellular components are broken down into more basic constituents ready for reuse by surrounding cells. It is a more environmentally friendly way for a cell to die because it protects neighboring cells from suffering damage due to the toxic molecules (e.g., hydrogen peroxide) that may reside within subcellular organelles and vesicles. The cell membrane stays intact until very late in apoptosis, and toxic molecules inside these cells are recycled into less toxic molecular forms before being released by apoptotic cells, typically within "apoptotic bodies." In general, we would always want to employ apoptosis in getting rid of diseased cells because it is planned, rather than unplanned. It also taps into the cells' natural programming and frees us from worrying about proper dose, as long as we are able to initiate the process of apoptosis in order to eliminate the diseased cell. As long as apoptosis is initiated, and unless we use huge doses that can adversely affect normal bystander cells, we really do not care if the dose is more than is needed.

Apoptosis is sometimes called programmed cell death. Once a cellular decision to undergo apoptosis has been made, the cell proceeds in an orderly fashion, first shutting itself down in terms of reproduction and then recycling its subcellular components for possible reuse by other cells. This process happens normally within our bodies in daily life. There are a number of well-studied signal transduction pathways in apoptosis. An ideal nanomedical system would try to take advantage of one or more of these pathways because, once triggered, the cell knows from its own genome programming exactly how to proceed in an orderly death process without injuring neighboring cells and without eliciting destructive inflammatory responses. The immune system is our friend until it is our foe! Unless we can be sure to engage it in a friendly mode, it is probably better to design our nanomedical systems to be independent of the immune system!

When testing for nanotoxicity, it is important to not only measure necrosis but also undesired or unintended apoptosis. We would like our nanomaterials to avoid injuring cells directly through necrosis as well as avoid injuring normal cells indirectly through apoptosis. For this reason, we need to design nanomedical systems that first target diseased cells, and not normal cells, before initiating apoptosis!

11.2.4   Other Forms of Toxicity besides Necrosis and Apoptosis

It should also be evident from the above discussions that there are many other forms of toxicity besides apoptosis and necrosis. A better way of thinking about nanotoxicity is that when *any* detrimental changes are made to a cell's basic properties (e.g., cell cycle and proliferation, differentiation process) beyond those under natural conditions, those deleterious changes should also be considered part of an overall nanotoxicity exposure. Most current single-cell nanotoxicity publications fail to look at any of these other variables.

11.2.5   Some Other Challenges in Measuring Toxicity of Nanomaterials on Specific Cell Subpopulations

In addition to all of the other preceding caveats about the complexity of nanotoxicity, the differential effect of nanotoxicity on specific cell subpopulations can be critical not only to those cells but also to other cell types not affected directly by the nanomaterials but instead indirectly affected by their nanotoxicity in mixed-cell situations, as in tissues.

11.3   Single-Cell Measures of Toxicity

Many direct or indirect measurements of toxicity can be made at the single-cell level either on isolated single cells in suspension using well-established methods involving flow cytometry or on attached cell monolayers or tissues using scanning image cytometry (Leary, 2014). Highly detailed protocols for single-cell measures of nanotoxicity are available for superparamagnetic iron oxide nanoparticles (Eustaquio and Leary, 2012a, 2012b) due to our extensive use of these nanoparticles in our research.

However, during dissociation of cells from monolayers or tissues into single-cell suspensions for flow cytometric analysis, we lose important information about the 2D and 3D tissue architecture involving cell–cell associations as well as individual cell morphology. Interestingly, one of the body's largest tissues is a complex liquid suspension of cells, peripheral blood, which we usually assume can be randomly sampled. For example, we assume that blood sampled from venipuncture in the arm would yield a result similar to that of blood sampled from another part of the anatomy. While such an assumption is reasonable, there is actually very little data in the literature proving this to be the case. I am unaware of any such published quantitative studies, and work in this area might yield valuable clues to some human diseases.

Analysis of blood cells for toxicity by flow cytometry allows us to assess as many as 17 fluorescence colors simultaneously at thousands of cells per second (Perfetto et al., 2004) as of 2004, with more colors in recent years. The level of detailed data for each single cell, as well as the vast numbers of cells that can be quickly measured, enables us to study the toxicity of nanomaterials simultaneously on many different blood cell types, including some rare cell types.

11.3.1   Why Are Single-Cell Measures Important?

Real biological situations involve the interaction of a number of many different cell types, each of which can have a different nanotoxicity when exposed to a given nanomaterial. "Bulk cell" measures of toxicity of artificial cell mixtures can be very misleading and sometimes completely incorrect and irrelevant predictions of overall toxicity. The importance of making measurements on individual cells is shown in Figure 11.1. Panel A seemingly shows a hypothetical collection of cells that all respond similarly in terms of toxicity to nanomaterials, but only under the incorrect assumption

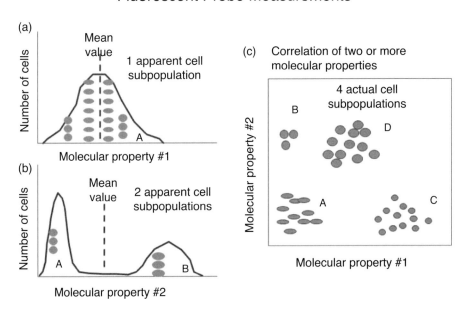

Quantitative and Correlated Single-Cell Fluorescent Probe Measurements

Figure 11.1 When single-cell measurements are not correlated on a cell-by-cell basis, and measurements taken as single-parameter histograms (Panels a and b), the number of cell subpopulations cannot be accurately determined. But when the measurements are correlated, we begin to see new cell subpopulations, previously hidden, as shown by the scattergram (Panel c). Source: Leary (2014)

that the variation in molecular property #1 is an error around a common. Bulk cell measurement (e.g., an MTT test with a colorimetric enzymatic assay for assessing cell metabolic activity; by reducing the tetrazolium dye MTT 3-(4,5-dimethylthiazol-2-yl)-2,5-diphenyltetrazolium bromide to its insoluble formazan, which has a purple color) and a single-cell test (e.g., a propidium iodide dye-exclusion assay by flow cytometry) will appear to yield similar data distributions. But the MTT assay has large errors in each measurement, whereas the flow cytometric measurement may have a measuring error on each cell of only 1 or 2 percent. For this reason, the flow cytometric measurement mainly represents "true biological variation," with only a very small instrument variation, whereas the MTT represents a large measurement error performed on many cells at once rather than on individual cells. In Panel B of Figure 11.1, we see another hypothetical case (which is much more typical than the first case) whereby two different cell subpopulations of different cell types 1 and 2 each respond differently in terms of cytotoxicity. The MTT bulk cell assay (and similar assays such as Western blots) yields results that are the weighted (by relative cell number or frequency) average of the two or more cell types. The resulting answer may prove to be poor, or even irrelevant, depending on the relative frequencies of different cell types and their relative cytotoxicities. A third example is shown in the hypothetical situation outlined in Panel C of Figure 11.1. In this case, two (or more) measurements are simultaneously made on the same individual cells. If one projects the data down onto the x- and y-axes, you can quickly see that this scattergram actually represents the same data shown in the two panels on the left. However, now we see that the one or two populations seen in the histograms are actually four cell subpopulations, as seen in a 2D scattergram. If one extrapolates this analogy to three or more simultaneous measurements, it is easy to see that there could be many other cell subpopulations difficult, or impossible, to see by direct visual inspection and these may need to be uncovered by multidimensional data-mining algorithms. A bulk cell measurement would completely miss the correct result (Leary, 2014) (Figure 11.1).

11.3.2   Measuring Cell Death at the Single-Cell Level

Many researchers try to directly measure cell death, thinking that it is a simple process. The cell must be either live or dead. Unfortunately, death at single-cell level is not a simple yes or no. Cells can be still living but starting to die. They can also die in many other ways.

Cells can die due to cell injury (i.e., necrosis), or due to programmed cell death (i.e., apoptosis). When cells die due to necrosis, either naturally or through drug-induced cell killing, they release many potentially toxic intracellular molecules that can injure neighboring cells. In nanomedicine, we try to avoid necrosis approaches to the elimination of diseased cells in favor of approaches that try to induce the diseased cell to undergo programmed cell death whereby the cell has a preprogrammed procedure for breaking down and safely disposing of all of its cellular components. When a cell is eliminated through apoptosis, it does not harm neighboring cells. For an excellent overview and discussion of the many differences between necrosis (accidental cell injury) and apoptosis (programmed cell death), see Darzynkiewicz et al. (1997) (Figure 11.2).

In fact, death is a surprisingly complex, multistep process. For this reason, simple measures of it can be quite difficult. Whatever single-cell assay we use for measuring cell death should be an absolute predictor of that cell's death. If we take a "late-stage" measure of cell death, we can be reasonably certain that the cell will not recover. A cell exposed to nanomaterials may be in an early stage of the death process and may or may not recover. It may be more accurate to say that there is a certain probability of a cell dying, or recovering, at a given point in the death process.

11.3.3   Are There Any Simple Measures of Cell Death?

One of the simplest measures of cell death is lack of cell membrane integrity. Most living cells must maintain an intact cell membrane to survive changes in its external environment. When that membrane integrity is compromised, the cell is unable to control the inflow or outflow of the many molecules essential to life and to buffer itself from sudden environmental changes.

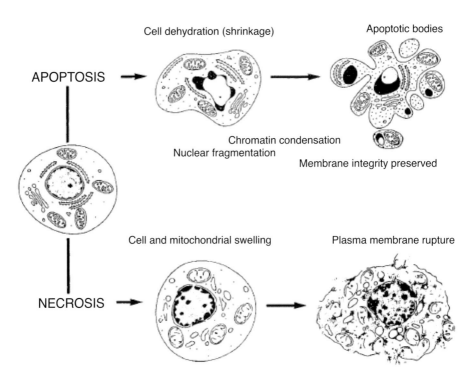

Figure 11.2 Cells can die due to cell injury or programmed cell death.
Source: Darzynkiewicz et al. (1997)

A number of assays have been developed involving the exclusion of fluorescent or nonfluorescent dyes, such as the exclusion of the dye trypan blue. Trypan blue exclusion is one of the most commonly used assays of cell death. But it is a poorly understood, and even more poorly applied, assay that is highly susceptible to serious errors. Actually, the correct concentration of trypan blue is highly dependent on cell type. If a too-high concentration of trypan blue is used for a given cell type, the dye will penetrate even intact cell membranes, giving a higher false-positive measure of cell death based on dye exclusion. If you apply trypan blue in excessive quantities, it will drive itself by concentration gradients even into living cells with supposedly intact cell membranes. Cell mixtures containing different cell types may contain cells of widely varying membrane integrity, meaning that the accuracy of viability studies by trypan blue may vary considerably by cell type. Devising a single concentration of trypan blue for an accurate measure of all cell types is difficult or impossible. A similar measure of cell death is the propidium iodide (PI) exclusion assay, which allows fluorescence measurements of cell death. The assay works because propidium iodide has very low quantum efficiency for fluorescence unless it is bound to DNA or RNA, whereupon its quantum efficiency increases by orders of magnitude. Cells with intact cell membranes exclude PI, and the quantum efficiency of the PI outside the cell is very low. But cells with leaky membranes allow propidium iodide to get through the cell membrane and bind to DNA or RNA within the cell, resulting in brightly stained red fluorescent cells that can be readily measured by the human eye looking through a fluorescence microscope, and much more rapidly measured by flow cytometry or scanning image cytometry. But some cell types (e.g., megakaryocytes in the blood) are inherently pretty leaky and yet survive and function. Other cell types have normally a very "tight" membrane structure such that they may be much farther along in the dying process before their membranes fail to exclude the dye.

All of these complications with simple one-color fluorescence measures of cell death have led to development of slightly more sophisticated multicolor measures that not only exclude one dye but also show the presence of a second dye, indicating activity of enzymes only present in living cells. Examples of the latter include the addition of a fluorogenic substrate (meaning it is a molecule that is nonfluorescent unless it is cleaved by another molecule) such as fluorescein diacetate in addition to propidium iodide. If the nonfluorescent, but otherwise membrane-permeable, fluorescein diacetate penetrated a living cell, it would have its diacetate group cleaved by intracellular esterases, leading to the green fluorescence of free fluorescein. The remaining cleaved molecule becomes fluorescent but with a charge that prevents it from easily diffusing out of the cell. A similar, and improved, assay is that of calcein AM (an acetoxymethyl ester of calcein) where the AM (acetomethoxy-) portion of the molecule confers easier permeability into cells with intact membranes. The calcein portion of the molecule is less pH sensitive than fluorescein and has a very high quantum efficiency, leading to intense green fluorescence if the AM group is cleaved by intracellular esterases present in the cytoplasm of nearly all living cells. The double requirement of excluding PI and also showing positive intracellular esterase activity gives a greater assurance that a cell is really live and at least partially functioning normally.

11.3.4   More Sophisticated Measures of Cell Death through Apoptosis

There are now more interesting and important measures of programmed cell death. We now understand that every normal cell, but not necessarily true for cancer cells or some cell lines, has a complicated and sophisticated "programmed cell death" agenda. Programmed cell death signal transduction pathways contain multistep processes. The cell goes through its death process in an orderly fashion, breaking down its components into subcomponents that are in some cases reused by surrounding living cells. Perhaps more sophisticated studies of apoptosis, looking at several apoptosis signal transduction pathways at the single-cell level, will show that apoptosis itself is multifaceted, leading to better predictions of whether a cell in the beginning stages of apoptosis can still recover.

11.4   Necrosis versus Apoptosis Assays

There are a number of assays used to distinguish between necrosis and apoptosis. The most obvious first step is to use membrane-impermeant fluorescent dyes to detect damaged membranes in necrotic cells. The harder task is to distinguish apoptotic cells from healthy cells, particularly in the early stages of apoptosis.

Though preceding discussions have pointed out the important differences between necrosis and apoptosis, it is not always possible to reliably distinguish late-stage apoptosis from necrosis. Since many nanotoxicity assays are taken at single time points, when such a measurement is taken can be crucial. Better assays should look at multiple time points on the same cells to see this difference. It is important to distinguish between necrosis and apoptosis because they are very different mechanisms, and we must test for them in different ways.

11.4.1   Annexin-V Assays for Early Apoptosis

The annexin-V early-apoptosis single-cell assay takes advantage of the fact that the phosphatidylserine molecule relocates from the inner portion of the cell membrane to the outer portion of the cell membrane during early apoptosis (van Engeland et al., 1998). Phosphatidylserine binds tightly to a ligand known as annexin-V, which can be fluorescently labeled. Since annexin-V is relatively inexpensive, this assay is much faster and less expensive than the TUNEL assay, which is only relevant to later-stage apoptosis. The annexin-V assay is also quite sensitive to the initiation of the apoptosis process, and for that reason it has become very popular for use by fluorescence microscopy or flow cytometry. It is commercially available with a wide variety of fluorescent labels, including FITC, PE, PE-Cy5, Cy3, Cy5, eGFP, and biotin (for subsequent detection by a fluorescently labeled avidin), that permit its detection in a wide variety of multicolor fluorescence experiments under various laser excitations by flow cytometry, image cytometry, confocal microscopy, and fluorescence microscopy.

An example of the assay showing both conceptual single-cell labeling and flow cytometric detection of data is shown in Figure 11.3 (Leary, 2014). It is necessary to add propidium iodide to the fluorescent annexin-V to distinguish apoptotic from necrotic cells. Since this is an early-stage apoptosis assay, cell membranes will be intact and will exclude PI during early apoptosis. So early-stage apoptotic cells will be annexin-V positive and propidium iodide negative (annexin-V+/PI−).

11.4.2   DNA Ladders and the Development of Single-Cell TUNEL Assays for Late Apoptosis

The ladder effect, observed on DNA gels before programmed cell death by apoptosis was understood, was due to DNA fragments of different sizes that occurred by intranucleosomal cleavage of the DNA at random intervals between the nucleosomes. This was then exploited in a late-apoptosis TUNEL assay that, instead of putting large numbers of cells and their contents on an electrophoresis gel, could then be used to identify single cells that were undergoing apoptosis.

Terminal deoxynucleotidyl transferase dUTP nick end labeling (TUNEL) is a late-stage apoptosis assay that is difficult to observe in vivo and is really much more of an

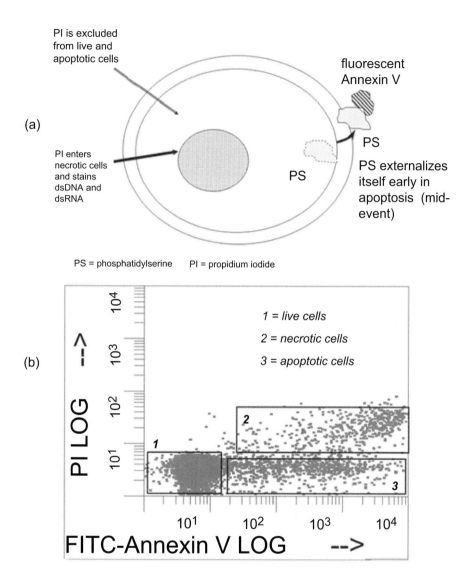

Figure 11.3 (a) The single-cell annexin-V assay labels the externalized phosphatidylserine with a fluorescent annexin-V ligand and also includes a propidium iodide dye exclusion assay to distinguish apoptosis from necrosis. (b) Single apoptotic cells can then be detected by flow cytometry, as shown in this example.
Source: Leary (2014)

in vitro assay. It also captures only late-stage apoptosis and does not reveal earlier stages. For this reason, there are better assays, such as annexin-V, that give a much more sensitive and complete picture of apoptosis occuring in cell populations. That said, it should be appreciated that any assay of apoptosis is only going to be sensitive to a "window in time" over the entire apoptosis process.

The TUNEL assay was adapted to flow cytometric detection (Darzynkiewicz, Galkowski, and Zhao, 2008; Gorczyca et al., 1992). Terminal deoxynucleotidyl transferase (TdT) is an enzyme that will catalyze the addition of deoxyuridine triphosphates (dUTPs) to the end of nicked DNA occurring in late apoptotic cells. The dUTPs chemically compete with the normal dTTP (deoxythymidine triphosphate is one of the four nucleoside triphosphates that are used in the in vivo synthesis of DNA) DNA precursor incorporating into the DNA sequence. But these incorporated dUTPs can be detected with a monoclonal antibody that can differentiate between dUTPs and the normal dTTPs. The single-cell TUNEL assay detects DNA fragmentation by labeling the terminal end of nucleic acids caused by DNA strand breaks. TdT is then used to add nucleotide analogues containing bromodeoxyuridine (BUdR). The BUdR containing cellular DNA is then recognized by a fluorescent anti-BUdR antibody. Development of an antibody capable of recognizing single base pairs of BUdR was a remarkable achievement at the time, and it has had numerous uses in biology and medicine, including in DNA repair (Gratzner et al., 1975). This assay, performed on brains of people during autopsy, revealed that supposedly reproductively dead neurons could actually repair themselves in some adults, which has led to a revolution in our understanding of brain injury and repair. In this way, nicked DNA within late-stage apoptotic cells can be marked with a fluorescently labeled antibody against BUdR of a specific color and measured by flow cytometry, as shown in Figure 11.4 (Leary, 2014).

Another characteristic of these programmed cell death (i.e., apoptotic) pathways is that in early apoptosis, the cell membrane remains intact and is able to exclude dyes such as propidium iodide and trypan blue. During apoptosis, the cell shuts down in a controlled fashion, preparing its molecular contents for recycling by neighboring cells. Only late in the process is the cell membrane permeable to PI, trypan blue, and other such dyes.

11.4.3   More Subtle Changes Not Resulting in Direct Cell Death

If we truly want to consider all of the potentially bad effects of nanomaterials on cells, we must also study how they might change the patterns of cell proliferation and differentiation. If a cell's proliferation or differentiation state is altered by exposure to nanomaterials, those changes should also be considered forms of nanotoxicity. Cells undergoing proliferation have different susceptibility to toxic agents. Likewise, cells undergoing unplanned cell differentiation can have different susceptibility to toxic agents. Particularly in vivo, these changes in proliferation or differentiation due to exposure to nanomaterials can affect the normal functioning of tissues and organs, which in turn can adversely affect the total organism.

11.5   Measuring Changes in Cell Proliferation and Cell Cycle

Changes in cell proliferation can be measured in a number of different, and sometimes complementary, ways. The first, most obvious measure of cell proliferation is studying

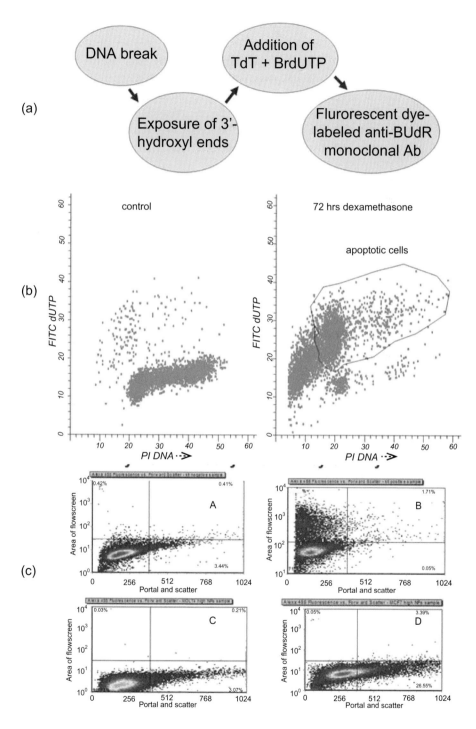

Figure 11.4 (a) Detection of DNA breaks by labeling with incorporated dUTP and subsequent detection of dUTP with fluorescent anti-BUdR antibodies. (b) Flow cytometric data of campothecin-treated positive control cells for terminal deoxynucleotidyl transferase dUTP nick end labeling (TUNEL) apoptosis. (c) Superparamagnetic iron oxide nanoparticle-treated cells show little TUNEL toxicity.

Source: Leary (2014)

how fast the cells are replicating by measuring absolute cell numbers using some method, manual or automated, to count cells. The cells may also not be dying, but they may be changing in their rate of proliferation by changing the fraction of cells that are actually cycling at any given point in time. Often when a cell is stressed by cytotoxic agents, but not killed or induced into apoptosis, the proliferation rate slows. The cell may go into a quiescent stage of the cell cycle called G_0. This can also be studied by measuring the total cell numbers at longer time intervals, allowing the cell to go into multiple cell divisions. However, this type of measurement is difficult and time-consuming.

11.5.1   Measuring Cell Cycle at the Single-Cell Level

Cells undergoing replication must synthesize enough DNA to be able to divide into two daughter cells. These measurements used to be performed through laborious tritiated thymidine (3TdR) uptake experiments using photographic film to be able to count the "grains" of radioactivity over each cell nucleus. In the early days, single-cell DNA synthesis was measured by counting photographic grains due to tritiated thymidine uptake in individual cells on exposed film. The counting statistics were poor and it was very labor intensive to measure just a few hundred cells. One of the early flow cytometry papers (Crissman and Tobey, 1974) used a DNA-specific fluorescence dye to measure thousands of cells per second and then perform cell cycle analyses to determine the fraction of cells in G1, S, and G2/M. This was a revolutionary advance in cell cycle analysis and is now performed routinely in labs and hospitals around the world. DNA-based flow cytometry data were easy to acquire but much more difficult to analyze, as they required a mathematical deconvolution to extract the cell cycle phases. One of the first such analyses came out from the same group at Los Alamos in the same year (Dean and Jett, 1974).

An alternative method that derived proliferating S-phase cells used an antibody against BUdR using simple gating methods, without the need for elaborate mathematical analyses. It has sometimes been called "the immunologist's version of cell cycle analysis." Many thousands of single cells can now be obtained in minutes by either flow or scanning image cytometry using BUdR (bromodeoxyuridine, a metabolic analog of 3TdR), which can be easily recognized by a fluorescently labeled monoclonal antibody. When co-labeled with PI, for DNA content per cell, the data appear for nanotoxicity applications (Leary, 2014) as shown in Figure 11.5.

11.5.2   Measuring DNA per Cell Using Fluorescent DNA-Specific Dyes

It is usually easier to measure the DNA synthesis process within single cells using fluorescent dyes that measure the amount of DNA within an individual cell. For example, Hoechst 33342 and recently available new dyes, such as the Vybrant® DyeCycle™ family of dyes (Invitrogen, Inc.), are actually cell permeable and can be used to label live cells with excitation at different non-UV wavelengths. DAPI (4',6-diamidino-2-phenylindole) binds strongly to AT-rich regions in DNA and can

(a)

Rate of DNA
synthesis

α

Rate of
Incorporation of
BUdR

α

FITC-anti-BUdR

α

Log green
fluorescence

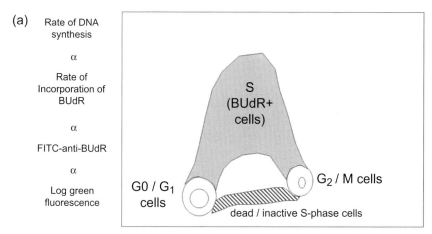

Red Fluorescence α Propidium iodide α DNA / cell

(b)

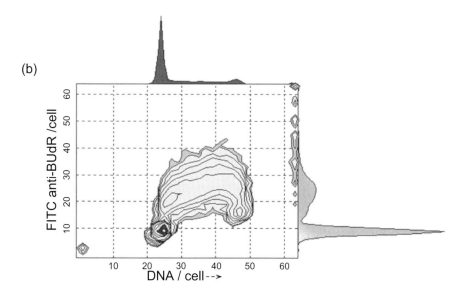

Figure 11.5 Rapid anti-BUdR assay by flow cytometry. (a) Flow cytometric measurement of anti-BUdR fluorescence is proportional to BUdR uptake during S-phase DNA synthesis. (b) Actual flow cytometric data are shown in a bivariate display with contour mapping. Source: Leary (2014)

pass through an intact cell membrane, staining both live and fixed cells. Since it passes through the membrane less efficiently in live cells, live and dead cells stain differentially (i.e., dead cells are brighter). DRAQ5™, a far-red fluorescent DNA dye, is a reagent for use in either live cells, dead cells, or fixed cells in combination with other common fluorophores (e.g., FITC, PE) in the visible part of the spectrum. A related dye, DRAQ7™, also emits in the far-red spectrumbut labels only dead or permeant (e.g., fixed) cells.

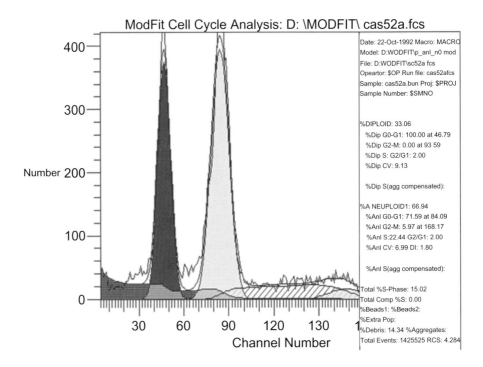

Figure 11.6 The toxic effects of materials on cells not only can cause perturbations of the normal cell cycle but can also, after some growth time in culture, potentially cause mutagenic aberrations in the numbers of chromosomes per cell. This aberrant number of chromosomes (usually an increase) appears as a hyperdiploid peak (aneuploidy) in a DNA content flow cytometry histogram. In addition to the number of cells in each cell cycle compartment, the number of diploid and aneuploidy cells can be determined by appropriate models and curve-fitting data analyses.
Source: Leary (2014)

There are other cytotoxic changes, including apoptosis, that can be measured on large numbers of cells treated with potentially cytotoxic or mutagenic agents. Aneuploid cell subpopulations with aberrant chromosome numbers appear as G0/G1 peaks beyond the normal diploid peak (Leary, 2014), as shown in Figure 11.6.

The simplest assay is to measure the amount of DNA per cell and then use "DNA cell cycle" curve-/model-fitting data analysis software (e.g., ModfitLT, Verity Software House, Topsham, Maine) to extract the relative numbers of G1, S, and G2 phase cells.

11.5.3   Disturbances in Cyclin Checkpoint Expression Patterns at the Single-Cell Level

DNA measurements alone do not always correctly indicate that proliferation compartment assignments (e.g., G1, S, G2, M) are correct. For example, highly perturbed or synchronized cells are not always really equivalent to normal cells in those same DNA

Expression of Several Cyclins around the Cell Cycle in Normal Cells

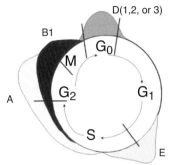

Figure 11.7 Cyclin expression can detect problems in cell cycle checkpoints due to drug treatments.
Source: Leary teaching

subcompartments. Checkpoint irregularities are important because they can indicate more subtle nanotoxicity effects. In those cases, the cyclin checkpoint expression patterns should also be studied (Gong, Traganos, and Darzynkiewicz, 1995) (Figure 11.7).

11.5.4   Changes in Cell Differentiation

If we are to measure the true toxic effects of agents on cells, we must also see whether the normal differentiation processes are proceeding or whether the cell is now on a different differentiation pathway. Monoclonal antibodies of different fluorescent colors label one or more cell surface molecules that are biomarkers of cell differentiation. Single-cell analysis by flow cytometry or scanning image cytometry is essential because only a subset of cells may have changed their differentiation states and, therefore, are difficult or impossible to detect in bulk cell measurements.

11.5.5   Changes in Cell Gene Expression

Sometimes, cellular changes are more subtle and cannot be easily measured with fluorescent monoclonal antibodies and flow or image scanning cytometry. Gene expression profiles can be easily measured on large numbers of cells, but this general approach is a good example of the potential pitfalls of a bulk cell assay approach. The gene expression profile will be the weighted average of the relative frequency of each of the cell subpopulations present. For this reason, it is necessary to purify different cell subpopulations by flow cytometry/cell sorting or other cell separation technologies. The purity of cells required to accurately measure gene expression profiles has been previously studied (Szaniszlo et al., 2004). It depends on a number of factors and has been discussed in some detail in that paper. The number of cells required and their relative purity, isolated from other cell types to avoid the pitfalls of measuring cell mixtures, vary, with the smallest frequency of cells and lowest purity required for those with strongly expressing genes. Single-cell gene expression analysis is more complicated and expensive, and the reliability of the results depends on the number of copies of genes expressed per cell.

11.5.6   Comet Assays for DNA Damage and Repair

Simply measuring the amount of DNA per cell does not always give an accurate measure of nanotoxicity. We need to know whether that DNA has been damaged. The comet assay (Klaude et al., 1996; McKelvey-Martin et al., 1993; Ostling and Johanson, 1984; Valencia et al., 2011) is a sensitive technique for the detection of DNA damage at the level of a single cell. It is a commonly used technique for evaluation of DNA damage/repair, biomonitoring, and genotoxicity testing. It involves the encapsulation of cells that are first encapsulated in a low-melting-point agarose suspension followed by lysis of those cells in neutral or alkaline (pH>13) conditions. Electrophoresis of the suspended lysed single cells, stained with a DNA dye, is then performed. The term "comet" refers to the pattern of DNA migration through the electrophoresis gel, often resembling a comet, with the comet's tail representing small molecular weight DNA fragments separating from the major mass of DNA of each cell. Semi-quantitative single-cell assays, either manually or by image analysis software, involve an analysis of the relative portion of DNA in the comet tail compared to the DNA in the comet head. An example of the comet assay used for detection of nanotoxicity, or in this case lack of toxicity, is shown in Figure 11.8. Hydrogen peroxide is a common positive control treatment that induces oxidative stress and DNA damage in virtually all cells (Prow, Salazar, et al., 2004).

11.5.7   A Simple Test for DNA Damage due to Oxidative Stress in Single Cells

Interestingly, there is a less well-known but even simpler assay for detecting nicks in DNA during oxidative stress toxicity to cells. It is a test for reactive oxygen species (ROS). A superoxide indicator, dihydroethidium, also called hydroethidine, exhibits blue fluorescence in the cytosol of a cell until oxidized, where it intercalates within the cell's DNA, staining its nucleus a bright fluorescent red. The dye dihydroethidium (DHE) is nonfluorescent and cell permeant. If the dye is hydrolyzed in the presence of reactive oxygen species within a cell, it becomes fluorescent and also binds to DNA (Haglund et al., 2009), as shown in Figure 11.9.

In this case, we see that QTracker quantum dot nanoparticles (Thermo Fisher Scientific, Inc.), used by many researchers as a supposedly nontoxic tracker of cells and also dividing cells, can be toxic to particular cell types under certain circumstances. There are at least two possible things happening in this instance. First, the cell may be degrading the biocoating of the nanoparticles, revealing more toxic nanomaterials beneath this biocoating layer. Second, exposure of these CdSe quantum dots to intense light can release elemental cadmium, which is much more toxic than the CdSe nanomaterial complex (Haglund et al., 2008).

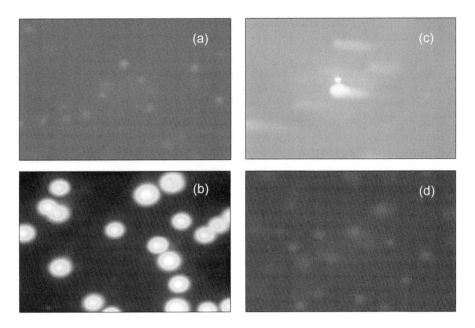

Figure 11.8 Comet assay performed on the following MOLT-4 cell samples: (a) untreated cells, (b) cells treated with 100 μM hydrogen peroxide for 18 hours, (c) cells exposed to 0.1 mg/mL ferric oxide nanoparticles, and (d) cells exposed to 0.2 mg/mL ferric oxide nanoparticles. Source: Adapted from Prow, Salazar, et al. (2004)

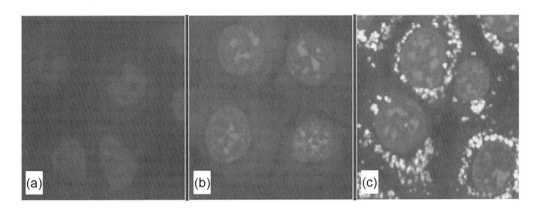

Figure 11.9 Production of reactive oxygen species (ROS) in MCF-7 breast cancer cells exposed to Qdots. Dihydroethidium is represented by the red and Qdots are represented with green. (a) Control: MCF-7 cells plus dihydroethidium. (b) Positive control: ROS induced with H_2O_2, plus dihydroethidium. (c) Experimental: QTracker plus dihydroethidium. The cells were exposed to the QTracker Qdots for 24 hours prior to application of dihydroethidium. The image illustrates the presence of dihydroethidium and therefore ROS in the nucleus. The cells are stressed by the presence of Qdots versus control. Source: Haglund et al. (2009)

11.5.8   Light Scatter Assays

In most flow cytometers (depending on the subtended angle of the obscuration bar or laser beam dump), the forward angle light scatter sensor will be sensitive to both cell size and refractive index. Side scatter (sometimes called 90-degree light scatter because it is collected roughly 90 degrees away from the direction of the exciting laser light) is sensitive to granularity, in this case the presence of higher refractive index nanoparticles within or on the cell as well as cell surface blebbing on stressed cells. These properties can be exploited to do rapid flow cytometric assays based on light scattering, as shown in Figure 11.10.

When cells die, their refractive index increases relative to that of living cells. This is why they appear darker in a phase contrast microscope. When refractive index increases for a given-sized cell, its forward scatter signal decreases (usually in the order of 50 percent). An important, but frequently ignored, fact to remember is that larger dead cells will look like smaller live cells in a flow cytometer, so good discrimination between live and dead cells only works if the cells are fairly homogeneously sized cell populations and will not distinguish well between live and dead cells in heterogeneously sized mixtures of cells. Since nanoparticles also have much higher refractive index than live cells (and probably most dead cells), their presence in large numbers will actually slightly increase the forward scatter signal since their diffraction will tend to scatter light beyond the angular range of the obscuration bar or beam dump. But that effect will only dominate over the live/dead cell effect if there are very large numbers of nanoparticles per cell. Side scatter will be increased by the presence of nanoparticles, but it will also be sensitive to granularity and cell membrane blebbing when cells become stressed.

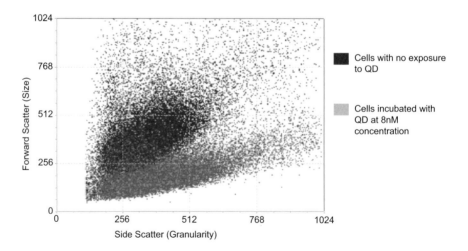

Figure 11.10 MCF-7 cells incubated with different concentrations (in this, case 8 nM) of amino group quantum dots (QD). The cells that have become dead or dying have lowered their forward scatter and slightly elevated their side scatter.
Source: Leary teaching

11.6   Measuring Nanotoxicity for Cell Systems In Vitro

It is important to measure the toxicity effects of nanomaterials on different cell systems that in some way approximates the presence of those cell types in the organ of interest. For this reason, most pharmaceutical companies now look at a variety of cell types, such as liver cells (for liver toxicity) and myocardial cells (for effects on the heart). Those tests can be performed in vitro with human cell strains or biopsies or in various animal systems.

11.6.1   In Vitro Measures of Nanotoxicity

In vitro measures of nanotoxicity can be used with a variety of animal and human cell types, including cell lines and primary cell strains. It is important to understand that cell lines are *not* normal. They are essentially cancer cells with unlimited growth, and they may test very differently than normal cells. Cell strains are normal cells and therefore grow for a finite number of times (i.e., the Hayflick limit) before the cell undergoes senescence, usually due to telomere shortening. Sometimes cell strains can behave and test differently depending on the number of times they have replicated in culture. It is also possible that cell strains may undergo spontaneous transformations such that they become abnormal cell lines, rather than cell strains.

11.6.2   Cell Lines

Cell lines have the advantage that they are easier to grow than cell strains because they have been "immortalized." This means that they tend to be mostly stable in genetic makeup (although this can sometimes change during cell tissue culture). However, the fact that immortalized cell lines are really cancer cells, rather than normal cells, means that we should be wary as to whether results from cell lines will be recapitulated in primary cell strains.

11.6.3   Primary Cell Strains

Primary cell strains are derived from actual normal animal or human tissue. Unless they become transformed or immortalized during cell culture, they have limited cell proliferation in culture and will die from senescence after 50–60 cell divisions, depending on their age at onset. They may also change as a function of their age in culture even if they do not ultimately transform into cancer cells. For these reasons, results should be compared to other cell strains of similar passage number.

11.6.4   2D versus 3D Cultures: A Need for 3D Tissue Engineering?

It is becoming increasingly clear that cells behave differently in their normal 3D conditions than they do in more artificial 2D cell culture conditions. This has led to new efforts in 3D tissue culture and even the construction of 3D cell complexes, including scaffolding by tissue-engineering methods. A problem with this approach, beyond the time and expense of construction, is that these 3D complexes are difficult to analyze and do not easily lend themselves to high-throughput screening methods. The sheer volume of 3D data and their complexity and the need for high-speed autofocusing methods means that high-throughput analysis of 3D tissue engineered complexes is still on the horizon in terms of practical use.

11.7   High-Throughput Methods for Measuring Nanotoxicity

As mentioned in the previous section, high-throughput methods for measuring nanotoxicity have thus far been limited to either flow cytometry of suspended single cells or to 2D cultures of cells. The limitation of flow cytometry is that cells may behave differently when they are suspended or when they are taken away from their neighboring cells. High-throughput 2D imaging requires high-speed imaging algorithms, particularly segmentation algorithms that "virtually" separate each cell from its neighbors.

11.7.1   Flow Cytometry

As long as cells behave similarly when they are in a single-cell suspension, flow cytometry is probably the technology of choice due to its considerable analytical power in multicolor (e.g., 14–18 fluorescence colors in four or five laser excitation instruments) fluorescence analyses of single cells at speeds of 10,000 cells per second or more. Cells of interest can also be captured and purified on the basis of these measured properties. This allows the simultaneous measurement of nanotoxicity by the preceding single-cell nanotoxicity assays on multiple cell subpopulations and also the detection of toxicity to small or even rare cell subpopulations that would be lost in bulk cell assays, such at MTT assays (Fotakis and Timbrell, 2006; Hamid et al., 2004).

11.7.2   Scanning Image Cytometry

High-speed scanning image cytometry offers the advantage of keeping cells attached to a surface, particularly in those cases of a cell type that changes its properties when put into single-cell suspensions. Scanning image cytometry usually performs image analysis on cells attached to surfaces, including in microwells, using an excitation light either as a cell-sized spot or a slit of light that can quickly excite a large number of attached cells and derive the fluorescence from each cell as well as its location. This storing of cell location information allows the researcher to return to a particular cell either to confirm a measurement or to perform additional measurements on the same cells, something not possible with flow cytometry.

11.7.3   3D Confocal Scanning: Can It Be Made High Speed?

When 3D information is required, either for 3D engineered tissue or for high-resolution 3D measurements within a single cell, confocal microscopy can provide that information and can measure different colors of fluorescence and confirm their spatial co-localization. However, there are two problems with confocal approaches. The first problem is photobleaching. In one-photon confocal microscopy systems, it is difficult to obtain quantitative information because the process of scanning a particular optical layer can cause photobleaching of the fluorescence and subsequent erosion of 3D boundaries. Two-photon or multiphoton confocal systems can improve the photo-bleaching situation by limiting the size of the excitation light beam, so that we are no longer photobleaching cells through the entire 3D structure, but such systems are much more complex and expensive. The second problem, as mentioned previously, is the problem of acquiring cell image information when cells are at different layers. This requires high-speed autofocusing, which is also compounding the preceding photo-bleaching problem. All of these problems make high-speed 3D image processing a continuing, but important, challenge.

11.8   Animal Models for Measures of Nanotoxicity

In vivo measurements in animals are challenging. Typically, animals are exposed to nanoparticles and then sacrificed so that their organs can be examined at the histopathological level. Nanoparticle biodistributions can be studied on living small animals, particularly in nude mice, with whole-animal in vivo imaging techniques using near-infrared imaging on near-infrared fluorescently labeled nanoparticles, as discussed previously.

11.8.1   What Are Good Animal Models for Human Disease?

While the vast majority of animal model systems used for nanotoxicity studies have been rodents (e.g., mice, nude mice, and rats), rodents are not particularly appropriate models for predicting nanotoxicity responses in humans. Their organs are different from human organs, leading to different biodistribution patterns. Their metabolism can also be quite different. One of the reasons they are used is because they are highly inbred genetic strains, leading to lower animal-to-animal variability due to genetic differences, but this very strength is also one of its greatest weaknesses since small differences in genetics can lead to potentially large differences in response to nano-materials and nanotoxicity. But even the process of scaling up the numbers of nanoparticles required to levels sufficient for human use can itself be challenging.

11.8.2   Some Common Animal Model Systems

What is needed in an animal system is an animal closer in size and metabolism to humans. Dogs and pigs are probably the two most important animal systems. For example, a bladder cancer in a Scottish terrier dog system has been developed that closely recapitulates human bladder cancer (Knapp et al., 2014). Because it is a naturally occurring bladder cancer in these dogs, it does not suffer from the potential artifacts of artificially induced animal model systems.

11.8.3   Nude Mice: Halfway between In Vitro and In Vivo

Nude mice are often a good system for starting a study for at least two reasons. First, the fact that it is "nude" means that it is immune-incompetent, so human tumor xenografts can be grown in these animals. Second, frequently not known by researchers, a side benefit of these "nude" mice is that they lack hair! This makes the detection of nanoparticles in whole-animal near-infrared imaging systems possible. The fur in most animals fluoresces red and interferes with the measurements of nanoparticle biodistributions. But it should also be remembered that nude mice are artificial animals and, as such, are only a bridge between in vitro assays and real animals with competent immune systems.

11.8.4   Immune-Competent Mice Add the Effects of the Immune System

The fact that nude mice, as discussed in the previous section, are immune-incompetent means that we get no information about the immunogenicity of the nanomaterials, which may be an important component of the overall nanotoxicity. Nanotoxic effects can lead to large disturbances in the immune system and need to be taken into account to obtain a more accurate picture of the effects of nanotoxicity on the animal or human.

11.9   Nanotoxicity In Vivo: Some Additional Challenges to Using Animal Models

Some additional challenges of measuring nanotoxicity in vivo concern the fact that animals represent "open" systems. Traditionally, we like to make measurements in closed-system laboratories whereby we control all of the inputs and outputs for a given environment. That is difficult to achieve in animals that move around in environments, breathe in and out, and excrete materials in their urine and feces.

11.9.1   Accumulations of Nanoparticles Can Change Toxicity Locally to Tissues and Organs

One of the main reasons we like to use animals to study nanotoxicity is that nanotoxicity depends on local concentrations of nanomaterials in specific organs. Depending on the size, zeta potential, targeting molecules, and other factors, nanoparticles can accumulate quite differently in specific organs of an animal.

Agglomerated nanoparticles can hugely change the overall nanotoxic effects of specific nanomaterials in tissues and organs.

11.9.2   Filtration Issues of Nanoparticles

There are a number of important issues concerning the filtration of nanoparticles in animals, according to the size and charge properties of these nanoparticles. There is also a so-called Goldilocks principle with respect to nanoparticle size. If the nanoparticles are too small, they are rapidly filtered out by the kidneys. If they get too large, they are captured either by the liver or the lungs. We need to design nanoparticles that are just the right size.

11.9.3   Removal of Nanoparticles by the Immune System

An important characteristic of nanoparticles is their residence circulation time in blood. If blood serum proteins are allowed to stick to the nanoparticles, then the immune system will rapidly clear them from blood in a process known as opsonification. Most nanoparticles developed for drug delivery use a "stealth layer" to prevent opsonification and thereby to help increase circulation time in blood. Longer circulation time allows more possibility for targeting to diseased cells, leading to lower doses needed for effective therapy. This is turn leads to fewer side effects and less toxicity for the patient.

11.9.4   Toxicity of PET Probes in Nanomedical Systems for Noninvasive Imaging

While this chapter has been concerned about the nanotoxicity of nanomaterials and chemicals used during synthesis of nanomedical systems, there is one additional area involving use of radioactive probes. Positron emission tomography (PET) probes contain radioactive materials that emit positrons, the antimatter counterpart to electrons. There are many different types of PET probes used for different purposes. The most common use of PET probes involves linking them to sugars to provide a sensitive detection of cancerous tumors. Cancer cells typically use a lot more glucose due to their heightened metabolism, so PET probes attached to glucose molecules accumulate more in cancer cells, causing those cancer cells to "light up" compared to the almost totally dark background of their normal counterpart cells. An exception to this situation is when patients have large-scale inflammatory responses such as Crohn's disease. The gut of those patients will light up due to the inflammatory cells rather than cancer. This is because antimatter positrons do not naturally occur in our matter world. They rapidly self-destruct when they combine with electrons. These PET probes are a radioactive analog to fluorescent probes, meaning that they appear as a positive signal over a totally negative background. These PET probes have very short half-lives and expose patients to approximately four times the background radiation levels, or about the same as spending one year living in higher-altitude places such as "mile-high" Denver or much-higher Santa Fe, where I live! Since PET probes usually have very short half-lives, they are not a major problem in terms of disposal in the environment, which is usually excretion from the patient in urine and feces. There is also minimal damage to the kidneys for the approximately 12 hours that it takes for the patient to excrete these probes after a PET scan.

11.10   "Organ-on-a-Chip" Approaches to Human Disease Modeling

Recently, new "organ-on-a-chip" approaches (van der Meer and van den Berg, 2012), whereby in vitro 3D engineered tissues approximate in vivo organs, allow for higher throughput testing techniques. Human organs-on-a-chip may provide better predictions of toxicity in humans than conventional animal testing. People opposed to the use of animals for drug testing should be delighted by the development, which should greatly reduce the number of animals used for this purpose. This advance is partially changing the paradigm of first testing in animals, then in humans. Since most animal model systems are not very good predictors of human responses to drugs, there was a push to develop artificial 3D human organs in vitro, whereby drugs could be tested on human cells and "pseudo-organs" without endangering humans. It also provides the advantage of being able to test on human cells of different races and ethnicities, even making use of human genome repositories. Artificial 3D tissue engineered organs-on-a-chip can provide a partial bridge between in vitro and in vivo testing platforms. These systems have many of the advantages of both in vitro and in vivo systems and can be a closed-system platform, allowing easier accounting of all experimental inputs and outputs. Some of these organs-on-a-chip are quite sophisticated, and multi-organ systems have been developed that even allow for trafficking of both molecules and cells between several different connected organs-on-a-chip. A good source of the latest information on the subject can be obtained by visiting the website of the Wyss Institute at Harvard University (http://wyss.harvard.edu/), which is currently the premier research institute in the world in this important new area.

Chapter 11 Study Questions

11.1 What are the important differences between necrosis and apoptosis?

11.2 Necrosis has some bad immune system side effects. What are some of them?

11.3 How do dye exclusion assays attempt to detect cell death by necrosis? What are some common dye exclusion assays?

11.4 Why are all but late apoptosis assays "dye exclusion negative"?

11.5 How does a TUNEL assay detect DNA damage? Why is TUNEL considered a late apoptosis assay?

11.6 How does an annexin-V assay work? Why is it considered to be an early apoptosis assay?

11.7 How does a comet assay detect DNA damage?

11.8 Light scatter can distinguish between live and dead cells but only under limited circumstances. How does it work, and what are those circumstances?

11.9 How does a dihydroethidium assay work? When is it a better assay compared to standard necrosis and apoptosis assays?

11.10 Nanotoxicity in vivo is far more complicated. How can we study the biodistribution of nanoparticles in live animals? In dead animals?

11.11 What advantages do organs-on-a-chip have when compared to conventional in vitro systems and animal in vivo systems?

12 Designing Nanodelivery Systems for In Vivo Use

12.1 Overview: The In Vitro to Ex Vivo to In Vivo Paradigm

The normal paradigm for developing a new nanomedical system is to start with an in vitro cell line (usually human) and then progress to excised or biopsied tissue from a human (ex vivo). Finally, to better simulate the effects of a total organism we begin in vivo studies, usually on an animal system. This has been the general paradigm for decades. Now we have new "organ-on-a-chip" in vitro models that generate organ-like human tissue on an in vitro format. There have even been more recent promising efforts at generating "human-on-a chip" technologies at the Wyss Institute at Harvard University, which is currently producing on the order of half of all the world's new organ-on-a-chip model systems. Multiple human organs have been simulated by organ-on-a-chip technology (https://news.harvard.edu/gazette/story/2020/01/human-body-on-chip-platform-may-speed-up-drug-development/) that connects several different organs-on-a-chip with trafficking of cells and molecules between different artificial organs. These organ-on-a-chip and human-on-a-chip technologies can even make use of human genomic information that includes race and ethnicity as well as models of human disease on a chip. More on this subject at the end of this chapter!

We may take part of the tissue out of an animal or a human and test on that excised tissue. The most common excised biopsy tissue is human blood. A human blood sample (sometimes referred to as a liquid biopsy) is the easiest thing to use because most other biopsy samples (e.g., liver or other internal organ) are usually only available as discarded human material from surgeries.

Most of our organs, other than the liver, do not regenerate very well, but blood regenerates all the time. The amount of blood in a venous blood sample is typically only about 10 mL or less. That is a small fraction of the seven to eight liters of blood in a typical adult human. Interestingly, one of the few occasions when this is a problem is in the case of newborns. For this reason, newborn blood biopsies are rarely more than a few drops of blood, typically from a heel pinprick. Geriatric patients represent a problem at the other end of the age spectrum because it is often difficult to find a good vein from which to draw blood.

Blood samples may be preferable to using human cell lines because the human genome is human, usually with not as many gene mutations as a typical human cell line. You can also continue to sample in the same human over time to eliminate transient artifacts and diurnal variations. Another advantage of using human blood samples is that there are many other components in the blood, including many enzymes that actually will break down, destroy, or otherwise change the things that worked great in vitro but will fail in blood. Most drugs, at least initially, will continue to be developed intravenously. There are many solutions to cancer in vitro and other diseases that need to change when you put them into blood. Drugs, and nanomedical systems, often do not work in blood because they become broken down or degraded or changed by the immune system, which is yet another reason why we want to protect the therapeutic molecules from the action of those enzymes and the immune system with stealth factors.

Once you get drugs or nanomedical systems inside cells, the environment is very different from what it is out in the bloodstream. We need to keep that in mind. What we need to get is a model system that mimics the complexity of in vivo with an ex vivo situation, adding as many of the active components of the real animal as possible. One of the things that people can do, although it is not often done, is test under ex vivo conditions a type of cell either from a cell line or from excised tissue or primary cells by putting them in blood added to tissue culture rather than just in tissue culture medium. Then you can get something closer to what the in vivo situation might be.

Both in vitro and ex vivo testing are done in a closed, or mostly closed, system, meaning we can control almost everything that is in the system. In vivo systems are really open systems. They may all be contained within the organism, but the complexity of that system is so large that many important variables cannot be accurately measured. In a closed system, we try to be able to measure everything important that is happening in that system. It is just vastly simpler to do this in closed systems.

For the in vivo system, which is no longer closed, the animal is breathing, metabolizing, and excreting things. It is truly an open system where things are changing and you cannot control the overall situation. If we actually have a perfused tissue infusing oxygen and nutrients into the tissue perhaps using a microfluidic chip, then we can create something that looks a little bit more like an actual organ.

I recommend you walk your way into a useful model system starting with a simple 2D in vitro system. Then get into something more sophisticated – either an ex vivo situation or some kind of organ-on-a-chip approach – so that you can test a lot more variables. If you cannot make critical measurements in your model system, you are really wasting your time and valuable resources because you will generate more questions than answers!

12.1.1   In Vitro Assays: Importance of Choosing Suitable Cell Lines

We need to start the discussion about in vitro testing by learning more about what a cell line is, and what it is not. A cell line is a human or animal cell that is removed from the organism and grown in vitro, in tissue culture, usually in flasks or plates containing cell media and serum. A cell strain is a cell grown in tissue culture without being altered for "immortal growth" characteristics. Hence, a cell strain ages and gradually dies due to senescence or programmed cell death. Human cell strains, similar to actual in vivo human cells, can replicate a finite number of times, on the order of 40–60 divisions, before they stop replicating and die (Hayflick, 1965). Elizabeth Blackburn, Jack Szostak, and Carol Greider received the Nobel Prize in Physiology or Medicine in 2009 for their work on genetic structures related to the "Hayflick limit," a term referring to the finite number of reproductive cycles of normal cells, as named after Leonard Hayflick, a researcher at the Wistar Institute. This "age limit" is caused by telomere shortening. Stem cells are an exception. Depending on how close they are to embryonic origin, they appear to divide indefinitely. Sometimes such cell strains turn into immortal cancer cells after a number of divisions and essentially become immortal. These immortalized cells are referred to as cell lines. Most human cell lines have been created using human tissue discarded during surgeries. Others have evolved from cell strains that became immortalized by a transformation, usually involving a transformation from a normal cell with senescence to a cancer cell where the senescence factor has been eliminated by random or induced gene mutations.

Why is all this important? Even human cell strains are not perfect predictors of response to drugs because they have adapted to growth in tissue culture. They are also usually grown in 2D cultures, which can behave very differently from 3D cultures or actual in vivo organizations of cells. Cell lines must be treated as cancer cells, meaning they have some, but not all, of the characteristics of their normal cell counterparts but have become immortalized in cell culture. But even cell lines grown in tissue culture can change over time because genes mutate. One of the most famous human cell lines, so-called HeLa cells (named after the patient Henrietta Lax, from whom the cancerous cells were removed during surgery), has mutated to the point where it can raise questions about whether it is even still "human" anymore. It is the result of a cell grown in tissue culture for so long that it has adapted to tissue culture so much that it can arguably be called a tissue culture cell. As such, while still often used due to its ease of growth in tissue culture, it is a poor model of a human cell line to be used with in vitro studies of drug treatments.

12.1.2   Ex Vivo Assays: Adding Complexity of In Vivo Background While Keeping the Simplicity of In Vitro

Organ-on-a-chip makes something a little bit more realistic than a two-dimensional tissue culture model. A three-dimensional situation eliminates at least some of the artificiality of a 2D situation. If we then perfuse that tissue with actual agents that would be similar to what it would encounter in an in vivo system, it becomes even more realistic. You can then have the advantage of making it more like an in vivo system but have it still semi-closed to allow for better control of a number of the variables. It is a more complicated system, but also a much more realistic one!

12.1.3   In Vivo Assays: All the Complexity of Ex Vivo Plus the "Active" Components of a Real Animal

There are good and bad things about doing animal experiments. The good is that you are coming up with something that comes a little bit closer toward the real in vivo situation. The bad is that you have to go through a lot of additional costs and regulations with animal studies. Animal studies quickly become expensive in terms of time, expertise, and money. What you want to do is to work out a lot of the details first in vitro and ex vivo before you go to an animal system.

That said, if you have not tested your system in some kind of an animal model system, then you are likely to get critiqued as to the relevance of your studies. Then you should also expect to be critiqued on the appropriateness of the animal system you chose. Although many studies are done on mice and rat model systems, they are not very good, or appropriate, models for human studies.

12.2   In Vivo Systems Are Open, "Active" Systems with Multiple Layers of Complexity

In vivo systems add several layers of complexity to the overall laboratory situation. The purpose of a laboratory is to be able to control as many things as you can so that experimental results are not washed out by things you either cannot control or about which you might even be unaware. But another issue is the added complexity of making meaningful measurements of all of the important variables. You can still do ex vivo sampling of blood, but in general you will need to make noninvasive in vivo measurements. For those of you who are engineers or nonbiology scientists, biology is messy! Once you add biology to any system, things will immediately go to at least an order of magnitude in complexity, so be prepared!

12.2.1   In Vitro and Ex Vivo Are Mostly "Closed" Systems, but Not Completely

Laboratory in vitro experiments are usually designed to be closed systems. Despite all of the above recommendations and caveats, you need to always remember that in vitro and ex vivo are still not perfectly closed systems.

12.2.2   What Is an "Open" System?

Animal systems are inherently open systems. Layers of complexity are added because it is usually difficult or impossible to be able to quantitate all of the variables. For example, for experiments done on animals, you would need to collect the excreted materials. Urine and feces are not too difficult to keep track of, but much is excreted through the lungs, which is more difficult to adequately sample.

12.3   Attempts to Reduce the Complexity of In Vivo Systems

There have been many attempts to reduce the complexity of in vivo systems, including restricting the genetic variations and initially ignoring the variations of the immune system. Since it is not ethical to do just anything to humans, attempts have been made to put human cells, particularly tumors, into animals. Normally, the animal's immune system would mount an immune response and try to reject the transplanted human cells. But by using "nude mice," we can simultaneously reduce the genetic variations and also eliminate most, but perhaps not all, of the immune response.

12.3.1   Human Cells in Nude Mice: A Mixture of In Vitro and In Vivo

Putting human cells into nude mice as "xenograft" transplants is one way to study human tumors in an animal without eliciting an immune response. Nude mice are "athymic," meaning they do not have a functioning thymus to mount certain types of immune responses. There is a joke in the field that says mice are "nude" for two reasons. First, the mice are nude, or lacking, in a normal immune response. But second, and far more visible, nude mice are indeed mostly nude, meaning they lack body hair. This secondary characteristic is surprisingly important if you ever decide to do NIRF imaging because hair has a confounding red fluorescence, as do some of the foods frequently fed to laboratory mice.

12.3.2   "Model" Small Animal Systems

There are a number of model small animal systems that are frequently used by researchers. These inbred mouse strains have little genetic variation, so each animal would be expected to have the same experimental response to drug doses. That is not vigorously true since there are always some variations from animal to animal, but it is certainly much less than the variations between humans, except perhaps in the case of identical human twins.

12.3.3   Nude Mice

There are other important things to consider when deciding whether to use thymic nude mice in your testing system. Many would argue that a nude mouse is not a normal animal because it lacks a fully functioning immune system. That is certainly true, but it gets you partway between ex vivo and in vivo systems. If you are using a nude mouse model, you are at least adding some dimensions to the ex vivo situation because when we inject tumors into a nude mouse, we are allowing the animal to actually undergo angiogenesis to the tumor to allow its growth through new blood vessels to that tumor. We can also test an intravenous approach of targeting through blood injections, which is at least a little closer to the real in vivo situation.

12.3.4   Larger Animal Model Systems

The following is due to the generous contributions of Dr. Deborah Knapp, a guest lecturer for many years of my Engineering Nanomedical Systems course at Purdue University. She has developed large animal dog models that are more like humans than you would imagine, considering that we coevolved over thousands of years, in many cases eating similar foods and living in similar environments. The dogs were also interbred more than humans are, but they are not the same as highly inbred mice with the genetic circuit quite simplified, so you are going to see a little bit more of the variations due to these greater genetic variations. With dog experiments, you would see more experimental variations than observed in highly inbred mouse experiments. In an additional complication to this research, most of these dogs are considered family members, so there are the many considerations in terms of concerns and permissions of dog owners similar to the concerns of parents when their children are being treated. The dogs are not just an experimental variable but are treated as something very close to human patients, at least in most Western societies. That said, many drug studies beginning in mice or rats work their way up to larger animals such as dogs and pigs. Because dogs are treated more like family members, we do not normally cause disease in dogs and then try to test cures. Instead, we try to find naturally occurring diseases that mimic human disease and then treat those dogs as patients. But, just as in humans, naturally occurring diseases do not occur at a very high rate, in general. For example, if that naturally occurring disease is bladder cancer, it only happens to one dog in a thousand on average. It is hard, and frequently impossible, to get enough dogs with naturally occurring disease to constitute a statistically meaningful study. However, Scottish terriers have high instances of naturally occurring bladder cancer, several hundred times greater than most dogs. It is an appropriate and important model of human bladder cancer (Knapp et al., 2014). Interestingly, dogs are one of the only large animals that develop bladder cancers, thought to be caused by their consumption of cooked, rather than raw, meat that they share in meals with humans.

12.3.5 Using Animal Data to Estimate Safe Doses of a Drug in Humans

To get a rough first estimate of the proper therapeutic dose in humans, the first approximation is to divide the dose by the weight of the dog versus the weight of the human. But as was discussed in Chapter 8, this is only a crude first estimate. However, it is probably better than attempting to estimate human doses based on mice experiments. It is also possible to get at least a first estimate and predictor of human response, such as the effective dose circulation time and excretion rate of the drugs. It can, therefore, be useful in evaluating the effectiveness of stealth factors that help protect the drugs from enzymatic degradation in blood and the immune system's attempt to eliminate the drug from the body.

12.3.6 Size and Shape of Nanomedical Systems Affect the "Safe Dose" Estimate

Size and shape are important in the ability of nanomedical systems to continue to circulate in blood. The body is set up to eliminate things of particular sizes and shapes and probably has evolved that way for obvious reasons. If you have bacteria in the blood, as in a serious medical condition known as septicemia, then the blood is probably tuned to try to eliminate rod-shaped objects a little faster than spherical objects. Rod-shaped objects tend to get eliminated from the blood unless you protect them from the immune system and the body's filtration (i.e., liver and kidney) systems by putting a good stealth layer on those rod-shaped particles. On the other hand, if they stay in the circulation system, the rod-shaped objects actually tend to get through the cell membrane more easily, promoting easier drug uptake. If you want to take advantage of improved cellular uptake of rod-shaped drug delivery systems, then first protect them with stealth layers so they will survive in the circulation long enough to get to the targeted cells.

In all of this, it is important to remember the Goldilocks principle. For those of you not familiar with the Goldilocks fairy tale, Goldilocks was a little girl who went into a bear family's cabin in the woods and tried three beds. One was too big, another was too small, and the third was just right. So the Goldilocks principle is about the concept of being "just right" in terms of size, temperature, and just about any other variable you can think of. In this case, to avoid the filtration of nanomedical systems by the kidneys or liver, we want the particles to obey the Goldilocks principle so they are big enough, but not too big, to have a maximum circulation time.

12.3.7   Stealth Factors Have a Major Impact on Drug Circulation Times

Polyethylene glycol (PEG) is currently the most common stealth factor component in most drug systems. PEG is a polymer with many possibilities of different lengths and branched structures that can behave differently in terms of improved circulation times. The main purpose of PEG as a stealth agent is to prevent the sticking of blood proteins to the nanomedical system. When certain blood proteins stick to the surface of an object in the blood, the immune system attacks that protein-coated object and attempts to eliminate it from the body.

Chitosan from the shells of crustaceans, including lobsters, crabs, and shrimp, as well as many other organisms, including insects. can function as a substitute for PEG. It is a deacetylated form of chitin composed of linear polysaccharide chains with randomly distributed β-(1→4)-linked D-glucosamine (deacetylated unit) and N-acetyl-D-glucosamine (acetylated unit). It is typically made by treating the chitin shells of shrimp and other crustaceans with an alkaline substance, such as sodium hydroxide. Some drug manufacturers have used liposomes to encapsulate drugs to protect them from enzymatic degradation and elimination by the immune system. Liposomes can also work, but care must be taken to prevent fusion of these liposomes randomly to cells and tissues, which can destroy desired cell subpopulation targeting.

You will also see small sequences of DNA used as a stealth factor. There is a lot of freely circulating DNA in blood due to the lysing of cells. For this reason, the body tolerates the presence of DNA without activating much of an immune response. This is one of the reasons why DNA aptamers have been used as targeting molecules since the aptamers simultaneously provide for targeting and serve as a stealth layer.

12.3.8   Importance of Delivering a "Therapeutic Dose"

Being able to improve the circulation time and getting enough nanoparticles to the targeted cells to have some kind of therapeutic response must be kept firmly in mind because if we do not get above a therapeutic threshold, then there is no point in pursuing a particular approach. Conventional medicine usually involves getting a certain concentration of small drug outside the cell so it can diffuse by a concentration gradient. That usually requires a high concentration of drug outside the cell in order to get enough inside by a concentration gradient. Unfortunately, this means that the amount of total untargeted drug administered to a patient to produce the required concentration gradient means exposing the patient to very high amounts of totally administered drug, which goes everywhere in the body unless it is targeted. This is the main problem of untargeted drug delivery, which results in adverse drug reactions and unpleasant side effects caused by these high overall doses. Any good nanomedical system with appropriate targeting and stealth factors should attempt to reduce the overall drug exposure to the patient by at least one and perhaps two orders of magnitude to eliminate most adverse drug events and side effects.

Many targeted drug delivery systems bind to biomarkers on specific cell subtypes, which then are subsequently taken into those cells by receptor-mediated uptake, bypassing the entire gradient approach and eliminating the need to have a high drug concentration gradient outside the cell. Alternatively, one can design a nanomedical system with a larger particle that does not actually go inside the cell, but targets the region just outside the cell and then does some kind of porous linear drug delivery in its vicinity.

12.4   Ex Vivo Analyses

Ex vivo analyses are the next most common ways of studying diseases. Some "tissues" (e.g., blood, skin samples) are easy to sample. Others are naturally expelled from the body (e.g., urine and feces).

12.4.1   Urine Samples

Urine samples and fecal samples are perhaps the easiest to obtain from both humans and animals because the process is totally noninvasive. Proteins in urine can serve as useful biomarkers for disease and are being used in a number of early disease detection strategies. Many of these types of studies begin in animals.

12.4.2   Feces Samples

Feces provide a good view into what is happening in terms of diet and disease. Wildlife experts probably have much more expertise in this area than most laboratory scientists. Interestingly, during the COVID pandemic they are being used to get rough measures of disease penetrance in metropolitan areas.

12.4.3   Tissue Biopsies

Tissue biopsies can be everything from very easy to very hard to obtain. Obviously, biopsies from most internal organs are more difficult to obtain either through endoscopy or through more invasive surgery, but it is possible for researchers to obtain access to these materials through proper human investigation approvals and appropriate arrangements with surgeons.

12.4.4   Blood Samples

Blood samples (sometimes referred to as liquid biopsies) can be relatively easy to obtain depending on the amount of blood. Finger pricks with a small lancet are relatively easy to obtain. Venous blood sampling of larger volumes of blood requires a trained phlebotomist and can be quite difficult to obtain in very young or very old patients.

12.4.5   Excess Biopsy Materials from Patients after Surgery

Another common source of cell and tissue samples is "excess materials" obtained from patients (after appropriate human investigation approval and patient permissions). These are cells and tissues in excess of what is needed by pathologists to make appropriate diagnoses.

12.5   Noninvasive In Vivo Imaging Techniques

The invention and widespread dissemination of noninvasive imaging modalities (e.g., CAT/CT scans, MRI, PET) has revolutionized not only medicine but also basic and clinical research. Nanoparticle-based nanomedical systems have harnessed these technologies due to the unusual and varied properties of nanomaterials, which are usually electron-dense, making them good X-ray contrast agents for 2D and 3D X-ray imaging. They are also MRI contrast agents. If tagged with PET probes, they can serve as PET contrast agents. A review of multimodal imaging methods and probes compares their resolutions and costs (Key and Leary, 2014).

12.5.1   X-Rays and CT Scans

Human bones are inherently electron-dense, which is why they give good x-ray and CAT images without the need for contrast agents. Since many nanomaterials are inherently electron-dense, they serve as excellent contrast agents for x-ray imaging of biological tissues, from simple 2D images to high-resolution 3D CAT/CT scans.

Nanoparticles can be T1 or T2 contrast agents depending on the nanomaterial. Their incorporation (e.g., barium contrast agent) or targeting to organs and soft tissues makes them electron-dense enough to be imaged by x-rays.

12.5.2   NIRF Optical Imaging

Near-infrared fluorescence (NIRF) imaging allows for noninvasive in vivo optical imaging, particularly in small animals such as mice. Though deep imaging is not possible by NIRF, it is sufficient to image xenografts of human tumors growing subcutaneously in nude mice. An interesting artifact is that hair on the bodies of regular non-nude mice autofluoresces red and can interfere with the NIRF imaging process. Hence, the "nude" in nude mice is important for the success of NIRF in these types of studies. Hair can be carefully removed from non-nude mice to reduce this artifact. Another interesting, and confounding, artifact is that some types of food that researchers feed mice and rats can itself autofluoresce red when in the mouse or rat intestines. Examples of NIRF imaging (Key, Cooper, et al., 2012; Key, Dhawan, et al., 2012) are shown in Figure 12.1.

When nanomedical systems are labeled with NIRF probes, the animals can be sacrificed and their individual organs scanned by NIRF for the amount (semi-quantitatively) of uptake of those nanomedical systems (Key et al., 2011) (Figure 12.2).

12.5.3   Magnetic Resonance Imaging

Magnetic resonance imaging (MRI) often uses just the naturally occurring protons in water in tissues to produce adequate-resolution in vivo images. If additional resolution is needed, there are a number of MRI contrast agents. But many nanomaterials are also natural T1 or T2 MRI contrast agents. Since many of them are also x-ray contrast agents, this provides for their use in multimodal imaging systems that can combine the best features of each imaging modality to yield additional useful information.

12.5.4   PET Imaging

Positron emission tomography (PET; sometimes called PET scans) uses PET probes that do not occur naturally in the human body or in animals. For this reason, these PET probes function in a way analogous to fluorescent probes, meaning they appear as a bright signal on a zero-background field. This makes them highly sensitive. That is why they are commonly used for detection of very small metastatic lesions in cancer and are the standard for detecting whether a patient is "cancer-free at the PET-scan level." Tumors as small as a few millimeters in diameter can be visualized in vivo by PET imaging. We typically do that with relatively long-lived PET probes, particularly if they must be shipped in from elsewhere, which is commonly the case since few research facilities or hospitals have their own cyclotrons to generate their own PET

Near-Infrared Fluorescence (NIRF) Imaging

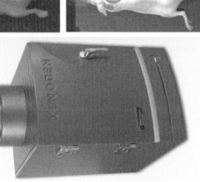
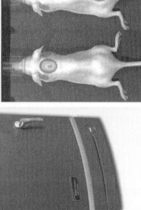
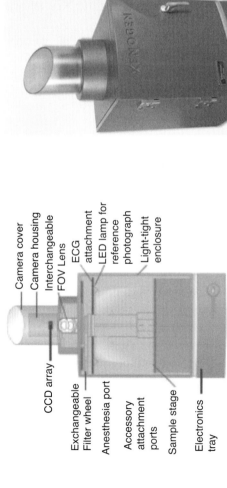
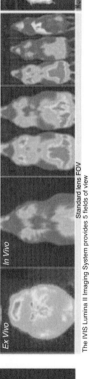

Camera cover
Camera housing
Interchangeable FOV Lens
ECG attachment
LED lamp for reference photograph
Light-tight enclosure

CCD array

Exchangeable Filter wheel

Anesthesia port

Accessory attachment ports

Sample stage

Electronics tray

Dorsal   Ventral

Xen 10

GFAP

Dual Reporter Imaging - High Resolution
EX Vivo Applications

Filels of View

Ex Vivo   In Vivo   Standard lens FOV

The IVIS Lumina II Imaging System provides 5 fields of view

Figure 12.1 NIRF imaging using appropriate near-infrared probes can provide useful in vivo information.
Source: Leary teaching

In Vivo and Ex Vivo NIRF Imaging

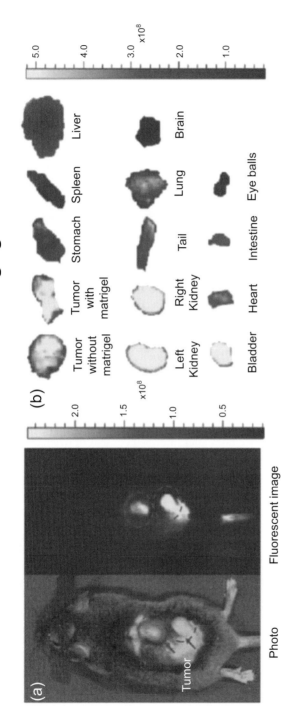

Figure 12.2 Ex vivo results in whole animals (a) and in surgically-removed individual organs (b) show that most nanoparticles (NPs) accumulated in cancer and liver cells. Accumulation in the liver is a problem that might be caused by large size or less flexibility of the NPs. However, when comparing current drugs available, it is still meaningful that the NPs mostly accumulated in cancer cells.

Source: Key et al. (2011)

probes. Cyclotron facilities can produce many different types of PET probes with a variety of lifetimes, from long to very short.

Since different PET probes provide different information, they are frequently used to detect tumors by labeling glucose and other molecules taken in by tumor cells. Since tumor cells are typically more active metabolically than normal cells, PET probes attached to glucose molecules will be taken up preferentially by cancer cells. One important exception to this is that inflammation diseases actively take up these PET probes and are a source of false positives. Since it is useful to have registration information with respect to the skeletal bones and various organs, PET is often combined with either x-ray or MRI. The dual-modality images can be overlaid (if they are properly registered) to provide useful location information.

12.6   Role of Animal Models in Translational Cancer Research

Despite their known limitations in predicting human doses and responses to drugs, animal model systems remain an important part of the overall drug-testing regimen. The important thing is to choose the best possible animal model for each human disease. One of the problems with animal testing is that inexpensive small animals such as mice and rats are really not particularly good models for predicting how drugs will behave in humans. It is also dangerous to try to scale up mice data to humans by simply compensating for size and weight. Pigs and dogs are usually much better models, but they are more costly to maintain. Also, in most Western countries, dogs are seen as family pets – effectively members of human families. For this reason, experimentation on dogs is much more limited in Western countries and often proceeds as animal clinical trials on dog patients with the permission of their owners, typically in veterinary clinics after informed consent of owners.

12.6.1   Steps from the "Bench" to the "Bedside"

The so-called bench-to-bedside paradigm attempts to streamline, as much as possible, the process of developing new techniques and drugs in the laboratory bench and then successfully translating them to use on human patients. It is a very long, arduous, and expensive process. Due to the large expenses and time scales, most of this bench-to-bedside process must be performed by large pharmaceutical companies with deep financial pockets. This developmental path often has smaller companies doing the early-stage research and then either licensing their technologies to the large pharmaceutical companies or partnering or even being bought out and incorporated within the larger companies.

12.6.2   The In Vivo Environment: 3D, Blood Supply, Microenvironment, and Immune System

Trying to model the human in terms of either animals or organ-on-a-chip technologies is complex, particularly in trying to include the human immune response component.

Many designs use microfluidics to provide pathways for cells and molecules to move around within these artificial organs. The current research leader in organ-on-a-chip technology is at the Wyss Institute at Harvard University, but there are also labs at many academic research centers doing this kind of research. The Wyss Institute has developed many different artificial human organs-on-a-chip and has even tried to develop a "human-on-a-chip" that tries to simulate real interactions between multiple organs. Although it has the potential for eliminating animal testing and has generated much excitement among groups opposed to animal testing research, it will probably be a number of years before that can happen. Whole organs-on-a-chip are still in early stages of development.

12.6.3 Studies Performed in Animal Models: Biodistribution, Pharmacokinetics, Toxicity, and Efficacy

Animals are perhaps most useful in obtaining the initial data about the biodistribution of drugs and nanoparticles; assessing cell, tissue, and organ toxicity; and having a basic system to test for drug efficacy as long as the animal model is appropriate for that human disease. These models allow simplified studies of biodistribution of nanomedical systems as well as some estimate of toxicity. These can also be used to study efficacy of targeting and treatment of diseased cells.

12.6.4 Expertise of the Team Needed to Take New Approaches from the Bench to the Bedside

The bench-to-bedside process, particularly with the current levels of multidisciplinary technologies, requires a challenging combination of talents in terms of research and clinical teams. The team members have very different perspectives and speak in different scientific and medical languages specific to their disciplines. Successful teams usually need a special leader, well versed across many disciplines, who can bring the team members together and bring the overall project to successful completion.

12.7 Types of Animal Models Available for Translational Cancer Research

The easiest and most common animal models are usually far from ideal. To minimize variables due to genetics and differing metabolism, the first animals of choice are usually inbred strains of mice with very specific genetics. But mice are really not good models for predicting human doses and response to drugs. A more appropriate model in terms of predicting scale-up are pig models since their organ structure is quite similar to that of humans. Dogs are another possible animal model, particularly useful in studying stomach and bladder cancer. But at least in Western countries, dogs are companion animals and must be treated more as research patients than research subjects.

Nude (athymic) mice permit the grafting of human tumors that will be nourished by their vasculature so that one can grow them and test drug treatments over time. The advantage is that human tumors can be grown inside these nude mice, which lack much of an immune response. The disadvantage is that these nude mice do not have a functioning immune system, so it is somewhat unrealistic eliminating something as major as an immune response.

12.7.1   Tumors Induced by Chemicals, Irritants, and Light/Radiation

Human tumors in nude mice can be created by chemical mutagens, irritants, and radiation. We can study the effects of chemical mutagens or irritants on the growth of tumors. Likewise, we can study the effects of drugs as well as radiation on these tumors.

12.7.2   Syngeneic Models

Preclinical research of new immunotherapies requires in vivo models that, unlike nude mice, have fully functional immune systems. Syngeneic tumor models are homografts derived from immortalized mouse cancer cell lines that originated from the same inbred strain of mice. This allows their study in animals with otherwise competent immune systems.

12.7.3   Immunocompromised Animals and "Foreign" Xenografts

Immunocompromised animals can have specific defects in their otherwise functioning immune systems. Foreign xenografts can be transplanted into these animals to study the rest of the immune system responses to these xenografts.

12.7.4   Transgenic Animals

Transgenic animals can be created that have some characteristics of human organs and allow studies of human diseases outside humans. In one particularly useful transgenic mice construct, researchers produced mice with human T-cell receptors, which are important in predicting human responses to infectious diseases (Cho et al., 2020).

12.7.5   Naturally Occurring Animal Models of Cancer

While the number of animals with naturally occurring cancers that mimic human disease is very small, they can serve as valuable ways to study human disease. One particular example is dogs that are excellent models for human bladder cancer. Scottish terriers have an 18–20 times higher risk for developing bladder cancer compared to mixed-breed dogs (Knapp et al., 2014). This is important for clinical trials involving dogs, as recruiting enough mixed-breed dogs for meaningful studies can be problematic. Scottish terriers can be studied in smaller total numbers to obtain enough naturally occurring bladder cancer (Fulkerson et al., 2017).

12.8 Naturally Occurring Cancers in Dogs as Models for Human Bladder Cancer

Much cancer research involves preclinical studies on animals before proceeding to human trials. The U.S. Food and Drug Administration (FDA) accepts some of this animal data as a rough indicator of safe levels of exposure to humans and also as evidence that such drugs may have beneficial effects in humans. The animal data can effectively change the benefit/risk ratio when under initial FDA reviews.

12.8.1 Nanomedicine Approaches in Dogs

Whereas conventional drug testing in animals has proceeded for over a century, nanomedicine approaches in animals are comparatively recent, including one study involving nanomedical diagnostics for bladder cancer in dogs. The nanomedical system was a multiple iron oxide superparamagnetic nanocube component core containing the chemotherapeutic agent, vinblastine, surrounded by a glycol chitosan stealth layer containing a NIRF probe Cy5.5 with a peptide targeting probe. In this case, we looked at intracellular uptake of multicomponent, peptide-targeted glycol chitosan nanoparticles (pMCNPs) by K9TCC bladder cancer cells (Key et al., 2016), as shown in Figure 12.3.

12.9 Organ-on-a-Chip and Human-on-a-Chip Technologies as a New Paradigm in Medicine

Organ-on-a-chip technology represents an exciting new advance in the testing of new drugs. It is an exciting new advance because it allows for rapid in vitro studies of human disease using human cells rather than animal cells. While it has not reached the stage of eliminating animal model testing, it is growing in importance as it becomes more sophisticated in terms of accurately modeling human organs. There have been recent attempts to model multiple organs interacting with each other as a system of interconnected organs, complete with trafficking of cells and molecules between the organs.

12.9.1 What Is Organ-on-a-Chip Technology?

Exactly what is organ-on-a-chip technology? It is the simulation of actual organs through artificial structures typically built with multiple cell types in 3D structures built on scaffolding. The latest advances include 3D printing of multiple cell types in patterned 3D structures. It represents a huge advance over 2D cell culture testing, which does not model in vivo situations very well.

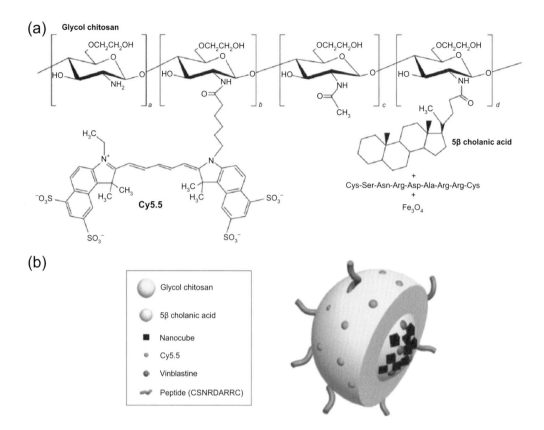

(a) Glycol chitosan

Cy5.5

5β cholanic acid

+

Cys-Ser-Asn-Arg-Asp-Ala-Arg-Arg-Cys

+

Fe$_3$O$_4$

(b)

- Glycol chitosan
- 5β cholanic acid
- Nanocube
- Cy5.5
- Vinblastine
- Peptide (CSNRDARRC)

Figure 12.3 "Chemical structures of glycol chitosan conjugated to hydrophobic 5β-cholanic acid, peptide (CS NRDARRC) and iron oxide NCs (a) and schematic diagram of pMCNPs containing iron oxide NCs and vinblastine (b). Notes: Glycol chitosan was modified with hydrophobic 5β-cholanic acid and a bladder cancer-targeting peptide, CS NRDARRC. The deacetylated free amine groups of glycol chitosan were conjugated with 5β-cholanic acids and the peptide. For NIRF imaging, Cy5.5 dyes were also conjugated to the free amines on glycol chitosan. Cy5.5 was chemically conjugated to the glycol chitosan (4:1 mole ratio, Cy5.5:Glycol chitosan). The oleic acid capped NCs (22 nm) were physically loaded with a probe-type sonicator and stabilized inside the glycol chitosan nanoparticles by hydrophobic interactions. Vinblastine was encapsulated by solvent evaporation. The schematic image indicates the surface conjugation of glycol chitosan with Cy5.5 and the peptide and the interaction of the core NCs and vinblastine with 5β-cholanic acid. Abbreviations: NCs, nanocubes; NIRF, near infrared fluorescent; pMCNPs, pCNPs loaded with 22 nm iron oxide NCs."
Source: Key et al. (2016, 4145)

Organ-on-a-chip technology uses a combination of 3D microfabrication techniques, including 3D bioprinting, to create 3D microstructures with microfluidic channels to permit drug delivery as well as trafficking of other cells, serum, and nutrients. An early formulation from our labs was a "breast-on-a-chip" model of human ductal breast cancer (Bischel, Beebe, and Sung, 2015; Grafton et al., 2011) (Figure 12.4).

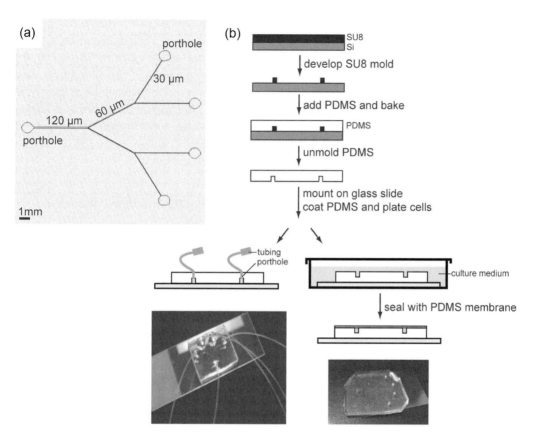

Figure 12.4 Example of the design and construction of a "breast-on-a-chip" artificial organ-on-a-chip technology for study of human ductal breast cancer. A microchannel system was molded in polydimethylsiloxane (PDMS), or dimethicone, which is is a polymer widely used for the fabrication and prototyping of microfluidic chips. In this case, PDMS is coated with laminin 111 and used as substrate for the culture of HMT-3522 S1 cells. (a) Schematic of the branched channel system. (b) Two independent approaches were developed. In the first approach (left, drawing and picture of the system), PDMS microchannels were sealed onto a glass coverslip, coated with dried or dripped laminin 111, and used for the culture of S1 cells in a closed environment. Cells were injected through tubing connected to the portholes using a syringe pump, and the medium was changed by immersion. In the second approach (right), cells were cultured in an open "hemichannel" system (top side of microchannel left open). The channels can be completed using a PDMS membrane on the day of the experiment.
Source: Grafton et al. (2011)

12.9.2   Concept of Disease-on-a-Chip

Many efforts in organ-on-a-chip technology focus on building artificial "normal" organs. This is a very important step in better understanding how normal organs work and can be modeled in vitro. But it is also important to extend these models into models of human disease, so-called disease-on-a-chip. This idea of artificial in vitro models of disease so treatments of human disease can be more rapidly and safely studied has been recently patented (Lelievre et al., 2018). This is particularly import-ant because it would be highly unethical to create human diseases in humans the way we sometimes induce or transplant disease in animal model systems. To this end, we decided to create "disease-on-a-chip" organ-on-a-chip technologies whereby we tried to re-create human disease, mimicking the pathology of human disease as closely as possible. As the logical extension of our earlier efforts to mimic an artificial human breast, we tried to mimic the pathology of human mammary cancers by designing and building a prototype of "human-disease-on-a-chip" (Grafton et al., 2011; Lelievre et al., 2018; Vidi et al., 2014).

12.9.3   What Are Some of the Challenges in Developing and Using These New Technologies?

There are many challenges in developing new organ-on-a-chip technologies. The first challenge is selection of appropriate molecules to mimic structures such as extracellu-lar matrixes and alternatives to bone and cartilage. The second challenge is to provide realistic patterns of nutrient delivery and waste disposal through appropriate perfusion technologies. Lastly, organs do not exist in isolation from other organs. This means that to be more realistic models of human disease, these systems must strive to create organism-on-a-chip levels, which are sophisticated and challenging to design and build.

12.9.4   What Are the Advantages and Disadvantages of Using Organ-on-a-Chip Technologies?

There are both advantages and disadvantages of organ-on-a-chip technologies. It partially bypasses animal testing, which is opposed by portions of the population. It also allows for testing potentially toxic substances directly on human cells without danger to actual human subjects. The organ-on-a-chip technology also allows for constructing human models of disease based on racial and ethnic differences, includ-ing rare genetic diseases and mutations.

Disadvantages include the inability to mimic the actual organ. It is also difficult to have artificial organs in isolation from the normal interactions with cells and mol-ecules from other organs. Attempts to create whole "human-on-a-chip" models con-tinue to be challenging, but it is important to develop better and more sophisticated systems.

12.9.5   What Is Human-on-a-Chip Technology, and Why Does It Represent an Important Advance?

It was perhaps inevitable that researchers would want to start connecting multiple organ-on-a-chip subsystems to represent larger systems and even extend them to human-on-a-chip technologies. This research was originally pushed by the Defense Advanced Research Projects Agency (DARPA), a branch of the US defense agency devoted to very high-risk research and development to create paradigm-shifting advances in science and engineering. The Wyss Institute at Harvard University won this initial large grant to produce a multi-organ simulation of the human body as a human-on-a-chip. While work on this idea is still at an early stage, it is becoming increasingly sophisticated. The importance of this cannot be underestimated since it allows for drug testing to be performed on a human organ–like structure in a total system that allows for trafficking of cells and molecules between organ-on-a-chip subsystems, at least partially similar to the human body. A preliminary human-on-a-chip system has been developed at the Wyss Institute using organ-on-a-chip technology, as shown in Figure 12.5.

12.9.6 Sophisticated Lung Organ-on-a-Chip That Simulates Breathing

Sometimes it is important to build an organ-on-a-chip model that involves not only organ structure but also function. A particularly good example is that of a lung-on-a-chip model developed at the Wyss Institute that incorporates the effects of breathing. Lung cells are sensitive to mechanical stretching during the breathing process (Huh et al., 2010) (Figure 12.6).

12.9.7 How Organ-on-a-Chip Is Accelerating Human Drug Studies

Organ-on-a-chip technologies are helping to accelerate human drug testing by providing a substitute for animal testing, which frequently is not very predictive for humans. There is a new NIH Institute, the National Center for Advancing Translational Sciences (NCATS) – one of 27 Institutes and Centers now at the National Institutes of Health. This new NIH Institute was largely set up in response to the advent of organ-on-a-chip systems, which can provide a convenient new human organ–testing platform prior to in-human clinical trials.

12.9.8 Will Organ-on-a-Chip Drug Research Supplant Use of Animals?

Will organ-on-a-chip drug research supplant the use of animals? This will not happen in the near future because the details of the circulatory system, metabolism, and excretion systems are hard to produce in artificial human-on-a-chip systems. But ultimately, I believe that the science will drive things in the direction of phasing out our use of animals for human drug testing due to the ability to test these human organ-on-a-chip systems with the variations of human cells of different human genetics and variations due to race and ethnicity, including the inclusion of human cells with genetic mutations making people with those mutations more susceptible to adverse drug reactions. It can also test those same variations rapidly and safely for prediction of drug efficacy, which varies according to these same variables. These artificial human-on-a-chip systems, while currently expensive to build, will eventually be mass produced, so thousands of drugs can be tested at different concentrations per day, leading to a revolution in the development of new drugs by the pharmaceutical industry.

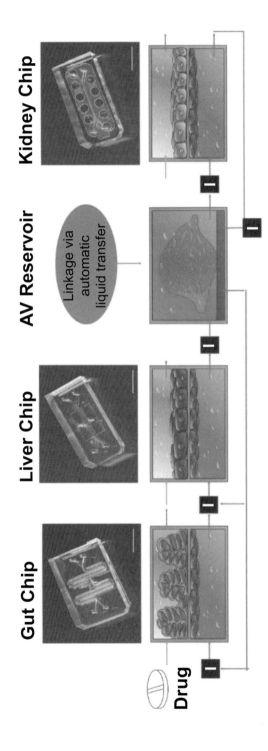

Figure 12.5 The Wyss Institute team designed a multi-human organ chip system with a human gut chip, liver chip, and kidney chip whose vascular channels are linked via a central arteriovenous (AV) fluid-mixing reservoir, and whose organ-specific channels are independently perfused. In drug-testing experiments, they added nicotine to the lumen of the gut chip's epithelial channel to mimic oral uptake of the drug, its first pass through the intestinal wall, and, via the vascular system, it journeys to the liver, where it is metabolized, and finally to the kidney, where it is excreted. Credit: Wyss Institute at Harvard University.

Source: https://wyss.harvard.edu/news/human-body-on-chip-platform-enables-in-vitro-prediction-of-drug-behaviors-in-humans/

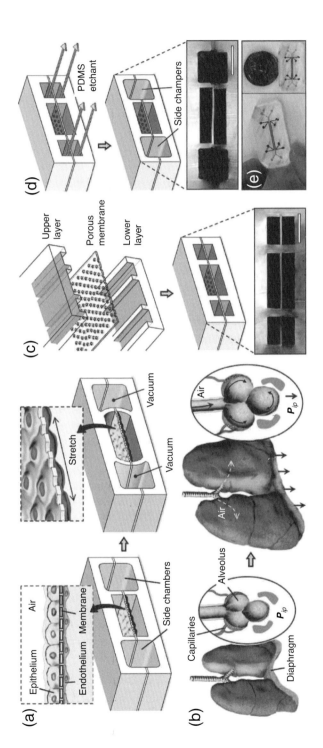

Figure 12.6 Biologically inspired design of a breathing human lung-on-a-chip microdevice. (a) The microfabricated lung mimic device uses compartmentalized PDMS microchannels to form an alveolar-capillary barrier on a thin, porous, flexible PDMS membrane coated with extracellular matric (ECM), a three-dimensional network consisting of extracellular macromolecules and minerals, including collagen, enzymes, glycoproteins, and hydroxyapatite, that provide both structural and biochemical support to surrounding cells. The device re-creates physiological breathing movements by applying vacuum to the side chambers and causing mechanical stretching of the PDMS membrane forming the alveolar-capillary barrier. (b) During inhalation in the living lung, contraction of the diaphragm causes a reduction in intrapleural pressure (Pip), leading to distension of the alveoli and physical stretching of the alveolar-capillary interface. (c) Three PDMS layers are aligned and irreversibly bonded to form two sets of three parallel microchannels separated by a 10 mm thick PDMS membrane containing an array of through-holes with an effective diameter of 10 mm. (d) After permanent bonding, PDMS etchant is flowed through the side channels. Selective etching of the membrane layers in these channels produces two large side chambers to which vacuum is applied to cause mechanical stretching. Scale bar, 200 mm. (e) Images of an actual lung-on-a-chip microfluidic device viewed from above.

Source: Huh et al. (2010)

Chapter 12 Study Questions

12.1   Why do we usually use a three-level system (i.e., in vitro to ex vivo to in vivo) to test nanomedical systems?

12.2   What are the advantages of a closed system in terms of testing? Why are such closed systems ultimately unrealistic?

12.3   Why do we often use a four-level (i.e., nude mouse to immune-competent mouse to larger animal to human) system to test nanomedical systems before human use?

12.4   What are the most important factors that affect nanomedical system circulation time in vivo?

12.5   Where (which organs) do nanomedical systems go in vivo?

12.6   How is the biodistribution of nanomedical systems measured? What are some of the problems in making these measurements?

12.7   What is an ideal therapeutic dose?

12.8   What are the most common modes of drug administration? What are the advantages and disadvantages of each mode?

12.9   How can we assess nanomedical system targeting in vivo?

12.10 Why is in vivo targeting almost always a case of rare-cell targeting?

12.11 What are some of the consequences of mistargeting in vivo?

12.12 Why is the balancing of dosing, therapeutics, and mistargeting usually an "engineering trade-off"?

12.13 What are some of the common measures of tumor load in tissues?

12.14 How can NIRF in vivo optical imaging be used to assess distributions of nanomedical systems in vivo? Ex vivo?

12.15 If the tumor is in blood, how can we get a better measure of tumor load using flow cytometry?

12.16 How is the nanobarcoding method useful in scanning over larger tissue areas to find out the distribution of nanomedical systems?

12.17 How do the MRI, CT, PET, and optical imaging systems compare in terms of (a) spatial resolution, (b) depth of penetration, and (c) sensitivity?

12.18 Why is a multimodal imaging design a good one for nanomedical systems?

12.19 What are some of the factors that make animal experiments time-consuming and expensive?

12.20 How do organ-on-a-chip systems provide a potentially faster way to test nanomedical systems for human use?

13 Designing and Testing Nanomedical Devices

13.1 Introduction to Integrated Designs

Taking our "onion skin" teaching approach, you will read things now that you read in earlier chapters, but with a little more sophistication as well as integrating concepts that could not be discussed at the beginning of the book without background knowledge. Hopefully all of what will be discussed in this chapter will be familiar to you now and be very clear and connected in your mind. You are now ready to design your own nanomedical system!

This chapter, drawing on what has been presented in Chapters 1–12, will guide you through the design steps as a series of decision trees about your exact objectives. Note that each design decision is influenced by the preceding design decisions. The construction, by its nature, is an inside-to-outside design, building from the innermost layers to the outermost layers to create a reverse-order process since the actual operation of these nanomedical systems (NMS) starts with the outermost layer and works its way back to the core. The outermost layers of your design are the first layers of interaction of your system with diseased cells, whereas the innermost layers constitute the final steps of the overall drug delivery process. These detailed reviews of the total design process may be helpful (Haglund et al., 2009; Seale-Goldsmith and Leary, 2009), as well as this comprehensive book on the fundamentals of drug delivery (Tekade, 2019).

13.1.1 "Total Design," but There Is Some Order in the Design Process

Total design of nanomedical systems means that you must take a total system-level design approach, but where the order in the design process matters – and each design step affects all preceding and subsequent design steps. The total design concept means that the essence of the design is a multilayered approach corresponding to a multistep targeting and drug delivery process (Haglund et al., 2009) (Figure 13.1).

The end product is the sum of all of the previous steps, which involves building layer-on-layer levels. Depending on the purpose of your design, you may or may not have the same exact steps. Some layers may be deliberately missing. Other layers may individually provide multiple steps in that one layer. For example, you may be running a diagnostic system that has no therapeutic step, so you can obviously skip the effect of the therapeutic step both in design and testing. There may or may not be

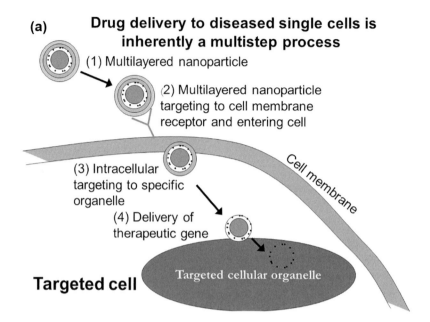

(a)   Drug delivery to diseased single cells is inherently a multistep process

(1) Multilayered nanoparticle

(2) Multilayered nanoparticle targeting to cell membrane receptor and entering cell

(3) Intracellular targeting to specific organelle

(4) Delivery of therapeutic gene

Cell membrane

Targeted cell

Targeted cellular organelle

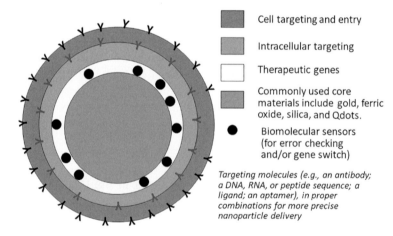

(b)   A multistep drug delivery process drives the design to a multilayered nanomedical device

Cell targeting and entry

Intracellular targeting

Therapeutic genes

Commonly used core materials include gold, ferric oxide, silica, and Qdots.

Biomolecular sensors (for error checking and/or gene switch)

Targeting molecules (e.g., an antibody; a DNA, RNA, or peptide sequence; a ligand; an aptamer), in proper combinations for more precise nanoparticle delivery

Figure 13.1 (a) The multistep targeting and delivery approach dictates a (b) multilayered design in reverse order, meaning that the outermost layers are first used and the innermost layers are for final drug delivery.

intracellular targeting. One layer, for example, may include initial targeting to the exterior of the cell as well as incorporate cell entry and retargeting within the cell.

13.1.2   A Brief Outline of the Total Design Process

It is important to go through a thorough "total design" process to avoid ugly surprises in the performance of the finished nanomedical system. We call such an approach a "systems design" because it tries to take into account the entire design operating as an integrated system. Think of the following as a checklist for constructing a NMS (with a good probability of success). As we go through this checklist, you will see the need for one or more decisions within each major design step.

The standard steps in the total design process for a NMS are as follows:

1. Choose autonomous or nonautonomous system.
2. If nonautonomous, choose type(s) of external modulation.
3. Choose core material(s), size(s), and shape(s).
4. Choose diseased single-cell molecule(s).
5. Choose type(s) of single-cell therapy.
6. Choose cell surface targeting system(s).
7. Choose zeta potential(s).
8. Choose stealth molecule(s).
9. Evaluate overall targeting system(s).
10. Evaluate single-cell therapeutic response(s).

To get you thinking about the overall total design as a multistep process, the above list can be thought of as a process flowchart with a number of decision points, as shown in Figure 13.2.

You will notice that the final steps of the total design involve the outermost layers of the NMS in a design where the last (in construction) shall be first (in operation). You should also notice that in some steps, multiple strategies may be employed. For example, there may be multiple targeting molecules, both extracellular and intracellular, as well as multiple therapeutic molecules. Even if some of these steps are not needed, it is good to at least go through this checklist in your overall design process to make sure you have not left something out of your total design!

13.2   Choose Autonomous or Nonautonomous Design

The first major question that you have, because it really fundamentally changes the design, is do you have an autonomous or nonautonomous system? Autonomous means that when you put your NMS in the body, there is no further intervention required, or possible. That means you are not going to try to manipulate it from

Process of "Total Design" of Nanomedical Systems

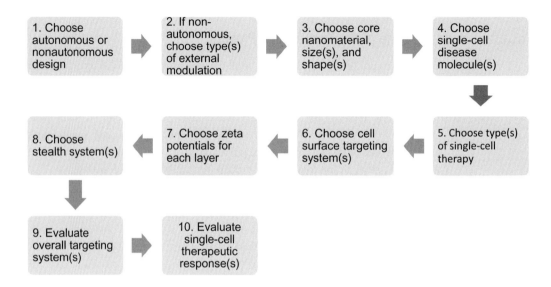

Figure 13.2 Each of the process steps in this diagram can have multiple decisions and multiple outcomes, as shown by the use(s) at each step.

outside the body with magnetic fields, or light, or by other means. That means it has to have everything that it needs in that structure to go through the multistep process, from initial targeting to the cells of interest all the way through to drug delivery and the therapeutic response.

13.2.1   If Autonomous, Will There Be Error Checking to Correct Mistargeting?

Autonomous systems tend to be a bit more sophisticated. They almost always require some kind of onboard error checking because you do not have the ability to error check that with external factors. This means that error checking must be built into the overall design because once the nanomedical system is initiated, there is no further means of controlling it externally. Often the error checking involves correcting mistargeting mistakes, including the consequences of mistargeting that were discussed in Chapter 3.

13.2.2   If Autonomous, Can the NMS Perform *All* of the Multistep Process?

Autonomous systems need to have objective measures as to whether the system is fully meeting its design goals. If not, it needs to takes alternative actions to redirect the system toward more desirable endpoints. Fully autonomous NMS require more sophistication to try to anticipate multiple decision points in the sensing and targeting layers as well as being able to make alternative decisions. In NASA speak, it is a case of "launch and forget" design, meaning that there will be no external midcourse corrections!

13.2.3 If Nonautonomous, What Form of External Modulation Will Be Used In Vivo?

If you have a nonautonomous system, you can have part of that turned on or off externally with, for example, magnetic fields or light or some other external directing agent. The advantage of systems that can be modulated externally is that the entire system can be turned off if the system is going in the wrong direction toward undesirable results. In a sense, the most valuable external factor not only turns the system on but also involves a "kill switch" to shut it off to prevent harm. You would rather kill the nanomedical system than kill the patient!

For example, if you are going to use a magnetic field, then you better have something that is superparamagnetic rather than just straight magnetic because just magnetic things will aggregate, they will remain magnetized, and you will end up with aggregation and agglomeration in vivo. When that happens, it can be life-threatening. Superparamagnetic means it is only going to be magnetic when the magnetic field is actually being applied. Another external trigger, for example, might be light of particular frequencies. An example might be gold nanoparticles that are spectrally tunable by their size and/or aspect ratio.

If you are going to have a nanomedical system deep inside the body, then you are not going to easily externally regulate it with an optical signal from outside due to the penetration depths required. These systems will then be limited except for maybe animals and nude mice. In a human being, you are definitely not going to be able to do that unless you use endoscopic approaches, for example, in a body cavity to be able to get the light or other external modulator directly there.

13.2.4 If Nonautonomous, Can the External Interaction Adequately Control the Entire Process?

We are going to assume that if a nonautonomous system is not externally triggered, the consequences of mistargeting are minimal. Only NMS by this externally modulated design will have consequences. There will be no consequences due to mistargeted NMS that are in a different bodily location outside the external modulation control system area. External modulators are like the so-called three laws of real estate in NMS design: location, location, and location. If the location is incorrect, the NMS will not be turned on and will therefore have no negative (or positive) consequences. If we know that we are trying to get the nanoparticles to a particular part of the body, the nonautonomous choice is particularly good because we can apply the external stimulus just to that area of the body and not to other body parts containing mistargeted NMS. Again, if the triggering step is necessary, then all of the other areas of the body where the nanoparticles may have gone will not be activated and therefore will have little or no mistargeting effects and little or no side effects.

13.2.5   Evaluate Reaction of NMS to External Intervention

Is the external interaction able to adequately control the functions? Is there typically going to be some threshold (e.g., you may be releasing a layer by external modulation by heat) whereby you actually melt off a layer by disrupting electrostatic interactions or actually breaking chemical bonds? How do you evaluate the response of the nanomedical system to that test? These tests are usually either in vitro or ex vivo. An ex vivo sample of blood obtained by liquid biopsy allows you to test your nanoparticle system outside the body using tests that would not be possible to use in vivo. For example, how long does the nanomedical system last before it might be enzymatically degraded within blood? How well is your stealth factor working, perhaps measured by the number of accumulating proteins on your nanoparticles? You can actually test the accumulation of proteins very easily by looking at changes in the zeta potential. There needs to be some testing endpoint of your NMS. It should typically be a functional endpoint of defined success. Otherwise, your nanomedical system might have made it to the targeted cell but did not actually do what you wanted it to do. You need quantitative measurements to determine what went wrong and to use those measurements to see whether your potential solutions are moving in the right, rather than the wrong, direction. Always try to have at least semi-quantitative measures so that you can walk your design toward a more optimal design.

13.2.6   Compare Actions of NMS with and without External Intervention

Do you truly need an external intervention to make your total design work? If you can, figure out a way to include the error-checking process in the total design without external modulators to make the design more robust. Sometimes this is possible; other times it is not.

13.2.7   Are the Actions of the NMS Linear or Nonlinear?

An external intervention in the design can cause either a linear or highly nonlinear response. If we stop the external intervention, can we instantly stop the reactions of the nanoparticles or is there some lag period? For example, if an oscillating magnetic field is used to cause localized hyperthermia at the site of the nanoparticles, will the heat continue to do something afterward and, if so, for how long?

13.3   Choose Core Material, Size, and Shape

The very next part of this really becomes the idea of choosing some kind of starting core material, size, and shape, as well as the fundamental questions of whether the core is to be used for diagnosis only, or diagnosis and therapeutics. It makes a huge difference if it's only for diagnosis. A diagnostic-only system typically has no negative consequences if it does not work properly other than the normal consequences of false negatives and false positives. But a therapeutic system can have consequences all the way from no therapeutic value to various degrees of therapeutic success. It can also have net detrimental therapeutic value, or even death. So diagnostic-only systems are much easier to design than diagnostic-therapeutic ("theranostic") systems.

13.3.1   How Will the Core Be Used for Diagnosis?

Will the diagnosis only be based on location? Medicine has always been focused on location, and for good reason. Will location at the organ level be good enough for overall diagnosis? What is the required location resolution sufficient for accurate diagnosis? For example, it may be used to locate metastatic lesions whereby cells have migrated to a part of the body where they should not be.

13.3.2   What Is the Therapeutic Radius?

By "therapeutic radius," we mean the effective radius around the nanomedical system that can deliver therapeutic doses of the drug. If the therapeutic radius is very small, one can effectively treat disease at the signal cell level. If the therapeutic radius is much larger, then cells far away from the NMS may be affected by the drug, both diseased cells and bystander normal cells. This is analogous to the differences between macrosurgery and microsurgery. If done properly, nanomedical solutions should be at the level of "nanosurgery" of diseased single cells.

13.3.3   Does This Dictate the Core Material Size?

Does this decision about diagnostics and therapeutics dictate the size of the core? Size consideration may be important if you are going to use an external magnetic field to guide these particles or cause a local concentration; then you are going to need at least a minimal diameter particle in order to have enough magnetizable material to be movable by a realistic external magnetic field. For example, to produce enough force with a reasonable magnetic field to a sufficient depth in the body, the superparamagnetic nanoparticle must be of sufficient size. A ferric oxide nanoparticle needs a strong enough external field to be able to move it or shake it in place to cause hyperthermia. If the magnetic nanoparticle is very small, it will require a very large, and perhaps unreasonable, magnetic field to be able to move it. The deeper the magnetic nanoparticle is in the human body, the larger the required external magnetic field required. With any kind of depth, you are going to need magnetic fields of at least 1–3 tesla. Researchers have made some high magnetic strength particles in the 50–100 nm range, but those are very special particles. If the nanoparticle exceeds a certain size, it starts to lose its superparamagnetism properties. An easier solution to this problem is to embed smaller superparamagnetic nanoparticles in a larger core material (Key and Leary, 2014). That allows an almost unlimited amount of superparamagnetism to be embedded in the NMS. A single region of superparamagnetic material cannot usually be made beyond a certain size before it is no longer superparamagnetic.

13.3.4   Does Shape Alter Circulation Time and Target Cell Penetration Efficiency?

In addition to size, shape can alter circulation time. It can also affect the penetration of cells in and of itself. The body is quite attuned to removing rod-shaped objects from the bloodstream to avoid septicemia. However, rod-shaped NMS can have desirable and useful properties. Gold nanorods allow you to tune them in terms of the energy absorbed by the aspect ratio of the length to the width. But you better make sure that you have a good stealth layer on it because in the absence of the stealth layer, those particles are going to be quickly taken out of the bloodstream by the immune system. The exact mechanism of nanoparticle shape in terms of penetrating cells is not well understood, but it is probably due to the radius of curvature of the shapes as they interact with single cells – perhaps a combination of potential for fusion in the case of lipid nanoparticles or electrostatics in the case of nonfusible nanoparticle materials. If you have a sharp edge on the end of a nanorod, electric fields can provide a way to make its penetration of a single cell much easier.

13.3.5   Evaluate Size and Shape of the NMS Core by Electron, Atomic Force, or Super-Resolution Microscopy

To evaluate the size and shape of the nanosized core, the technology of choice has been either transmission electron microscopy or scanning electron microscopy. Atomic force microscopy (AFM), a technology allowing measurements in an aqueous environment, is both tedious and demanding in terms of expertise and skill. Some nanoparticles do not change their shape much in aqueous environments. For example, ferric oxide, whether it is in water or whether it is being processed for electron microscopy, tends to retain its size and shape. But other core materials, particularly those with organic components, could drastically change size and/or shape, depending on whether they are in a vacuum environment for electron microscopy, in air, or in an aqueous environment. With bio-AFM, we are actually doing the tapping in a liquid environment. If you are using polymers, you might need to include some electron-dense material in order to be detectable by electron microscopy.

13.3.6   Evaluate Size of the Complete NMS

Dynamic light scattering (DLS) is really a measure of hydrodynamic size. It is necessary to put polymers, or otherwise hydrophilic molecules, on the outside of the nanomedical system so that it will stay in aqueous solution, which is the in vivo situation not only in blood but also in interstitial spaces between cells. DLS measures hydrodynamic radius, which includes the part of the surrounding fluid that travels with the nanomedical system. Typically, in an electric field that is not going to be the same as the physical size of the core. This can lead to some rather interesting differences compared to other size measurements, such as electron microscopy, which only sees the electron-dense part of the nanomedical system. It also depends on the viscosity of the fluid in which the DLS measurements are made. The hydrodynamic radius will always be bigger than the radius of the nanomedical system because it includes molecules from the surrounding fluid that travel with the nanoparticle.

13.3.7   Evaluate Materials Present at Each Layer of Construction

When we start attaching molecules to the core, we are altering surface chemistry. There is always the question of whether the layering step actually happened. If you have any doubts about whether the layering process was successful, you can use X-ray photo electron spectroscopy (XPS) to look for the presence of specific atoms or the presence of very specific chemical bonds (Seale, Haglund, et al., 2007). For example, did you really accomplish that cysteine linkage to the peptide? Use XPS analysis to find out!

13.4   Design NMS Targeting and Evaluate Its Effectiveness

Back in Chapter 3, we discussed many ways to use small molecules to help target NMS to diseased cells. We have also previously discussed many NMS that people have made without targeting molecules, using only the leaky vasculature with the enhanced permeability and retention factor (EPR) effect around the tumor and the fact that nanoparticles will naturally tend to cluster in areas of leaky vasculature even without active targeting. This EPR method is particularly important if there are no known diseased cell biomarkers with which to design a targeting system.

There are a number of diseases for which we really do not have a unique disease biomarker on the surface of that cell type. Enhanced permeability and retention (EPR) may suffice in the absence of identifying biomarkers, particularly in the case of tumors with leaky vasculature. In these cases, it is a perfectly valid design for you to proceed without a specific targeting strategy, as long as you recognize those limitations. It may still be "good enough" either diagnostically or therapeutically. You may have to deal with the mistargeting consequences. Many of these untargeted NMS are releasing drugs in the vicinity of a cell where there is a local gradient due to the buildup of NMS at that site. This gradient then pushes the drugs across by a local gradient without any kind of specific biomarker or even, in any case, a receptor-mediated uptake.

13.4.1   Choose Cell Surface Biomarkers

First, you should choose cell surface biomarkers on the diseased cells. Are the biomarkers unique or just varying in quantity? The most desirable situation is having a unique biomarker cell surface molecule on the diseased cell that is simply not expressed on normal cells. More commonly, the biomarker on diseased cells is expressed more or less than on normal cells. In the case of upregulated biomarkers on diseased cells (e.g., folate receptors on cancer cells), one can use that to target selectively, if not exclusively, to those diseased cells (Frigerio et al., 2019; Lu and Low, 2002). The consequence of this approach is that typically you are going to injure a number of normal cells in the process. That may or may not be important in terms of clinical consequences.

Folate is a small molecule found on many cell types. But it is expressed in different amounts on different cells and particularly on cancer cells. This is a biomarker that is not simply present or absent. Fortunately, different types of cancers can have elevated levels of folate receptors that help distinguish themselves from normal cells. Folate biomarkers work, but it also means that they are going to be sensitive to NMS concentration. If you flood the body with too many folate-targeted nanoparticles, many normal cells in your body will eventually take up those nanoparticles. That said, folate targeting to cancer cells has been demonstrated to do the job well enough to serve as biomarkers useful in determining the margins of a tumor during surgery (Low et al., 2018).

13.4.2   Choose Targeting Molecule Type

There are a variety of different ways that you can make targeting molecules. In Chapter 3, we talked about how people quite often use antibodies to initially prototype their NMS. Part of this is the availability factor. There are many thousands of antibodies commercially available. But most of them are made in animals such as mice, rabbits, or goats, which can cause severe allergic reactions if put into humans, including anaphylactic shock, which can even lead to death. For example, if it is a mouse monoclonal antibody, the human body's immune system will recognize the mouse protein sequences on that antibody and either attack the antibody or, worse yet, the patient. For this reason, we no longer put mouse monoclonal antibodies into human beings. Instead, we first "humanize" the structure, meaning we change the amino acid sequences slightly so that the body does not perceive them as foreign proteins. Therefore, while it is fine to do ex vivo and in vitro targeting experiments with mouse monoclonal antibodies, you must first "humanize" those mouse monoclonal antibodies before putting them into a patient!

The other problem with antibodies is their large physical structure. Most whole antibodies are approximately 200,000 daltons in size. That is very big compared to the size of some nanoparticles! Instead, we could use a biomimetic peptide, meaning a peptide sequence that will recognize the same epitope in the recognition sequence that the antibody is finding on the cell surface. These peptides can be rather small (e.g., six to eight amino acids in length) and fit much more easily on the limited surface area of a nanoparticle without steric hindrance problems. Another useful feature about some peptides is that not only can you find recognition of epitopes, but you can actually also have sequences on the same peptide that caused them to be taken rapidly across the cell membrane – so-called membrane-penetrating peptides. We have used as few as six amino acids in a peptide sequence that not only recognized a receptor on human breast cancer cells, but also served as a cell membrane–penetrating peptide – effectively accomplishing two steps of a multistep targeting process with a single nanoparticle layer (Haglund et al., 2008)!

Another type of molecule to use for targeting is an aptamer. Aptamers are typically DNA but can also be RNA or even DNA:RNA hybrid aptamers. DNA aptamers have many advantages for cell targeting. DNA aptamers, unlike RNA aptamers, are quite stable in blood because free DNA is floating around in your blood all the time. This also means that DNA aptamers are not very immunogenic and are therefore unlikely to cause allergic reactions. This allows you to not have to worry about "humanizing" aptamers, as you would with mouse monoclonal antibodies.

Your body is well equipped to not attack DNA aptamers. As we learned earlier, DNA barcoding (Eustaquio and Leary, 2011; Eustaquio and Leary, 2012b) can be used to construct targeting molecules that also contain a DNA barcode. DNA aptamers with high negative charges can also be used as a stealth factor to increase circulation time. This allows us to find the DNA-barcoded NMS more easily in a tissue. It also allows for interesting possibilities like conducting multiple experiments in the same animal (e.g., nanobarcoded to signify site of administration, time points of different infections, different doses), thereby reducing the need for larger numbers of experimental animals. However, aptamers can sometimes have a very high charge due to the large number of negative charges on a DNA molecule, so you may end up with a very high zeta potential that might need to be modulated downward by adding some positively charged molecules in the overall nanomedical device structure. Otherwise, the zeta potential may be so high that it cannot bridge the electrical repulsion when it comes close enough to a negatively charged cell to permit receptor-mediated uptake of the nanomedical system.

13.4.3 Use Flow or Image Cytometry to Evaluate Correctness of Targeting to Single Diseased Cells

The proper method for evaluation of targeting and mistargeting is largely determined by a few factors. If the cells are able to be in a single-cell suspension, as in blood samples, the most common and powerful way to evaluate the targeting process is by flow cytometry. More details about the use of flow cytometry were discussed in Chapter 3. If the cells are not suspended as single cells, then some kind of image cytometry analysis must be used, but this image analysis still needs to be at the single-cell image analysis level using single-cell image segmentation algorithms. Since nanomedicine is, by definition, single-cell medicine, evaluation methods also need to be at the single-cell level. If the targeting is intracellular, it is possible to analyze it using flow cytometry, but, in general, flow cytometry is a zero-resolution method with only crude capability of determining *where* the nanoparticles are inside a cell. Since conventional image analysis, or even confocal image analysis, does not have the resolution to see single nanoparticles, what you are going to see are aggregated (i.e., agglomerated), rather than single, nanoparticles. Currently, the only way to "see" individual nanoparticles inside a cell is super-resolution microscopy (Betzig et al.,

2006), whereby a "calculated image" is able to obtain an image below the normal diffraction limit of resolution.

13.4.4   Finding Diseased Cells in Human Blood

If you are targeting to diseased cells in blood, such as leukemia or lymphoma, you can just take a blood sample (i.e., liquid biopsy). A typical blood draw takes approximately 7 milliliters per tube. That 7 milliliters of blood is out of approximately 7 liters of total blood in a normal adult, depending on size and weight. The only cases for which these liquid biopsies are not easy are in geriatric patients, who often have difficult-to-find venous blood veins that may have many internal valves, or in newborns and infants, for whom 7 milliliters of blood represents a much more serious fraction of their total blood volume. These "liquid biopsies" are routinely performed on cancer patients whereby clinicians analyze by flow cytometry the immunophenotyping of tumors to figure out what the tumor load is in a patient. This information is then used to guide and evaluate therapy for that patient. There are now many thousands of papers in the literature using flow cytometry for this clinical approach. Once the treatment reaches a certain degree of success, it becomes difficult to detect these now rare tumor cells in a sea of normal cells in blood. At this point, "rare-event" sampling statistics determines how many total cells must be analyzed to have, for example, 95 percent statistical certainty that the tumor load estimate is accurate. To detect one rare tumor cell in a million normal cells, one must sample 2.99 million cells to have a 95 percent confidence level in the estimate of tumor cell frequency! ALthough most people assume that the sampling obeys Poisson statistics, that is actually not the case. It is really combinatorial statistics, involving gamma functions, which are very difficult to compute. During our work, an expert statistician collaborator came up with a log transform calculation method that permits simple calculation of confidence limit sampling of very rare cells (Rosenblatt et al., 1997). Monitoring so-called minimal residual disease in patients can be quite challenging. This need led to the invention of high-speed flow cytometry technology (Leary, 2005) and rare-event targeting and analysis methods (Leary, 1994), as well as techniques to molecularly characterize rare sorted tumor cells obtained by high-speed single-cell sorting, whereby these papers demonstrated the capability of detecting, and isolating by cell sorting, a single rare tumor cell in a sea of more than a million normal cells.

13.4.5   Not All Monoclonal Antibodies Are the Same, and Some Are Unsuitable for Flow Cytometry

Most monoclonal antibodies are made using a single epitope of a receptor. But sometimes that same epitope is shared by very different receptors. This is a surprise to many people when they discover a lot of false-positive targeting to different cell types that may have that same epitope on a different receptor!

Remember that a monoclonal antibody or biomimetic peptide can actually recognize a peptide sequence as small as five or six amino acids. It should be obvious that a sequence of five or six amino acids may appear in many different proteins. Many antibodies, which are quite sufficient for Western blots in electrophoresed gels, have this problem but are still usable because these shared epitopes appear at different molecular weight regions of the gel and can therefore be eliminated on the basis of gel location, which separates out different proteins on the basis of their differing molecular weights. However, in the case of a cell, you have 20,000–30,000 different proteins expressed in a single cell, each containing multiple epitopes. Unlike in locational proteomics where you have a situation analogous to the Western blot whereby the protein labeled with the antibodies may be in the right or wrong place in the cell, flow cytometry as a zeroth-order resolution technology cannot, in general, distinguish the location on or within a cell. Other than use of time-of-flight techniques whereby it is possible to distinguish whether the fluorescence is coming from the nucleus only (Leary et al., 1979), the cell surface, or intracellular molecules, it is not possible by flow cytometry to use antibody location as a way to eliminate false-positive epitope labeling. However, it is possible to use "imaging in flow" techniques (Barteneva and Vorobjev, 2016) or "slit-scan" techniques (Wheeless, Reeder, and O'Connell, 1990) to obtain information about the location of fluorescence within a cell, provided that the instrument has sufficient sensitivity for detection. A commercial instrument originally developed and sold by AMNIS Corporation, and now sold by Luminex Corporation, built on the work of Wheeless to produce a robust imaging in flow instrument that has been quite popular. It was challenging to develop to get the fluorescence sensitivity to form an image, as well as the extreme requirements of "real-time" image processing. It does require that the cells are in suspension as opposed to lying on a surface. For the cells lying on a surface, there are a variety of techniques discussed earlier in Chapter 3 that are better in terms of resolution as well as for tissues where cells are physically joined and must be separated by "image segmentation" techniques to obtain single-cell information.

13.4.6 How Much Mistargeting Is Anticipated?

Sometimes you can anticipate that a particular choice of targeting molecule is going to lead to some mistargeting. If a better targeting molecule cannot be found, then you should consider a more complicated design that requires that two different targets be simultaneously present on the cell of interest. Such targeting designs are not simple to achieve, but the requirement that more than one targeted molecule be present is a powerful solution to the problem of mistargeting.

13.4.7 Determine Degree of Mistargeting and Consider the Costs of Misclassification

If mistargeting is inevitable with your design, then the next question to ask is the *cost*, or consequences, of such mistargeting. In the case of targeting to rare cells, for example, rare cancer cells in tissue containing a huge excess of normal cells, what are the consequences of losing a few normal cells for each cancer cell eliminated? If the organ will survive and function normally, then the costs of mistargeting may be acceptable. Indeed, this strategy is practiced all the time with chemotherapy whereby doses of chemotherapy are used to try to either eliminate or reduce the tumor load of cancer cells to a level that can be handled by the patient's immune system. Some "collateral damage" to normal cells, either in proximity to the cancer cells or elsewhere in the body, may occur. An example might be some collateral destruction of the bone marrow by systemic chemotherapy. While this is certainly far from desirable, it is a case of the good not being precluded by absence of the perfect. We should always try to improve upon what is currently done, even if the revised treatment is not perfect. Science is incremental, and even incremental progress represents an advance!

13.4.8   Based on Costs of Misclassification, Reconsider Diseased Cell Biomarkers?

Other times, the collateral damage (i.e., unintended damage) to normal cells is simply unacceptable. Then some protection for normal cells may be needed. This is a powerful case for using a nonautonomous system with an external trigger. For example, a magnetic field triggered design may allow treatment only at the site of the cancer and not at the bone marrow, where the NMS may have mistargeted but then become inactive and caused little or no damage.

13.5   The Need to Evaluate Intracellular Targeting

Just getting the nanomedical system to the cell, even if binding to that cell, is not sufficient. The NMS is tiny compared to the cell. It is frequently necessary to deliver the nanomedical system to at least an approximate vicinity inside the cell. Sometimes it is necessary to target precisely to an intracellular organelle. All of this requires ways to evaluate intracellular targeting, if possible, to the single nanoparticle level.

13.5.1   Evaluate Intracellular Targeting by TEM If the NMS Is Not Fluorescent

Since these NMS are almost always below the optical limit, namely, less than the wavelength of light in the visible spectrum, it is necessary to use wavelengths below visible wavelength. One technology to use is electron microscopy, whereby the wavelengths used are far below the size of the nanoparticle. While transmission electron microscopy is very useful, it has an important limitation. Since electron microscopy requires a medium, or at least most of it, surrounding the cell to be a vacuum, the process can only work on chemically fixed, rather than alive, cells. There have been clever attempts (e.g., ultrafast electron microscopy; Flannigan, Barwick, and Zewail, 2010) to image live cells, but there is also skepticism that definitions of "live" are being stretched, particularly that of reproduction since the electron beam densities are probably at a lethal dose (de Jonge and Peckys, 2016). The cell may stay intact but might have many of its cellular functions ended. But we have probably not heard the last of efforts or claims of electron microscopy being performed on living cells.

13.5.2   Evaluate Intracellular Targeting by 3D Confocal Fluorescence Microscopy

If the nanomedical system has a fluorescent tag, for example as part of the core materials, then fluorescence microscopy can be used. Fluorescence detection is powerful because it is a technique with potentially "zero background," particularly if cellular autofluorescence is minimal. But again, optical detection of a suboptical nanoparticle is not really possible due to the diffraction limit. Most of the time, the fluorescence detected is that of agglomerated nanoparticles, sometimes very large, as earlier illustrated in the case of quantum dots within the cell.

13.5.3   Evaluate Intracellular Targeting by 2D Microscopy If Confocal Microscopy Is Unavailable

If you do not have ready access to a confocal fluorescence microscope, you can use your eyes and your brain to focus up and down and use human pattern recognition to effectively construct a confocal image of that cell in your head. Recognizing that you do not have much time to do it if it is photobleaching rights, you must do it rapidly. But we get at least get an idea whether it is within the cell or not by focusing rapidly up and down. For very rapid prototyping, 2D fluorescence microscopy is cheap, easy, and fast. Anti-fading solutions can largely eliminate the rapid photobleaching problem, but at a price. Anti-fade solutions, in general, can only be used on chemically fixed cells.

13.5.4   Evaluate Intracellular Targeting by Super-Resolution Microscopy

A recent advance to address the problem of detecting suboptically sized nanoparticles with visible light is super-resolution microscopy (Betzig et al., 2006). The way this method tries to get around the standard optical diffraction limit is to reconstruct an image of a suboptically sized object by taking multiple blurry images caused by Brownian motion of that nanoparticle and then computing a clear image based on the multiple blurry images. It is a "computed image" rather than a real image. That said, this technology can be performed on living cells, so it represents an important technological advance.

13.6   Choose Zeta Potential

If the zeta potential is horribly wrong, one of two things is going to happen. First of all, if the zeta potential is not sufficiently negative or positive, the nanoparticles will rapidly agglomerate. Agglomerated NMS in vivo can lead to possible dangerous embolisms and therefore should be avoided at all costs! If your NMS are agglomerating, you should probably focus on zeta potential problems!

In general, since you do not want your NMS to stick all over everything, and since most cells and tissue have a net negative charge, you will want a net negative zeta potential, typically in the range of −10 to −30 millivolts. If you have a net negative charge beyond −50 or −60 mV on your NMS, then you may find out that your NMS can never get across that potential barrier of the NMS and the cell to link it to a cell surface receptor. If you have a peptide or an antibody trying to grab hold of that receptor across too large a negative potential barrier, you may never be able to bridge the gap and bind that receptor!

The zeta potential of your NMS in the bottle, in order to keep it a nice, monodisperse suspension, may totally change in different ionic strength or different pH environments. The zeta potential also changes when it is in a microenvironment that gets diluted with water. If the NMS is sensitive to pH, then it might vastly change its zeta potential in the vicinity of a tumor that might have a more acidic pH microenvironment. Whereas the peripheral blood has a pH of about 7.2, the often-acidic microenvironment of a tumor may have a pH closer to 6.5–7.0. Since pH is on a log scale, that is a huge difference, and it is going to make a huge difference in the zeta potential resulting in NMS that may behave very differently!

13.6.1　Determine Required Zeta Potential for Outer/Inner Layers

If you are doing a layer-by-layer construction of your NMS, it is probably a good idea to look at the zeta potential after each layer step. You may find that you are going to have to add positively or negatively charged molecules back to your NMS or to get and keep the NMS total zeta potential in the zone where you need it to be. If you do a layer-by-layer assembly consisting of alternating positive and negative charge layers, you must guard against your NMS sticking nonspecifically over everything when it has an outermost positive layer exposed during its delayering, multistep targeting journey toward its destination.

That same process has to take place within a cell. Once inside the cell, the NMS must navigate through different microenvironments of subcellular organelles. While one common strategy is to avoid uptake of NMS by endosomes, the opposite philosophy is to deliberately get endosomes to take up the NMS and then use endosome acid pH microenvironments to release drugs.

Injecting NMS into the blood may cause a major shift in the ionic strength microenvironment of the NMS. If care is not taken, the NMS may agglomerate at or near the site of injection. Agglomeration is your enemy. Almost never would you want to have any kind of a situation where you are going to get agglomeration in vivo. If you are having any problems with your NMS sticking to cell or cell-like surfaces, that is a clear sign that you have charge (i.e., zeta potential) problems. There are many different surfaces in the bloodstream and organs!

13.6.2　Determine pH of Encountered Microenvironments

Since zeta potential very much depends on the pH of the microenvironment of the nanomedical system, it is important to try to anticipate the pH of the multiple microenvironments of the system as it traverses the human body to find its final target.

Sometimes the vicinity of the diseased cells may have a quite different local pH. An example of this is tumor cells, which frequently produce a more acidic local microenvironment. This acidic pH situation is often exploited as a way to release either a layer of the nanomedical system or to directly cause release of the therapeutic drug. You should first test your NMS in a variety of different in vitro pH environments to better predict what is going to happen in vivo.

13.6.3 Determine Ionic Strength of Encountered Microenvironments

Zeta potential is also very much affected by the ionic strength of the microenvironment, which can vary greatly not only in different parts of the body but even within different intracellar subcompartments and organelles of a cell. If you can simulate the ionic microenvironment of an in vivo portion of the body, you can check how your NMS is going to respond if you try to replicate that ionic strength in vitro.

13.6.4 Evaluate Suitability of Zeta Potential

In all of the above instances, it is important to evaluate the suitability of the overall design of the nanomedical system. Here is where the idea of using multiple layers is particularly important. As layers peel off in a multistep targeting process, the new layer can be designed to have an appropriate zeta potential for that new microenvironment.

13.6.5 If Signs of Agglomeration, Modify Zeta Potential of NMS

During all of the multistep targeting process, the zeta potential needs to stay away from conditions whereby aggregation becomes a problem. A general strategy should be that the zeta potential of the remaining nanomedical structure never approaches zero. Although a temporary positive zeta potential is generally undesirable, it is better than a situation where the NMS zeta potential comes close to zero, even transiently.

13.6.6 Are the NMS Sticking to Any Surfaces or Cell Types?

It is possible that the nanomedical system, while staying away from zero, may be briefly positive rather than negative. In such instances, it is important to prevent the system from sticking nonspecifically to cell surfaces within the human body that are almost always negatively charged.

13.6.7 Are the NMS Being Rapidly Filtered by the Kidneys In Vivo?

It is important that the NMS not be rapidly filtered out by the kidneys in vivo. In addition to the removal mechanisms of the immune system through opsonification, there is also an active filtering process by size going on through the kidneys.

13.6.8   Are the NMS Becoming Concentrated in the Liver over Time?

The function of the liver is to remove toxic things from the blood. If your NMS are sensed as potentially toxic objects by the liver, then the liver will trap them and attempt to break them down. If the liver is unable to break down toxic things, they will accumulate in the liver to the point where the organ becomes damaged. You should assume that some of your NMS will go to the liver. This means that as a part of your overall design there are no potentially toxic components of your NMS. Liver toxicity is one of the major reasons some drugs cannot be used on patients. There is considerable variability in the ability of patients to break down toxic things with their livers.

13.7   Choose Stealth Molecule and Test in Different Environments In Vitro

The biggest reason for stealth molecules is that you want to increase the circulation time of your NMS. Drugs or NMS that do not have some kind of stealth layer can be rapidly taken out of the blood by the immune system. The process of opsonification involves the deposition of serum proteins on particulate things like NMS. What you want to do is to try to avoid or at least limit the ability of proteins to stick to the surface of your NMS. Again, this is something you can test in vitro by measuring what happens when your NMS encounters those types of proteins. A mistake that many people make is to develop their entire NMS in something like saline and then get surprised by how differently those same NMS behave in the complex environment of human blood, which contains many different components not found in saline. A good place to start is to add albumin to your saline to see what happens when your NMS encounters proteins in general. The zeta potential of your NMS will change in this environment. Finally, before you inject your NMS into the human bloodstream, you should first test it in vitro using a sample of human blood. Then you should sample peripheral blood at periodic intervals to measure the concentration of NMS over time. These measurements are not easy to make, but they will give you at least an estimate of how long the NMS stay in blood and are available to deliver drugs to cells. The main purpose of using stealth molecule layer(s) is to avoid the common exponential or biexponential decay of drugs/NMS in the blood. Then you can estimate how much you can lower the overall dose to the patient. A general design goal I made for my students was to try to be able to lower the total dose by 90 percent or more. If you can do that, you will be able to eliminate most of the side effects in most patients. Side effects are a very serious problem in medicine. Either the patients get injured, directly or indirectly, by the drug or the patients suffer side effects severe enough to cause them to stop using the drug.

13.7.1   Determine Required Time of Circulation

The required time of circulation is determined by a number of different factors. Perhaps the most important of these is the effect on overall dose to the patient. The faster the NMS are cleared from the body, the more overall numbers of NMS required. If we want to keep the overall drug exposure to the patient at a minimum, we should try to maximize the circulation time.

13.7.2   How Can We Determine Circulation Time in Peripheral Blood?

The circulation time, a drug concentration versus time measurement, can be most easily obtained by taking blood samples at appropriate time intervals and measuring the drug concentration at each time point. How one determines a circulation time depends on the definition of the measurement. It might be determined by half-life as the time intervals between successive half-concentrations, for example, ½, ¼, or ⅛ of the original concentration. Or it may be determined by the 1/e points, which are useful in a logarithmic analysis if the concentration range of interest is very large.

13.7.3   Evaluate Effectiveness of Stealth Molecules

The effectiveness of stealth molecules can be evaluated according to the previous circulation time measurements. The more effective the stealth factor, the better the stealth molecule performance.

13.7.4   Do the NMS Show Signs of Protein Deposition In Vitro or In Vivo?

The presence of serum protein deposition in blood as part of the opsonification process can perhaps best be measured by changes in the zeta potential. Initially, bovine serum albumin (BSA) can be used for initial studies of protein depositions on your nanomedical systems.

13.7.5   Are the Circulation Times of the NMS Adequate to Target the Diseased Cells In Vivo?

All that really matters is that the circulation times are "good enough" to accomplish the therapeutic goal, with dose exposures to the patient below those where there is significant bystander normal cell damage or cause of side effects that may make the treatment not worth it in terms of quality-of-life factors.

13.8   Choose Type and Intracellular Target of Therapy

You may or may not have a therapeutic nanoparticle, but is your purpose to eliminate the cells or try to fix them? I do not mean chemically fix the cells. Rather, I mean to "fix them" in terms of repair them from a diseased state to a more normal state. Most of conventional medicine right now, including most of nanomedicine, is still very much focused on killing diseased cells, rather than repairing them.

If your strategy is to eliminate the diseased cells, then you should try to do so in an environmentally friendly way, by inducing programmed cell death (i.e., apoptosis), because you want to avoid injury to neighboring healthy cells either through direct injury from the contents of the diseased cells or through the immune response, which may lead to the immune system attacking not only diseased cells but also neighboring healthy cells.

Removal of diseased cells by conventional surgery, chemotherapy, or radiation therapy is a pretty blunt-edged sword in terms of spatial resolution that can create additional injury. Injured cells are going to become necrotic cells, and necrotic cells are going to cause inflammatory response, as well as releasing all of the nasty things (e.g., hydrogen peroxide) that are meant to stay inside a cell within intracellular compartments and vesicles. If instead you activate things like apoptotic pathways or related signaling transduction pathways, you can use the process of apoptosis to break down and dispose of these molecules in ways that prevent necrosis. Once the apoptotic process is triggered, the cell can proceed without further outside stimulation to dispose of itself. It is essentially an on-off switch. If your goal is to repair diseased cells, it is usually not a simple on-off switch but rather one that needs to be continuously controlled. This usually requires either an external controlling system or a self-regulating process controlled by a combination of biosensing and feedback control.

It is important to understand what part of the diseased cell state is potentially dangerous. For example, a benign tumor may be of concern only if it turns malignant; a malignant tumor probably has certain molecules that cause it to invade neighboring tissue or surface molecules that cause it to want to home to specific organs during metastasis. For example, benign prostatic hypertrophy (BPH) is a common occurrence in older men. But men can have fairly normal lives over many years unless it becomes malignant, in which case it wants to home to bone and then grow approximately 20 times faster! This is why clinicians try to get older men's prostatic antigens (PSAs) tested periodically to see whether their PSA value is increasing. A modern "sensitive PSA" test, unlike earlier, less sensitive versions, is highly predictive of prostate cancer stage. The point is that diseased cells, even cancer cells, are not necessarily life threatening. But it is important to have biomarkers that predict their metastatic and proliferation capabilities. This means that one strategy of nanomedicine can be to prevent the overexpression of certain molecules that can lead to invasion or metastasis. You do not need to change the entire gene expression profile (GEP) of the diseased cells to treat it with your NMS. You just need to control the production of certain molecules in diseased cells if those molecules are what makes the diseased cell dangerous.

13.8.1 Eliminate or Fix the Diseased Cells?

Most medical treatments define success by elimination of the diseased cells. This is an appropriate medical goal as long as the disease involvement is not so extensive that killing or eliminating all of the diseased cells would not lead to organ failure and patient death. Systemic disease, or chronic diseases involving a large percentage of cells, may require a more sophisticated and challenging strategy whereby we try to fix, or at least reduce, the defect at the single-cell level. The promise of CRISPR gene editing is that it will provide a way to treat, and even fix, chronic diseases if they are due to mistakes in the DNA.

13.8.2 If the Choice Is Elimination, Choose Appropriate Therapeutic Molecule

If the medical choice is to eliminate the diseased cells, then it is important to consider how that elimination process is to be made. Literally killing cells, whereby killing is defined as such severe injury that the cell breaks up, is often not a very good idea. Death of cells by injury (i.e., necrosis) was described previously. A more elegant solution to protect normal bystander cells from injury by nearby necrotic cells is to have the therapeutic strategy involve apoptosis, in which the cell is instructed to undergo programmed cell death. Apoptosis allows the elimination of diseased cells whereby the contents of the cell, including molecules that would be harmful to neighboring normal cells, are recycled and either used by neighboring normal cells or eliminated from the body as harmless waste material.

13.8.3 How to Control a Therapeutic Molecule That "Fixes" a Cell

The choice to try to fix, rather than kill, a diseased cell to make it a normal phenotype creates a more complex therapeutic strategy whereby the diseased cell is at least partially "reprogrammed" to try to change it from a diseased state to a normal state. This may appear to be a daunting task, but as described in Chapter 9, it may not require more than a modest number of genes to be reprogrammed by gene-editing technologies.

13.8.4 Choose Molecular Measure of Effectiveness of Therapy

In all of the preceding therapeutic strategies, it is important to come up with some quantitative, or at least semi-quantitative, measurements to determine whether the strategy is headed in the desired direction. As I used to instruct my students, if you have no measurements to guide you, then you are proceeding blindly and may quickly run off course from your objective. The measurement need not be perfect. It may be measurement of a surrogate variable. But as long as you are receiving useful feedback about the effectiveness of your strategy, then the measurements are useful and will, in most cases, save you overall time of development as well as improving the overall probability of success.

13.8.5 Use Flow Cytometry for Molecular Measurement of Cells in Suspension

Since these measurements need to be performed at the single-cell level (nanomedicine is, after all, single-cell medicine!), technologies with the sensitivity to make measurements on a single cell, rather than on bulk cells, need to be made. The measurements must be made on enough cells to evaluate the overall effectiveness of the therapeutic strategy. If it is at all possible to obtain single cells that can be maintained in a cell suspension, then flow cytometry is probably the preferred technology. It is highly quantitative at the single-cell level, it is very fast (at the rate of thousands of cells per second), and many different correlated measurements can be made simultaneously on the same cell. A disadvantage of flow cytometry is that it is a whole-cell, zero-resolution method, so you cannot determine the location of your NMS in or on a cell. It is possible, in some circumstances, to determine whether the drug is in the whole cell or just the nucleus using time-of-flight flow cytometry measurements (Leary et al., 1979). More sophisticated measurements using fluorescence energy transfer techniques and flow cytometry can look at direct molecule–molecule interactions to see, for example, whether a drug is binding its receptor within a single cell. These flow cytometric measurements require expert-level flow cytometry. The best scientists in the world for making these measurements have published a number of papers on this subject, including basic science studies (Tron, Szollosi, and Damjanovich, 1984) and clinical-level studies (Szollosi, Damjanovich, and Laszlo, 1998).

13.8.6   Use Scanning Image Cytometry to Measure Attached Cells

If single cells in suspension are not possible, then some form of image analysis must be performed. Scanning image analysis techniques allow the virtual separation of a tissue into single cells such that quantitative or semi-quantitative measurements can be made on single cells within the tissue. Major advantages of an imaging approach are that the location within a cell can be determined and that the tissue architecture, frequently a good diagnostic of disease, can be obtained. Flow cytometry loses that tissue spatial architecture information if the cells are removed individually into a cell suspension. There are *many* different types of image analysis, both 2D and 3D. Each of them has advantages and disadvantages, as were described in Chapters 3 and 4.

13.9   A Few Final Words on Design of Integrated NMS

Remember that we are still in the very early days of designing these NMS. Some people will criticize that the technology is overhyped; I would argue that nanomedicine approaches are not overhyped (Leary, 2013). Some will claim that no nanomedical system has ever cured a patient. That may or may not be true in the absolute sense. But the "first cured patient" is always next. Currently, we are PEGylating drugs and making pseudo-nanoparticles for increased circulation time, so-called time-release drugs. I would argue that those are very early, simple forms of NMS. We already have antibody therapies for cancer, which are really forms of nanomedicine. Monoclonal antibody approaches are a pseudo-nanomedicine approach right now and can eliminate or reduce a lot of the really bad side effects of untargeted chemotherapy, providing a much better alternative to killing all proliferating cells in the patient (including many in the gut or bone marrow) and then depressing the immune system. Destroying the gut is not a good way to encourage that patient to get well! Fluorescence-guided surgery uses fluorescent nanoparticles targeted to cancer cells to make the cancer cells easily distinguished from normal cells, providing guidance to the surgeon about the true location of the cancer cells within a sea of normal cells (Frigerio et al., 2019; Low et al., 2018; Lu, 2002).

13.9.1   We Are Still in the Early Days of Designing NMS

We are still early in the design of NMS. Although there was some early hype concerning nanomedicine, the discipline is now working its way step-by-step through the difficult implementation process (Leary, 2013). Some of the feedback we need for better designs awaits early clinical trials on human patients and volunteers. Only when we are able to test NMS on real human patients will we know for sure how well they will work. There may be some surprises, but hopefully not ugly ones. While we can gain valuable information from testing NMS on animals and on organs-on-a-chip, we still need clinical trials to more fully understand the actions and interactions of NMS in the human body. Nanomedicine is almost certain to have a major impact on healthcare in the near future (Leary, 2010)!

13.9.2   How Good Must Our Understanding Be to Make a Major Improvement in Disease Treatment?

We do not currently understand some of the processes well enough to fully control their design. It is important to know what actually is important, even if you cannot yet control it! We do not understand some of the processes of this well enough to fully control the design and have tried to indicate that you do not have to be perfect with your nanomedical system. Your NMS does not need to be perfect; it just has to be better than the existing therapy. That is what we do in science: We do not ever achieve perfection. We just strive to make things better. Science is usually an iterative process of improvements, just doing things a little better than they were done previously. To know whether you are heading toward improvements, always try to find a way to take some kind of measurements, even semi-quantitative, on what you are doing so that you know when you are making the situation better or worse. We have a lot of highly imperfect things in this world, but they are "good enough" to at least improve over conventional processes, which are themselves imperfect. The difficulty of accomplishing any new progress is that the old processes are grandfathered in with all their imperfections. For example, there are many far from perfect drugs that were approved that would probably not be approved if reevaluated today. They serve as a "gold standard" even if they are really "fool's gold." Fool's gold looks like gold but is not real gold! It is very frustrating to create the next generation of nanomedicine while being compared to hopelessly outmoded medical techniques just because they were the state of the art many years ago. In many cases, the techniques can be so different that it is impossible to compare the new method to the old one. Yet the new method is forced to try to come up with some way to measure itself against the old, even if that comparison does not really make any sense! I say all this so you will not be intimidated in designing a perfect NMS. Just design your NMS and then try to make the next version even better.

Chapter 13 Study Questions

13.1   Why is the order of assembly so important in the design of NMS?

13.2   Why is it more difficult to design an autonomous nanomedical system compared to a nonautonomous one?

13.3   What are some of the ways an external control can be built into a nonautonomous nanomedical system?

13.4   Why does choice of core material tend to drive the rest of the design of the nanomedical system?

13.5   How does overall size affect the performance of NMS?

13.6   What are some of the ways we can evaluate targeting in vitro? In vivo?

13.7   Why is zeta potential such an important design factor? Why is zeta potential more complicated in multilayered systems?

13.8   Why is the stealth layer such an important design factor?

13.9   Why is the design of regenerative systems that reprogram cells usually more complicated than designs to either kill the cells or induce them into programmed cell death (apoptosis)?

13.10   Why are feedback-controlled NMS more challenging to design?

14 Quality Assurance and Regulatory Issues of Nanomedicine for the Pharmaceutical Industry

14.1 Nanotechnology Task Force

The thing that we do in the United States – usually when we don't really know what we're doing – is to first form a committee and study the problem. Knowing what you don't know is the first important step in the process of gaining better understanding. Nanotechnology had been simmering in the scientific background for a number of years and especially since the year 2000. It was evident to many people that we did not understand the full implications of nanotechnology and its coming effects on science, medicine, the workplace, and the environment. In 2006 a committee, the Nanotechnology Task Force, was formed. It reported its preliminary findings and gave recommendations to Dr. Andrew Eschenbach, then commissioner of the FDA, who promised to embrace its findings and recommendations.

14.1.1 Findings and Recommendations of the Nanotechnology Task Force

When a potentially disruptive technology appears on the scene, it is important to think through its effects on society. As with any new science or technology with potentially vast implications, it is important to try to make sure that it is used for good purposes and not bad. It is also important to try to get out ahead of the new technology with some rules regarding its use, anticipating potentially harmful situations while not destroying the future good. It requires regulation with a light touch! That is exactly what the FDA, in particular, decided to do – to not kill the technology at an early stage in its development. I was a very small part of this process when I was invited in 2007 to the FDA to give a talk on nanomedicine and its implications for the FDA. At the time, the general approach of the FDA was to approve a process in toto. In other words, if nanomedicine was a "nanocar," if you changed the windshield wipers you would need to reapprove the entire nanocar all over again. Clearly, such an approach would kill the field of nanomedicine in terms of big pharma developing new nanodrug delivery systems. There actually is some validity to the total nanocar reapproval process, as it assumes that the behavior of a complex, multicomponent vehicle sometimes involves more than just the sum of its parts. Sometimes the parts do not behave independently of one another, such that the behavior of the overall vehicle is unexpectedly different from just the behavior predicted by the sum of its parts, which assumes that there are not synergistic or nonlinear dependencies between those parts.

That said, it makes sense to assume that if you use the same car just with different windshield wipers (representing perhaps the targeting molecules), along with the same cargo you have already studied and approved (repackaging the same drugs), that most of the time you can predict the performance of the nanodrug delivery system. For example, you can fairly accurately predict the homing of nanomedical systems to diseased cells by first testing the ability of the free-targeting molecules to target the diseased cells. This builds on a great deal of previous research in terms of seeing the specificity and sensitivity of those targeting molecules to find the diseased cells of interest. Depending on how the targeting molecules are attached to the nanomedical system, they might be slightly more or less effective but probably not radically different. The drugs, many in untargeted formulations already approved for human use, would still be safe – presumably a lot safer, due to vastly smaller doses! The nanomedical vehicle independent of targeting is fairly predictable based on its size and biocoatings.

All of these arguments are important because big pharma will not expend the tremendous effort, expense, and time to reapprove a nanodrug delivery system if they have to start all over each time. They will want to settle for a limited number of nanovehicles, use well-established targeting molecules, and just slightly adjust drug doses for optimal therapeutics. The safe doses of a given drug are already well studied, so using far lower drug doses should be safe. With a predictable manufacturing pathway to regulatory approval, big pharma will embrace nanomedicine as a way to extend patent protections to existing drugs repackaged into nanodrug delivery systems. This would represent a revolutionary path for modern medicine. Untargeted drugs expose patients to totally unnecessary and dangerous side effects. This can be largely eliminated by using targeted drug delivery systems. Regulation should not kill the good because it is not perfect. Good regulation protects against really bad things happening while encouraging potentially good new things. It is always a delicate balancing act.

14.1.2   The Need to Reexamine the Existing Regulations to Check Their Relevance

Sometimes new research with old things reveals important new considerations. Many things were "grandfathered in" because they were in use for many years without apparent bad effects. Indeed, the FDA has a whole area of such things listed in the category of generally recognized as safe (GRAS), such as products realted to food additives, dietary supplements, and color additives. Cosmetics were generally never considered by the FDA because they are things you put on the outside of your body, and they presumably stay there. That was true until the FDA realized that nanoparticles were being put into cosmetics. New findings showed that nanoparticles, if they are small enough (typically less than 10 nm in diameter), go right through the skin (Liang et al., 2013) and may even go on to cross the blood–brain barrier (Zhou et al., 2018)! This means the FDA needs to seriously reconsider their policies on cosmetics containing nanoparticles, particularly those smaller than 10 nm in diameter.

14.1.3   The Transition from the Lab to the Workplace and Environment

Technology in general, and nanomedicine in particular, started out small – if you would excuse the pun! By small, I mean that it was created in small quantities in the laboratories of a few researchers and subject to all the difficulties of being able to take something from the lab and replicate those results elsewhere. It also only required small amounts of material, so researcher exposure, while not without incidence, was less of a problem. As nanotechnology started scaling up, exposure risk to workers rather than researchers started becoming a potential problem. Nanomaterials were also being released into the environment, making it a potential problem for both the workplace and the environment. Manufacturing of nanomaterial products – in consumer applications as diverse as water-shedding jeans (nanomaterials tend to be hydrophobic!) and microbe-resistant kitchenware (silver nanoparticles are antimicrobial) – has put large amounts of nanoparticle-laden products in the hands of consumers. Although most nanomaterials are probably harmless, it is worrisome that so few are thoroughly tested. As discussed in Chapter 11, nanomaterials need to be tested in multiple ways to see not only toxicity due to necrosis, but also induced apoptosis, loss of normal function, and so on. But to put all this in perspective, more than 5,000 new chemicals per year are put into the hands of consumers or in the environment, most without thorough testing.

14.1.4   The Need for Some Regulation Even during the Learning Process

While governmental agencies wisely strove to not kill commercial applications of nanotechnology and nanomedicine, most people could see the wisdom in having at least a few guidelines and preliminary regulations. The FDA decided to treat regulations with a "light touch," a mostly wait-and-see attitude while nanotechnology matured and both the problems and opportunities became more visible. Although the FDA has taken some criticism for proceeding in regulation baby steps (Bawa, 2011, 2013), it was probably a wise policy, particularly in the early days of nanomedicine when so little was known or understood. Hopefully, as our knowledge and experience grow, we will be able to move faster and more decisively in terms of appropriate and selective regulations on aspects that are problems.

14.1.5   The Need to Determine Some Parameters for Nanomanufacturing

When scaling up a process and trying to make it reproducible, it is usually wise to determine what underlying principles make those processes reproducible and reliable. This means that nanotechnology is perhaps only now ready to try to determine good manufacturing practices that have been in place for other technologies since they matured. There is a well-determined pathway for "graduation" of practices suitable for the laboratory, so-called good laboratory practices (GLP) to those suitable for general manufacturing practices (GMP) for the workplace.

14.2   GMP-Level Manufacturing Compared to GLP

Good laboratory practices (GLP) can be thought of as a precursor to good manufacturing practices (GMP), but at a laboratory level rather than at a manufacturing level. GLP helps with the process of replicating processes to produce high-quality products at the laboratory level. But it does not necessarily solve the inconsistencies that occur during the scale-up from laboratory levels to true manufacturing. To make that leap from GLP to GMP, we need to understand the role of process control and testing in the overall manufacturing process.

14.2.1   Predictable Methods Lead to Predictable Products

Process control will beat your best sample-testing statistics every time! Random sampling may seem like a good way to find defects in a manufacturing process, but even random sampling is sometimes not possible. Since most of the testing on product samples is destructive, by definition those samples are destroyed, leading to lower overall yields. To get accurate predictions, for example, 95 percent confidence limits may require a significant, and unacceptable, fraction of the total sample. Process control has been around for a long enough time to be a relatively well-understood science studied by professional process engineers.

Controlling a process involves trying to create more predictable outcomes by doing the same process over and over in the same exact way. That is far more difficult than it would appear. Frequently, tiny and even invisible changes to a process can lead to very different outcomes!

Early on, the FDA understood the importance of process control to improving the safety of products. Rather than testing a large fraction of the total product sample, the total product is divided up into aliquot subsets called "product lots." Within each lot, a statistically significant number of products are tested (usually destructively) to have a 95 percent confidence limit (or some other limit, depending on the needs of the user and the consequences of a defective product to that user) that the result accurately predicts the number of defective products in a given lot. If a lot does not pass a certain level of performance, the entire lot may be rejected. This is the reason that the FDA is so obsessive about procedures and documentation. The main information submitted to the FDA for approval concerns precise documentation of the process.

That said, process control in nanodrug delivery systems for nanomedicine is at the very early stages. One can imagine process control test points for nanomedical systems being measures of size, zeta potential, targeting, and other measurements, perhaps after each step in constructing multilayered nanomedical systems. There are new possibilities for contamination to be considered. For example, our lab got very different zeta potential measurements in two laboratories at two different locations on our campus using similar equipment on a lot of nanomedical systems we synthesized. The mystery was solved only after we realized that the two results happened due to other nanoparticles and chemicals leaching out of the container walls during the transport process across campus!

14.2.2 The Code of Federal Regulations Sections on GMPs

There is a Code of Federal Regulations (CFR) that describes in detail the regulations regarding GMP practices and many other things. The CFR is the codification of the general and permanent rules and regulations published in the Federal Register by each of the executive departments and agencies of the federal government of the United States. There is a CFR annual edition of the general and permanent rules published by the Office of the Federal Register (part of the National Archives and Records Administration) and the Government Publishing Office. But there is, in addition to this annual edition, a version of the CFR that is published and updated daily in an unofficial format online on the Electronic CFR website (www.ecfr.gov/cgi-bin/ECFR?page=browse). The online version can give a good sense of where the regulatory agencies are headed. In many cases, they also use this online mechanism to solicit responses to proposed new regulations, providing an opportunity to companies and the general public to point out potential problems that may not be visible to regulators.

14.2.3 What Is Covered under cGMP?

The term "cGMP" is the abbreviation for current GMP, which means it is the most currently updated version of GMP. It is a reflection that best practices can, and should, constantly change. When a product is manufactured under cGMP, it is recognized that the manufacturer is following the latest (i.e., current) and best good manufacturing practices. The implications for the term are that it is a moving target because things are always changing. What was good enough a year or two ago may not be good enough now. It is called progress, and progress should be embraced to improve any product or process. Standards need to change to keep up with the latest, and presumably better, current standards.

The process details need to be highly detailed, and even voluminous, so that anybody anywhere in the world could actually reproduce that manufacturing process to construct a specific product. We know how hard that is to do even in the same lab. When a technician changes in the same lab, or we try to actually move the process from one lab to another lab, we sometimes get very different results. This is a well-known variable, whereby we can have different results from each technician. Often the results vary little by an individual technician but vary greatly between technicians. This is particularly the case when the test requires a technician interpreting results that may not be highly quantitative (Corsetti et al., 1987). Process details should never depend solely on the prior expertise of the technician. Prior expertise, if required, is a reflection that process details are missing steps that may or may not be filled in by a given technician based on his or her personal expertise. The semiconductor industry is an excellent example of highly controlled processes, with greater reproducibility than just about any other manufacturing process. Their success did not happen by accident, and it took them many years to reach that point!

14.2.4   Enforcement

There is no sense in making rules if there is not going to be rule enforcement. GMP rules regarding food and drug delivery are enforced in the United States by the Food and Drug Association (FDA). Many countries around the world have their own regulatory systems. The problem is that the regulatory agencies of different countries may have very different rules. The FDA of the United States has been doing this for many decades and serves as a model for many other countries. The World Health Organization (WHO) has also become involved in determining acceptable practices in this area (https://apps.who.int/iris/bitstream/handle/10665/64465/WHO_VSQ_97.01 .pdf;sequence=1; see also www.who.int/medicines/areas/quality_safety/quality_assur ance/TRS986annex2.pdf).

14.2.5   What Can Be Learned from the Semiconductor Industry Clean Room and Manufacturing?

We can learn from the semiconductor industry and clean room manufacturing because, believe it or not, the computer industry probably manufactures things with greater reproducibility than just about any other manufacturing process. It took them years to be able to do this. That's why not everybody goes out and creates the latest Intel® chips in their basements!

14.2.6   Why Not Just Sample the Product in "Lots" and Analyze Statistically?

While one can indeed sample the product in lots and perform statistical analyses, the issue is really understanding how small steps in process control can have large effects on outcome. You really want to understand where in a multistep process the sources of product quality variation are. To make any given level of product performance even reproducible requires a much greater understanding of what variables in the overall process are most important and focusing on those steps to reduce the larger sources of variation. Only after that is done does one have any hope to improve the product.

Most sample testing methods involve destructive sampling, meaning that you examine one product and then you cannot use that product again after testing. Usually, tested samples are just thrown away, unless the product is very expensive, making it worth the time and expense to try to fix the problem. Destructive testing is the enemy of yield, which is an important variable in any manufacturing process. The synthesis of nanomedical systems is at the scale of many billions of nanodevices, yet it is difficult to find defective nanomedical devices in an ocean of normal nanodevices using bulk testing methods, and single nanodevice testing is difficult or impossible. As the father of rare-event analysis and high-speed flow cytometry and cell sorting, I have often thought about trying to build a very high-speed nanoparticle "cytometer" and sorting device that could individually analyze every nanoparticle to make sure each one has every layer that is supposed to be there and remove the defective ones by sorting. Given the complexity of the multiple measurements per nanoparticle and the vast number (in the many billions) required for a typical patient dose, I am not sure it is even possible. I hope one of my readers proves me wrong, but if you decide to tackle this problem, rest assured it will require a powerful solution and probably a lot of research and development! If the number of defects is very small, such as one defect in a total of a million nanoparticles, then you would have to sample almost 3 million nanoparticles to reach a 95 percent confidence level of certainty that your analyses correctly estimate the number of defects in the total sample (Leary, 1994, 2005). Hopefully, we can have a much higher level of defects but still have acceptable consequences. I think the secret to good manufacturing of nanomedical systems will have built-in error checking for defects and a design that will mitigate the consequences of those defects.

14.2.7   Why SOPs Are So Important

Why are standard operating procedures (SOPs) so important to cGMP manufacturing? An SOP is a standard operating procedure written in great detail that you use over and over again. It is one of the most critical documents in a GMP manufacturing process and an essential part of an application for product approval to the FDA. A good SOP really has two parts. The first is a detailed description of the procedures used to produce a product. The second part tells how you can test your product to make sure that what you were trying to produce actually happened. This second part must include some tests that prove the results of the first part are actually working according to design. How do we actually know a bioconjugation bond is really there? Did we really add that layer? How do we know that if we didn't successfully add a given layer?

14.3   Bionanomanufacturing

Most people think that bionanomanufacturing will follow some variation of semi-conductor device manufacturing. We certainly can and should learn from the semiconductor manufacturing industry. First and foremost, we need to have ultra-clean air and water and protections against sources of contamination. We also need to pay attention to the gases that our nanomedical systems are exposed to since those atoms in the gas, including nitrogen and oxygen, can produce unplanned chemical reactions and bonds.

But bionanomanufacturing is *not* the same as semiconductor electronics manufacturing. Biological requirements of water-based microenvironments introduce many additional complications to address the question of whether bionanomanufacturing can be performed in an environment that ensures cGMP conditions (Figure 14.1).

Can Nanomedical Systems Be Bionanomanufactured under cGMP Principles?

Figure 14.1 Bionanomanufacturing under cGMP conditions has many of the requirements of semiconductor electronic plus additional complications once water and biology are added to the mix.

Source: Leary teaching

14.3.1   What Is So Special about Bionanomanufacturing?

There are a number of things that make bionanomanufacturing special in its require-ments. Someday, it may be possible to use some kind of not-yet-invented nano 3D printing process. But in the meantime, we need to use thermodynamically self-assembling structures. This means that we need to precisely control a number of variables such as temperature and pressure, pH, ionic strength, and other variables we have discussed earlier in this book.

14.3.2   Nano-Clean Water Necessary for Nanopharmaceuticals

What most people consider adequate water is really not adequate, particularly if you are making nanoparticles on the scale of a virus. Then you have to remove all other particulates down to under 10 nm in size. Ordinary water, even if 0.2 μm filtered, still allows particulates on the order of 200 nm to pass through, meaning that the particulate contamination may be larger or even comparable in size to what we are trying to manufacture. "Nano-clean water" used in semiconductor device fabrication removes not only most particulates but also many ions. For this reason, it is probably good for bionanomanufacturing but very bad for a person to drink. It represents the same problem as marathon athletes consuming too much water at the expense of electrolytes, but much, much worse. What you get from a drinking fountain has "nanoboulders" in it!

However, during the bionanomanufacturing process, adding biomolecules, DNA, RNA, and other molecules may introduce further problems and additional requirements. For example, if the nanomedical system uses RNA, then all water must be RNase-free and DNase-free, something that is not easy to provide. These RNases, even in small concentrations, can cause RNA components of the nanomedical system to degrade, and in some cases rapidly degrade!

14.3.3   Contaminants at the Nanolevel

We usually remove contaminants by size filtration, but what if the contaminants are on the scale of the nanomedical system? For example, viruses in the environment are on the size scale of nanomedical devices. How can we remove similar-sized object contaminants? Obviously, we need to use techniques more sophisticated than simple size filtration. Electrical charge is an obvious parameter, as objects of similar size, but different charge, can be removed by electrophoresis. If the nanomedical systems have a magnetic component, then they can be separated from other nonmagnetic object contaminants with magnetic fields. This is one reason why I keep recommending superparamagnetic core materials. It is then relatively simple to separate NMS with superparamagnetic cores from all other similarsized contaminants.

14.3.4   Can You Scale Up the Process?

It is time to discuss some of the issues faced when you try to move from laboratory scale to industrial manufacturing scale. Sometimes the scale-up is simple, but in other cases it is not possible to simply scale up a small-scale process to a large-scale process. We initially faced this problem when we moved from small mouse experiments to much larger-scale experiments in dogs. Even going from dogs to humans represents additionally required scale-up, although it is much more in the same size regime than trying to scale up from mice. In addition to the possibility that synthesis procedures need to be vastly scaled up, material costs can start to be a significant factor. This is where good targeting and stealth factors can at least partially compensate for the expense of vastly scaling up for use in large animals or humans. Those two factors may reduce the need for scale-up at least by an order of magnitude and perhaps more!

14.4   Some Quality Control Issues: How to Test

In order to do GMP manufacture of NMS, it is important to be able to perform tests on manufactured lots to ensure they are built as intended. Some of these tests will undoubtedly be bulk tests, but bulk tests will miss a small percentage (in number) of defective nanomedical devices. But considering that there may be billions of nanodevices, it will probably be impossible to test more than a small subset of the total nanodevices produced. Some tests will be more sensitive than others.

14.4.1   Correctness of Size: Size Matters!

The size of nanomedical systems has been much discussed because size really is a critical parameter. Size determines where these nanodevices will travel, how they get filtered out of the body by the liver and kidneys, and how easily they can travel to tumor sites using enhanced permeability and retention (EPR) mechanisms that take place even in the absence of targeting. The size has to be not too small and not too large, but "just right." There is a "correctness of size" because nanoparticles of the same material can behave very differently in their interaction with cells based on their size.

14.4.2   Composition: Atomic-Level Analyses

It is important to check the composition of nanomedical devices, perhaps after each layer. A technology that can examine what is in the outermost layer is X-ray photoelectron spectroscopy (XPS). An example of XPS analysis of the layers of an NMS to detect bonds in the outermost layer is given in Seale, Haglund, et al. (2007).

14.4.3   Monodispersity versus Agglomeration

Monodispersity is an important parameter. Agglomerated nanoparticles not only behave differently in vivo, but they can also lead to potentially serious problems such as embolisms. Agglomeration is caused mainly by an incorrect zeta potential, in particular when the zeta potential goes close to zero. Most of the time, we want a negative zeta potential of about –15 mV to –20 mV, but it is possible to have positive zeta potentials of approximately the same magnitude as long as they do not cause the nanoparticles to nonspecifically stick to cells and tissues. Monodispersity measurements were discussed in detail in Chapter 7.

14.4.4   Order of Layers and Correctness of Layers

How would you test, and how would you know, what fraction of the particles that you made actually had every layer in the right order? Remember, if you are going to control a sequence of events and make a "programmable nanoparticle"(Seale, Haglund, et al., 2007) using a layer-by-layer assembly, then you have to know that each layer was in the right order. While programming is being used as a controlled order of events, just like in computer code, lines of code cannot be incorrect or missing if we want to get to our desired results. Nanomedical systems missing a layer may not be able to correct for the error. But one missing layer in a multilayered process may or may not have serious consequences.

14.4.5   Correctness of Zeta Potentials

I have repeatedly emphasized the importance of correct zeta potentials. I urge you to check the zeta potential at every stage of the manufacturing process, particularly after each layer of a multilayered nanomedical device. Zeta potential may prove to be one of the most important measurements in a quality control process of bionanomanufacturing of nanomedical devices.

14.4.6   Does the Nanomedical System Contain the Correct Payload?

We know that getting drugs into nano drug delivery systems, even liposomes, has only a certain amount of drug uptake and is not a very efficient process. This means you may have considerable variation in the amount of drug per nanoparticle. That may or may no, turn out to be a critical factor, as has been discussed previously.

14.4.7   Targeting (and Mistargeting) Specificity and Sensitivity

Targeting and mistargeting specificity and sensitivity are obviously very important, not only in the process of therapy but also in terms of whether scale-up is physically or financially feasible. This targeting process may be multistage, in which case each stage must be tested separately. This multistage testing means that partially completed nanomedical stages will be needed to test for intracellular targeting since the outer layers will be used in the initial targeting stages. The construction of the nanomedical device is in reverse order, so testing must create late-stage targeting structures by using nanomedical devices that may have been accidentally stopped in their construction in the stage with the inner layers.

14.5   Role of US Agencies Involved in the Regulation of Nanotechnology

As nanotechnology has evolved and matured, it has become evident that it was a technology that touched almost every aspect of life and function. For this reason, a number of government agencies saw the need to become involved in the regulation of nanotechnology to protect the health and safety of citizens. While the involvement of the NIH was central to the intersection of nanotechnology and the life sciences, and the FDA was essential due to nanomedicine's role in drug delivery, the initial involvement of other agencies is less obvious. My initial involvement with nanomedicine was through a very early joint project between the National Cancer Institute and NASA, whereby I was asked to help invent some nanomedicine for the Mars mission. There were no guidelines at that time, and I had to devise my own strategies to try to protect my students and technicians as they synthesized NMS.

It was perhaps inevitable that one US agency got involved in nanotechnology at an early stage. Other US agencies, unfamiliar with nanotechnology, sought out the help of the National Institute of Standards and Technology (NIST) to even define some of

**US Government Agencies Involved
in the Regulation of Nanotechnology**

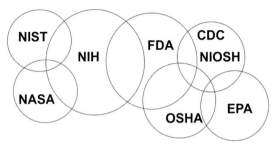

Figure 14.2 Government agencies involved in the regulation of nanotechnology and nanomedicine in the United States.
Source: Leary teaching

the fundamental parameters of this new technology. NIST is the premier US agency whose expertise is measuring things. NIST, working with the Nanotechnology Task Force, started out by trying to define what was nanotechnology and what was not. They came up with a number that has both helped define nanomedicine and also caused many problems. NIST scientists are primarily physicists and engineers with little or no knowledge of biology and medicine. As could be expected, based on their expertise, there was an initial standard of a physical dimension whereby nanotechnology was defined as objects 100 nm or less. While initially useful, that measurement was simplistic and has caused some continuing problems, primarily by people obsessed with what was just supposed to be a rough starting place for helping determine what was nanotechnology and what was not. As we will see, that 100 nm definition is useful, but too restrictive.

Once nanomaterials were being produced and used to make products, the issue of worker safety became important. The National Institute of Occupational Safety and Health (NIOSH), an agency of NIH under the direction of the Centers for Disease Control (CDC) and Occupational Safety and Health (OSHA), became concerned about worker exposure to nanomaterials. Soon thereafter, the Environmental Protection Agency (EPA) became concerned about the effect of nanomaterials on the environment.

There are many intersections between the missions and responsibilities of these agencies. While the total interactions between these agencies is much more complex, the basic interactions are shown in Figure 14.2.

14.6   FDA and Nanomedicine Regulations

It is important to try to understand how the FDA thinks about nanotechnology in general and nanomedicine in particular in order to approach the process of FDA approval in a realistic way. My biggest advice is to engage the FDA early on in your overall process to seek their guidance. They have been trying very hard to be helpful in working with manufacturers, and particularly so in the case of nanomedical devices. That said, it is also important to understand how the FDA approaches the general problem and what they are looking for in their approval process.

14.6.1 How Does the FDA Think about Nanomedical Systems?

Is it a drug or a device? Most products that the FDA reviews are one of those categories or the other, not both. A third "combo" drug-device category exists, but most pharmaceutical companies have strongly avoided that third category, thinking that their product will fall between the cracks of device and drug and therefore not get approved. Unfortunately, many nanomedical systems really are combo devices. The FDA is going to need to prepare for a much higher number of combo device approvals, and manufacturing companies are going to need to learn to be more comfortable operating in the combo device regime. Both the FDA and companies will become more efficient with practice on much higher numbers of combo devices so that there are not regulatory decision delays. Since most nanomedical systems are inherently combo drug-devices, pharma is going to need to learn to navigate those regulatory waters!

Drug-eluting stents were one of the first combo devices. Interestingly, they represent macro versions of miniature drug-eluting nanomedical systems. But these drug-eluting stents were placed in a particular part of a patient's anatomy. Nanomedical devices will be free to roam the body and home in on the diseased cells using multistep targeting systems, as has been discussed earlier in this book. For these reasons, drug-eluting stent experiences as a model for nanoscale drug-eluting nanomedical devices are only partially useful.

That said, the economic potential for repackaging current FDA-approved drugs into a nano combo device at much lower overall patient drug exposure levels is very attractive. It should also buy big pharma another 10 years of patent protection. Given the benefits of targeted nanodevices to both the patient in terms of reduction or elimination of side effects and to pharma in terms of cost of materials to manufacture, it should be a clear win-win situation for everyone. In several talks to big pharma, I tried to make this case. I have joked in invited talks to big pharma that they could end up using as little as 10 percent of the normal dose (where the drug itself may be very expensive) per patient and generously give the patient a 50 percent discounted price – a clear win-win situation for everyone!

The packaging of the drug itself is part of the NMS device because the packaging may be the targeting process and stealth layers. For example, PEGylation of a drug to prevent it from being taken out by the immune system and to improve circulation time is now becoming common even in over-the-counter drugs. I think what you are going to see is a lot of combination products using existing drugs with targeting and stealth biomolecules attached. This will represent an attractive new marketing area for big pharma as well as a straightforward way to extend the patents on drugs that had a high cost of initial development.

14.6.2   CDER and CBER Centers within the FDA

As described from their own websites, the Center for Drug Evaluation and Research (CDER) is the part of the US Food and Drug Administration (FDA) that regulates over-the-counter and prescription drugs, including biological therapeutics and generic drugs. This work covers more than just medicines; it also covers dental products, hair products, sunscreens, and many other common items (www.fda.gov/about-fda/fda-organization/center-drug-evaluation-and-research-cder). CDER is responsible for testing chemical-based drugs, although this distinction is not absolute. There is much back-and-forth between CDER and CBER.

The Center for Biologics Evaluation and Research (CBER) is the center within the FDA that regulates biological products for human use under applicable federal laws, including the Public Health Service Act and the Federal Food, Drug, and Cosmetic Act. CBER protects and advances public health by ensuring that biological products are safe and effective and available to those who need them. CBER also provides the public with information to promote the safe and appropriate use of biological products. CBER is responsible for the regulation of biological and related products, including blood, vaccines, allergenics, tissues, and cellular and gene therapies. CBER was set up largely in response to the development of antibody treatments, such as anticancer antibodies. Biologics, in contrast to drugs that are chemically synthesized, are derived from living sources (e.g., humans, animals, and microorganisms), they are not easily identified or characterized, and many are manufactured using biotechnology (www.fda.gov/about-fda/fda-organization/center-biologics-evaluation-and-research-cber).

Nanotechnology is a cross-cutting technology with products evaluated in several regulatory offices within the Center for Devices and Radiological Health (CDRH) of the FDA. CDRH develops nonclinical in vitro, in vivo, and in silico tools and models using a multidisciplinary (i.e., cell biology, toxicology, biomedical engineering, analytical chemistry) approach to more readily assess and improve upon the prediction of the safety and efficacy of FDA-regulated products incorporating such technology. Projects are focused on four broad areas of investigation: (1) medical devices containing discrete nanoparticles, such as nanosilver; (2) medical devices with immobilized surface nanostructures and topographies for orthopedic and dental implant applications; (3) genotoxicity assessment of nanomaterials using standard and alternative methods; and (4) developing provisional tolerable intake values for nanomaterials using toxicological risk assessment approaches (www.fda.gov/about-fda/fda-organization/center-devices-and-radiological-health).

14.6.3   Where the FDA May Need to Meet the EPA on Nanoscale Materials

The Environmental Protection Agency (EPA) is responsible for chemicals being put into the environment. Environmental exposures affect plants and animals, both wild and domestic, as well as humans who travel into those. The endpoint of the total life cycle development of a product is its safe disposal. As we develop our NMS, we should be conscious of this total life cycle and plan ahead through green chemistry to minimize any negative impact on the environment.

14.6.4   Will the FDA Revisit GRAS Products Containing Nanomaterials?

The FDA is revisiting so-called generally recognized as safe (GRAS) products. GRAS products have been used for a very long time (many years and decades, even centuries). The FDA has been in existence for less than 100 years, so it inherited a lot of things, such as tobacco and alcohol, that it might never have approved. This also includes many natural products, home remedies, and herbal medicines. GRAS products may or may not be safe. Most of them were never scientifically tested. Just because humans have been using foods and drugs for a long time does not mean those products are really safe or good for us. But it is difficult to get people to think differently about familiar things in their lives. It took many years to convince people that smoking, and even more so for secondhand smoke inhalation, was bad for human health. This is particularly true for products produced by companies that strongly lobby our politicians who make policy. In America, we have the "best government that money can buy" – a sad statement about the way many policies are decided and implemented. Sometimes the only strategy is to convince the consumers themselves to simply stop buying and using particular products even in the absence of proper regulation.

The FDA's position, and my own, is that many of these GRAS products may not actually be effective, but many of them are probably harmless. It is a fact that Americans spend an almost equal amount of money on these unproven and unregulated remedies as they do on regulated prescription drugs. The FDA has enough on their hands just dealing with new chemicals and biologicals. That said, many doctors now want to know about their patients' use of these remedies as well as over-the-counter drugs and "nutraceuticals," all of which can interact, sometimes dangerously, with prescription drugs. Few, if any, drugs are tested in combination with other drugs. Mostly, bad interactions are found by unexpected medical incidents. It is in large part due to the efforts of pharmacists, who understand this problem, to help steer us around potentially dangerous drug interactions. Unlike prescription medicines, and over-the-counter generic versions of them that are under the purview of the FDA, these natural products and remedies are almost totally unregulated. Personally, I would much rather have well-defined high-performance liquid chromatography (HPLC)–defined chemicals than natural products that may be contaminated with all kinds of other things. Even "organically grown" items may contain toxic chemicals that were already in the soil. Just because plants are grown without the use of toxic chemicals does not mean that preexisting chemicals in the soil have not contaminated the produce. You will not see what you do not measure!

14.7   How Does the FDA Consider Nanomedical Systems?

Traditional views are that new products are either drugs or devices. That worked until drug-eluting stents came up for approval. The FDA was forced to consider the possibility that a product could be *both* a drug and a device. For this reason, the third category "combo device" was initiated. Potential products are normally screened through either the CBER or CDER or back and forth, as deemed appropriate. The combo category has presented a bit of a problem for both the FDA and companies seeking FDA approval. While it may or may not be true that a combo designation requires additional regulatory hurdles to be jumped, this is at least a perception by companies seeking approval of their products. Many companies are loath to having their products put into the combo category. This is an unfortunate and unreasonable situation. The FDA has another agency, the Center for Devices and Radiological Health (CDRH), that functions as a catchall for things using nanotechnology. Perhaps nanomedical devices deserve a center of their own devoted to nanomedicine?

14.7.1 Nanomedical Systems Are Integrated Nanoscale Drug and Drug Delivery Devices

Nanomedical systems are really totally integrated drug and drug delivery devices. The packaging of the drug contains targeting and stealth components that help deliver the drug to the site(s) of the diseased cells. As such, they are inherently combo devices.

14.7.2 Either a Drug or a Device? How about a "Combination Product"?

There is little doubt that many nanomedicine products will naturally fall into the category of combination products. But it is important that this category not become a regulatory black hole to be avoided at all costs by companies hoping to get new nanomedicine products approved.

14.7.3 Drug-Biologic Combination Products

Further complicating matters is the fact that many targeting molecules will come from already approved anticancer monoclonal antibodies whereby instead of being a direct therapeutic, the same antibodies may be used within nanomedical devices as targeting molecules for delivering drugs. This means that we now potentially have a chemical drug inside a complex delivery device guided by molecules that have already been designated or even approved as direct biologics. All of this suggests the inherent problem of trying to fit new situations into old categories.

14.8 Types of Human Clinical Trials

Human studies normally progress from laboratory studies to limited tests of human safety without consideration of efficacy, to expanded tests of human safety, and finally to studies that include both safety and efficacy (Figure 14.3).

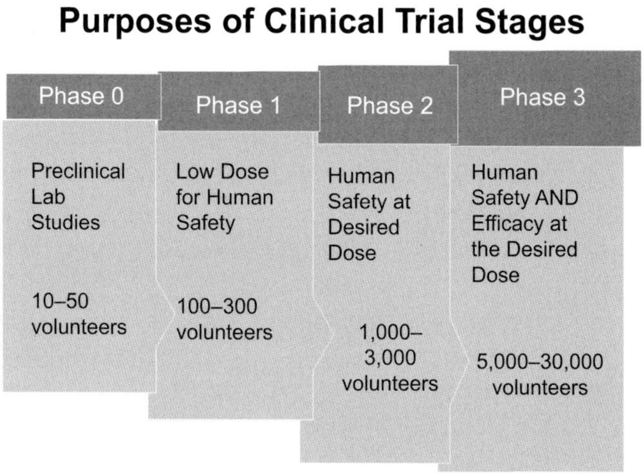

Purposes of Clinical Trial Stages

Phase 0	Phase 1	Phase 2	Phase 3
Preclinical Lab Studies	Low Dose for Human Safety	Human Safety at Desired Dose	Human Safety AND Efficacy at the Desired Dose
10–50 volunteers	100–300 volunteers	1,000–3,000 volunteers	5,000–30,000 volunteers

Figure 14.3 Clinical trials of new medicines or medical devices are done in phases. These phases have different purposes and help researchers answer different questions. For example, Phase 1 clinical trials test new treatments in small groups of people for safety and side effects. Phase 2 clinical trials look at how well treatments work and further review these treatments for safety. Phase 3 clinical trials use larger groups of people to confirm how well treatments work (i.e., efficacy), further examine side effects, and compare new treatments with other available treatments.
Source: Leary teaching

These stages of clinical trials have been formalized into structures called "Phases," which progress from 1 to 4. These stages start with proving safety before attempting to prove their effectiveness (Figure 14.4).

What do each of these phase designations actually mean? Let's discuss the purposes of each phase and see how they proceed in a logical order to first prove safety and then effectiveness.

14.8.1   IND

The first and earliest designation is the Investigational New Drug Application (IND) classification, or preclinical stage, if you are developing something new. You would request an investigational new drug application. There are at least three common ways this process is done. Guidance to potential IND investigators was given in an FDA document (21CFR312) in 2006 (www.accessdata.fda.gov/scripts/cdrh/cfdocs/cfCFR/CFRSearch.cfm?CFRPart=312).

One common investigator-initiated IND can be done by a physician wanting to treat a patient using a drug that has been proven safe, but has not necessarily been proven effective, and may not have been used for a different disease therapy. There are always reports out in the literature about how a drug that was made to cure a certain disease seems to have some beneficial effects toward another disease but has not been firmly tested. It is at least known to be safe, but it may or may not be effective. The

Clinical Trial Phases

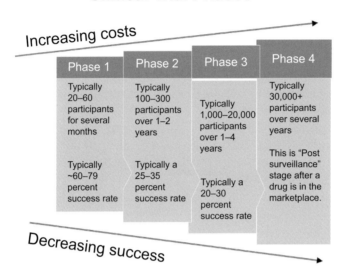

Figure 14.4 The differences between Phase 1, Phase 2, Phase 3, and Phase 4 by relative numbers of participants, time to completion, costs, and probability of success.
Source: Leary teaching

drug is unapproved in the sense of being applied to that second disease, but it has already been approved for other uses. A physician can request IND use and prescribe that drug for a patient. IND approval can allow studies to go forward on a very limited basis. The physicians are supposed to take information about patient outcomes and pass that information on to the FDA. IND drugs can reveal multiple uses that were not initially envisioned. Another use of the IND category is one where a patient is dying and all other known treatments have been tried unsuccessfully. The physician wants to try a drug that has been proven at least reasonably safe, but he or she does not have time to get FDA approval because the patient is going to die before an approval process could be achieved. The physician can use a drug on that patient for "emergency use." A third IND use is for immediate, serious, or life-threatening situations where there can be no delays in trying to save a patient's life. It makes little sense to protect patients from unknown but potentially dangerous side effects from drugs if the patient is going to die anyway. The risk factor for that dying patient is different from that of a normal patient. Use of a drug is considered on the basis of benefit versus risk. A dying patient has a more serious risk than potential side effects or even medical incidents. We would not do this to people who are healthy for whom these uses present a significant risk with less potential benefit.

The requirements for a more typical, non-emergency use IND application usually include at least some preclinical data, typically on animals, for safe testing. Now that organ-on-a-chip technology is increasingly available, I believe we will start to see some IND applications with organ-on-a-chip safety data, even without animal data.

There are many challenges regarding how to scale up from animal data to human data. This is already being done with what are considered appropriate animal models. Will it be possible to estimate safe doses from organ-on-a-chip data? You can actually scale up the animal studies and proceed, at least for very limited or emergency use, for a patient to be treated by an individual physician. You are going to base proposed doses on data that looks like it is reasonably safe. You had better have some proof that your manufacturing process is at least similar because otherwise you might be manufacturing in a way that comes out 10 times more potent per dose by weight (i.e., bioeffectiveness). That kind of variation can get you into some dangerous territory! You need to be able to be credible that you can manufacture it in a similar way so that the predicted dose seems very safe.

The IND classification is sometimes used interchangeably with a Phase 0 study, but Phase 0 studies deserve their own discussion.

14.8.2　Phase 0

A Phase 0 study is basically an IND with a new, untested drug as a first in-human trial system, and it is usually done with microdosing (i.e., using very small doses). Sometimes Phase 0 studies will put in a very small amount and then gradually increase doses, all of which are far below what would be considered a dangerous dose.

A Phase 0 study does not require data on safety or efficacy, although anything that will help make the case for approval is a good idea. A Phase 0 study is just the first time putting a drug in humans, usually in microdoses, which may be, and probably are, less than fully effective therapeutic doses. The purpose is to gather some very preliminary data for a subsequent Phase 1 application.

Typically, a small group on the order of 20–50 healthy volunteers is recruited. Phase 0 is tried on healthy volunteers because it is just safety testing. If it adversely affects normal, healthy volunteers, it will probably cause more problems in unhealthy patients with more serious underlying conditions.

14.8.3　Phase 1

Phase 1 studies get closer to actual therapeutic dose levels. It is also the first stage that allows studies of diseased populations, but it is still more concerned with safety than efficacy. It helps provide data on pharmacokinetics and dynamics of a drug as well as getting closer to what might be an actual therapeutic level of drug to treat the disease and still be at safe enough doses for the patient. Small groups of subjects are given a single dose and then observed. If there are no bad effects, they are given ascending doses, with slightly bigger doses until some volunteers start having mild adverse effects. The dose where volunteers start to have significant, but not life- or even health-threatening side effects, is considered the "maximum tolerated dose." Although most drugs are initially developed for intravenous use because it is simpler, they need to be further adapted so they can be taken orally. Drugs taken orally can be affected by whether the person has eaten or taken the drug on an empty stomach. Food effects are often studied in Phase 1. A lot of medicines will stipulate taking the drug an hour before a meal or two hours after a meal (because food in the stomach is normally passed through that digestive stage about two hours after eating). Taking some drugs on an empty stomach can sometimes drastically change their potency at a given dose! In the absence of evidence to the contrary, we should suspect that similar effects may change the potency of nanomedicines taken orally. There is still not much data on oral delivery of nanomedical devices. Many people are familiar with the phenomenon of feeling more affected by drinking alcohol on an empty stomach as opposed to drinking it mixed with food during a meal.

14.8.4   Phase 2

Phase 2 studies start to get more serious. They involve much larger groups to reveal the true side effects. The number of patients studied must be high enough to provide statistically significant data on the tails of a presumed Gaussian curve (we really do not know whether the curve is Gaussian in the case of nanomedicines). Many people blindly assume that most distributions are Gaussian. That is simply not the case! As you increase the number of patients in a study, you start seeing that some people are less, or more, normal than others. When you have more and more patients, typically on the order of 100, 200, or 300 patients, you begin getting an idea of the true variations from person to person. For this reason, Phase 2 clinical trials start to get a little more expensive. In addition, patients tend to be followed longer to make sure that there are no longer-term adverse effects. Phase 2 also tries to establish the clinically effective drug dose. People metabolize drugs differently, so this can become quite a source of variation, which you will not see until you get enough patients to see the variation among outliers who may be much more sensitive to a drug.

Since humans are greatly affected in their response to a drug by their expectations, there is a well-known "placebo effect" that is not normally studied during Phase 2 studies but becomes an important part of Phase 3 studies. Placebo effects can constitute up to two-thirds of the response to a drug! A patient's positive expectations may also be a reason why home remedies are perceived as effective even if there is no actual biochemical basis for the home remedy.

14.8.5   Phase 3

Phase 3 studies involve doing randomized studies with very large numbers of patients (perhaps 5,000–30,000). This is also when studies must obtain "placebo effect data" that help distinguish the real therapeutic effects of a drug from patient-anticipated placebo effects whereby the patient can have what appears to be a therapeutic response in the absence of being dosed with an actual drug. Most Phase 3 trials are conducted as a "double-blind, placebo-controlled" study whereby neither the patient nor the doctor knows whether a certain patient has been given the drug or a placebo. The placebo should not be easily seen as different from the actual drug, so pills should be the same color and size. Liquids should be indistinguishable from the real drug in terms of being the same color and in similar vials.

Phase 3 clinical trials become very expensive (typically more than $1,000,000,000) due to the large number of patients who need to be recruited and followed, as well as the costs of the drug and the labor to administer it to the patients. Recovering Phase 3 clinical trial costs, as well as the preceding research and development costs, is the primary reason why new medicines are so expensive, even if the cost of drug materials is very cheap! On top of all of these costs is the high probability that the Phase 3 trials will result in failure. Another negative factor is the time that it takes from the filing of a patent to obtaining FDA approval. The new drug is protected from competition for only 20 years, and the entire process from basic research and development to FDA approval may take 10–15 years. After the patent expires, generic drugs that are "biosimilar" can be made for a small fraction of these costs.

There are a number of ways whereby the risk of engaging in Phase 3 clinical trials resulting in failure can be reduced. The first way of reducing risks of Phase 3 clinical trials is using personal genomic information so that genetic variations that result in a patient being more adversely sensitive to a drug and its side effects can be anticipated a priori. This changes the situation from guesswork to a point where these known adversely sensitive patients are not only excluded from a study but also excluded from being the type of patient who will ever actually receive a given drug. Individual genome sequencing is becoming sufficiently inexpensive that it is really not a cost issue at this point. Rather, it is an information issue whereby we do not yet know the critical gene variations for adverse responses to drugs. Once those sequences are known for particular drugs, I predict that most drugs will be designed to steer clear of patients who are adversely sensitive and will target their markets to the patients who will not be adversely affected. The effect of decreased market size will, in most cases, be tiny compared to the costs of medical incidents and lawsuits. The second way of reducing risk of Phase 3 clinical trials is to use nanomedical methods to reduce overall patient dose by increasing circulation time and using targeted drug delivery. Drugs that are untargeted and unprotected from rapid degradation will become a smaller and smaller fraction of newly developed drugs once big pharma realizes the wisdom of nanodrug delivery systems. We *must* use these and other ideas to reduce the cost of bringing new, affordable drugs to market. Otherwise, there will be fewer and fewer new drugs. The present path of drug development is simply unsustainable!

14.8.6   Phase 4

Phase 4 studies are not something you normally hear of. They are sometimes called "postmarketing surveillance." After marketing and selling a new drug, you are still required to send in new data to the FDA every time there is a severe side reaction, an injury, or a death. While rare, probably due to the required rigor of Phase 3 trials, a drug can still be pulled from the market by the FDA. This action can lead to some really serious consequences, including massive lawsuits!

Figure 14.5 Integrated decision-making process for evaluating environmental impacts. This figure shows how the life-cycle assessment (LCA) is one of several tools to consider in the integrated decision-making process. LCAs are useful for identifying major system impacts on society and the environment.
Source: Leary teaching

14.9   EPA and Other Regulatory Agency Issues

Let's shift over to environmental aspects, as the Environmental Protection Agency (EPA) tries to assess the impact of nanotechnologies on the environment. Integrated decision making is an important part of the overall process of evaluating environmental impacts as done by the International Council of Chemical Associations (ICCA) (Figure 14.5).

14.9.1   Concept of Life-Cycle Assessment

The life-cycle assessment (LCA) is a relatively new concept that says, if you are going to create something new, you need to know the actual true costs of the product, including its proper disposal, as shown in Figure 14.6.

When I use the term "true costs," I mean the true total costs, not just the parts that people and companies are willing to pay for. Otherwise, those additional costs will either be paid by the taxpayer or left unfixed to cause damage to the environment and to society. Indeed, much of our past environmental transgressions, including "Superfund" toxic waste cleanups, usually were picked up by the taxpayer long after the companies responsible for the mess had declared bankruptcy – having realized that

True Costs in Life-Cycle Analysis

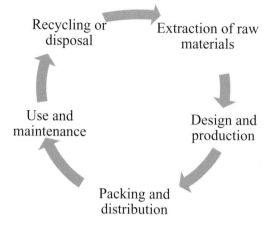

Figure 14.6 The true costs of producing any product must take into account many variables, not the least of which includes the costs of its safe disposal/recycling.
Source: Leary teaching

their profits on a subset of the true costs were about to end – before they would be forced to pay for the rest of the true costs. The idea of a life-cycle assessment is an important part of this overall process (Figure 14.7).

Good life-cycle analysis can be proactive and prevent the creation and disposal of products that could have serious environmental consequences. In the USA, there were serious consequences from companies producing and disposing of toxic substances that resulted in massive Superfund toxic waste sites that took many years and lots of taxpayers' money to try to remediate these disasters long after the responsible companies went bankrupt and just walked away from their responsibilities.

14.9.2   Assessing the Environmental Impact of Emerging Nanotechnologies

There may be a potential impact of nanomaterials on plants and animals in the environment. As humans interact with that contaminated environment, there is also a human risk. Responsible development of new nanotechnologies involves excluding as much as possible the introduction of toxic intermediates during the manufacturing process as well as anticipation of how the products decay over time after disposal.

14.9.3   Toxicity of Nanomaterials

By 2005, industry was already producing more than $30 billion of nanomaterials per year. Much of the early waste, by-products, and disposed end products were not well

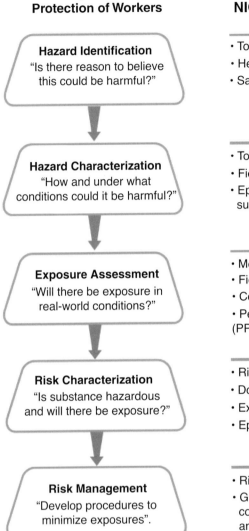

Gaps in the Protection of Workers	NIOSH Focus
Hazard Identification "Is there reason to believe this could be harmful?"	• Toxicologic research • Health effects assessment • Safety research
Hazard Characterization "How and under what conditions could it be harmful?"	• Toxicologic research • Field assessment • Epidemiologic and hazard surveillance research
Exposure Assessment "Will there be exposure in real-world conditions?"	• Metrology research • Field assessment • Control technology research • Personal protective equipment (PPE) reasearch
Risk Characterization "Is substance hazardous and will there be exposure?"	• Risk assessment • Dose–response modeling • Exposure characterization • Epidemiologic research
Risk Management "Develop procedures to minimize exposures".	• Risk communication • Guidance development for controls, exposure limits, PPE, and medical surveillance • Information dissemination • Adherence investigation

Figure 14.7 There are still many gaps in nanomaterial health and safety knowledge, even though nano-enabled products continue to increase in commerce and worker exposure to engineered nanomaterials (ENMs) is likely occurring. A challenge for the NIOSH NTRC program is to determine how to conduct timely research that addresses the elements of hazard identification through risk management, as the introduction of ENMs to the workplace continues and the workforce exposed to them becomes larger and more diverse. The NTRC approach has been to conduct concurrent, focused research to address knowledge gaps in each step of the risk management process. This enables the NTRC to provide occupational safety and health guidance across the life cycle of ENMs in a predictive manner to avoid health and safety impacts.

Source: www.cdc.gov/niosh/docs/2019-116/pdfs/2019-116.pdf

documented. The Canadian government, at a very early stage, required detailed documentation about the quantities and locations of nanomaterials. The US government was a bit slow getting started on this path. The EPA eventually became concerned and started to at least require some reporting of the amounts produced and disposed. By 2014, up to $2.6 trillion was produced in an exponential growth of the industry using nanotechnology, according to the textile industries, one of the biggest users of nanomaterials for water-resistant fabrics whereby nanomaterials are embedded in those fabrics. Additionally, the nanomaterials leach out of these fabrics as they are washed and nanomaterials end up in wastewater and groundwaters Although a laboratory setting is a closed environment where we can pretty much measure everything, once these same materials are released to the environment – an open rather than a closed environment – it is difficult or impossible to accurately account for the true quantities and movement of these materials through that environment.

Lastly, while this chapter is devoted to the toxicity of nanomaterials and chemicals used in the manufacture of NMS, we must not ignore the safety and environmental hazards of radioactive probes that are used in some NMS. The major ones with regard to nanomedicine are PET probes used for noninvasive imaging during PET scans. The patient is "radioactive" for about 12 hours afterward, usually the time it takes pass the PET probes via urination. In general, PET probes are not very toxic to the kidney and have short half-lives. Their environmental effects are usually minimal, but it is important to confirm that assumption.

14.9.4   Use of Green Chemistry to Lower Toxicity to the Environment

Whereas many nanomaterials are not in themselves very toxic, other chemicals involved in their synthesis can be quite toxic. In particular, organic solvents, used during synthesis can be difficult to impossible to remove. Even a few parts per million of some of these solvents can lead to subsequent disposal problems whereby microbes and other life-forms may interact with those toxic disposed by-products.

14.10   Nanotechnologies and the Workplace

Nanomaterials are already becoming an increasingly large and important part of the workplace. It is vital that we have a realistic assessment of their potential dangers before we have serious worker safety problems in the immediate workplace as well as problems in the environment. In the workplace, we can at least provide personal protective equipment (PPE) as we do in many industries where workers handle hazardous substances. Once nanomaterials are dumped into the general environment, they can travel through the air and groundwater, causing potentially larger and more serious problems.

While my professional experience tells me that many nanomaterials are relatively safe, it is a lot of work proving that to be the case, and that work has simply not been done yet.

14.10.1 NIOSH: Formulating Workplace Safety Standards for Nanotechnology

The National Institute for Occupational Safety and Health (NIOSH) is responsible for formulating workspace safety standards for nanotechnology, in addition to their other duties. NIOSH continues to try to fill gaps in their knowledge of nanotechnology workplace risk factors, and recently they have published new guidelines for this process for 2018–2025 (Figure 14.8).

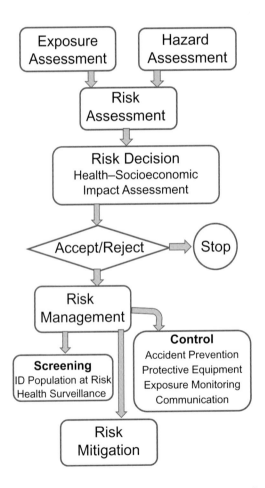

Figure 14.8 Decision tree of risk management procedures.
Source: Adapted from www.azonano.com/article.aspx?ArticleID=2611. Courtesy AzoNano

14.10.2   Protecting Workers in the Workplace

The Occupational Safety and Health Administration (OSHA) is the main formulating body that tries to establish safety rules for workplace safety. In regard to nanotechnology, there are no really firm rules yet. Whereas OSHA always had jurisdiction in academia, they only recently started exercising that authority. For almost 15 years, I had to largely make up my own laboratory safety rules in order to protect my students and technicians. This is why my lab became involved at an early stage in nanotoxicity studies, as discussed in Chapter 11. While there have been comparatively few accidents and exposures, my lab experienced a potentially serious one when personnel not from my lab used my laboratory on a weekend without proper supervision. The accident, an improper use of an incorrect oil to reach higher temperatures for a synthesis of nanomaterial, spilled and almost incinerated the laboratory workers! While my own laboratory workers were well trained and performing syntheses in a safe workplace environment, these untrained workers were carrying out unsafe practices without my knowledge or control. This incident led me to literally move my laboratories to another building and center, which better supported my concerns about worker safety. I cite this example because it is not only the nanomaterials themselves that can be dangerous, but also the synthesis and handling methods of those nanomaterials.

Proper protection of workers, both in the laboratory and the general workplace, requires that hazards and potential exposures be properly assessed in terms of the risks. No workers should have to endanger themselves or their families, and rules are necessary to guard against accidents caused by ignorance and irresponsible behaviors. A simple flow chart of this kind of risk assessment (Oberdorster, Oberdorster, and Oberdorster, 2005) is shown in Figure 14.9.

14.10.3   Assessing Hazards in the Workplace

Since we are still in the early days of understanding the true risks of nanomaterials, and their proper handling, in the laboratory and workplace, it is important to try to develop an appropriate response, erring not only against excessive behaviors but also excessive caution. There is risk to doing anything, and humans are notoriously bad at assessing risks and changing their behaviors accordingly. To say that one behavior has three times the risk as another, it is important to know whether that compared risk, in absolute rather than relative terms, is 1 in 10 or 1 in a million. Relative risks, as opposed to absolute risks, are a poor way to assess risks. As in the joke, there is a risk in getting out of bed in the morning. There is also a risk in not getting out of that same bed in the morning! As human beings, we live our lives amid constant risks. But we still need to live our lives and do our work, conscious but not paralyzed by risks that are difficult to evaluate. We need to learn how to balance risks in a reasonable way. This is what "risk management" is all about. This is obvious and common sense, but difficult to apply.

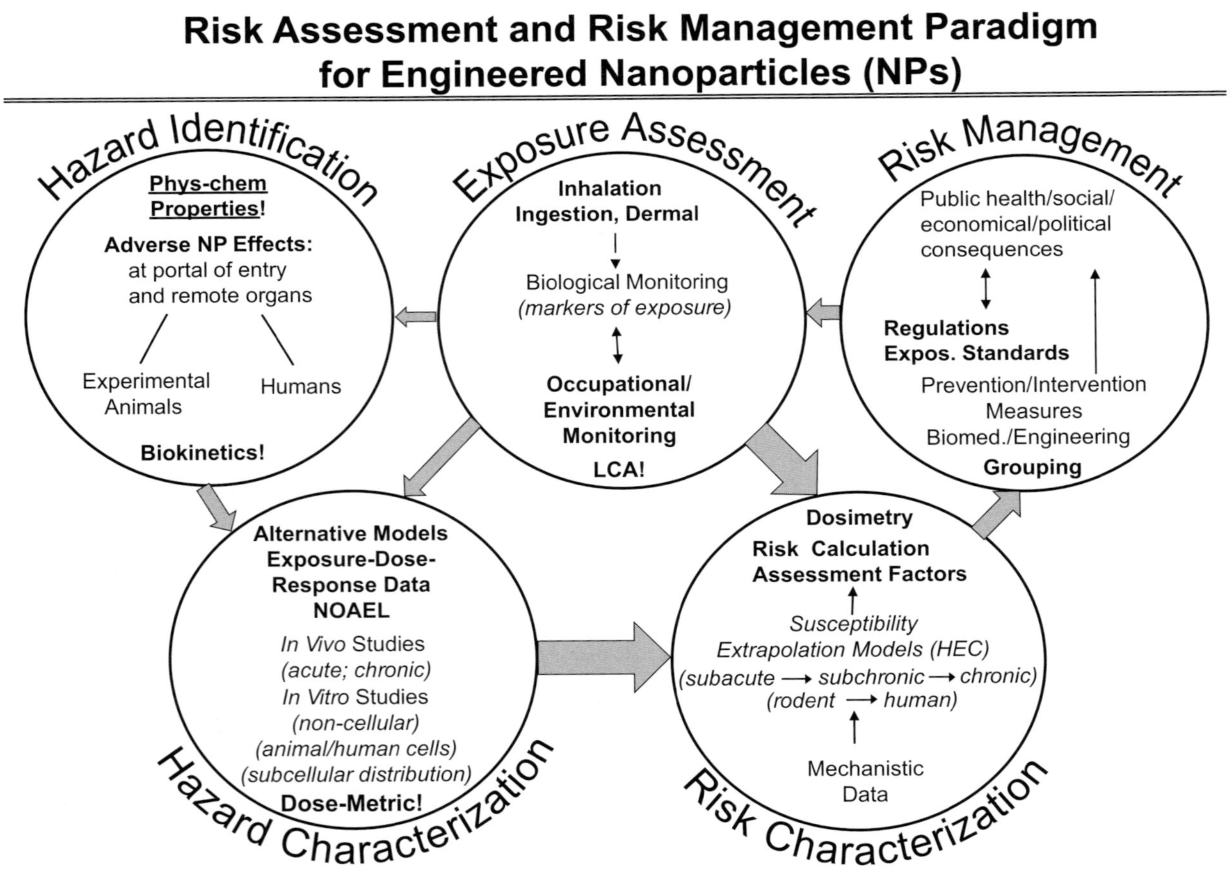

Risk Assessment and Risk Management Paradigm for Engineered Nanoparticles (NPs)

Figure 14.9 Risk assessment and risk management paradigm for nanoparticles.

Source: Adapted from Oberdörster et al. (2005)

Reducing Risks

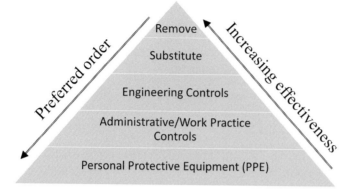

Remove
Substitute
Engineering Controls
Administrative/Work Practice Controls
Personal Protective Equipment (PPE)

Preferred order
Increasing effectiveness

Figure 14.10 The traditional hierarchy of risk control.
Source: Adapted from Oksel et al. (2016)

14.10.4   Establishing a Nanotechnology Safety System

Once hazards and exposures have been established, then a hierarchy of risk controls can be formulated in order to reduce both risk and exposure (Oksel et al., 2016) (Figure 14.10).

14.11   The Future of Nano Healthcare

In 2007, I gave an invited talk at the FDA (Leary, 2007) concerning nanotoxicity as well as making recommendations about approving new nano combo drug-device applications. At the time, the FDA treated every new combo drug-device as totally new. If anything changed in the structure of the nano drug delivery device, the total approval process needed to be repeated from the beginning. It was like approving a new car, whereby if a windshield wiper blade changed, the entire car had to be reapproved. I warned against that strategy because I felt that it would kill the future of nanomedicine since pharma was unlikely to want to go through an entire reapproval process if one component changed. For example, a robust drug delivery process could be identical except for the targeting molecule. As long as the targeting molecule was assessed in terms of its targeting effectiveness and mistargeting issues, the toxicity of the rest of the structure was unlikely to change. Pharma would want to settle on a few general nano drug delivery system designs and then swap out the targeting molecules. This approach would make it relatively straightforward to package new drugs in nanodelivery systems or repackage existing approved drugs in these nanodelivery systems with the reward of a patent extension – a win-win situation for everyone!

Chapter 14 Study Questions

14.1 Why do we usually use a three-level system (in vitro to ex vivo to in vivo) to test nanomedical systems?

14.2 What are the advantages of a closed system in terms of testing? Why are such closed systems ultimately unrealistic?

14.3 Why do we often use a four-level (nude mouse to immune-competent mouse to larger animal to human) system to test nanomedical systems before human use?

14.4 What are the most important factors that affect nanomedical system circulation time in vivo?

14.5 Where (which organs) do nanomedical systems go in vivo?

14.6 How is the biodistribution of nanomedical systems measured? What are some of the problems in making these measurements?

14.7 What is an ideal therapeutic dose?

14.8 What are the most common modes of drug administration? What are the advantages and disadvantages of each mode?

14.9 How can we assess nanomedical system targeting in vivo?

14.10 Why is in vivo targeting almost always a case of rare-cell targeting?

14.11 What are some of the consequences of mistargeting in vivo?

14.12 Why is the balancing of dosing, therapeutics, and mistargeting usually an engineering trade-off?

14.13 What are some of the common measures of tumor load in tissues?

14.14 How can NIRF in vivo optical imaging be used to assess distributions of nanomedical systems in vivo? Ex vivo?

14.15 If the tumor is in blood, how can we get a better measure of tumor load using flow cytometry?

14.16 How is the nanobarcoding method useful in scanning over larger tissue areas to find out the distribution of nanomedical systems?

14.17 How do MRI, CT, PET, and optical imaging systems compare in terms of (a) spatial resolution, (b) depth of penetration, and (c) sensitivity?

14.18 Why is a multimodal imaging design a good one for nanomedical systems?

14.19 What are some of the factors that make animal experiments time-consuming and expensive?

14.20 How do organ-on-a-chip systems provide a potentially faster way to test nanomedical systems for human use?

14.21 Why is it important for different countries to agree on regulations for nanomedical devices? What is the most likely international agency for countries to use?

14.22 Why are nanomedical systems not being treated as simply the collection of chemicals comprising the particular nanomedical system?

14.23 Why is the whole nanomedical device not necessarily the "sum of the parts"?

14.24 What is the meaning of the term "GRAS" (generally recognized as safe)?

14.25 Under which FDA classifications is a nanomedical system most likely to be a drug? A device? A combination or "combo" device?

14.26 What are some of the advantages of a combo device for the pharmaceutical industry?

14.27 What is an IND study?

14.28 What is a Phase 0 study?

14.29 What is a Phase 1 study? What is the approximate size of a Phase 1 study?

14.30 What is a Phase 2 study? What is the approximate size of a Phase 2 study?

14.31 What is a Phase 3 study? What is the approximate size of a Phase 3 study?

14.32 What is a Phase 4 study? Why is it fundamentally different from Phases 1–3?

14.33 What is the concept of "life-cycle assessment"? Why is it important to keep in mind as we create different nanomaterials and attempt to manufacture them?

14.34 What are some of the problems in assessing the true effects of nanomaterials on the environment?

14.35 Why is it important to continuously assess nanomaterial hazards in the workplace?

References

ADEREM, A., and UNDERHILL, D. M. 1999. Mechanisms of phagocytosis in macrophages. *Annual Review of Immunology*, **17**, 593–623.

ANASTAS, P. T., and WARNER, J. C. 2000. *Green Chemistry: Theory and Practice.* New York: Oxford University Press.

ARYAL, S., PARK, H., LEARY, J. F., and KEY, J. 2019. Top-down fabrication-based nano/microparticles for molecular imaging and drug delivery. *International Journal of Nanomedicine*, **14**, 6631–6644.

BABER, N., and PRITCHARD, D. 2003. Dose estimation for children. *British Journal of Clinical Pharmacology*, **56**, 489–493.

BAEZ, A. 1967. *The New College Physics: A Spiral Approach.* New York: Freeman.

BAGASRA, O. 2007. Protocols for the in situ PCR-amplification and detection of mRNA and DNA sequences. *Nature Protocols*, **2**, 2782–2795.

BAGASRA, O. 2008. In situ polymerase chain reaction and hybridization to detect low-abundance nucleic acid targets. *Current Protocols in Molecular Biology*, **82**, 14.1–14.28.

BAILEY, J. M., and HADDAD, W. M. 2005. Drug dosing control in clinical pharmacology. *IEEE Control Systems Magazine*, **25**, 35–51.

BANDURA, D. M., BARANOV, V. I., ORNATSKY, O. I., et al. 2009. Mass cytometry: Technique for real time single cell multitarget immunoassay based on inductively coupled plasma time-of-flight mass spectrometry. *Analytical Chemistry*, **81**, 6813–6822.

BARRAS, D., and WIDMANN, C. 2011. Promises of apoptosis-inducing peptides in cancer therapeutics. *Current Pharmaceutical Biotechnology*, **12**, 1153–1165.

BARTENEVA, N. S., and VOROBJEV, I. A. 2016. *Imaging Flow Cytometry: Methods and Protocols.* In *Methods in Molecular Biology*, vol. 1389. New York: Springer.

BATES, M., HUANG, B., and ZHUANG, X. 2008. Super-resolution microscopy by nanoscale localization of photo-switchable fluorescent probes. *Current Opinion in Chemical Biology*, **12**, 505–514.

BAWA, R. 2011. Regulating nanomedicine: Can the FDA handle it? *Current Drug Delivery*, **8**, 227–234.

BAWA, R. 2013. FDA and nanotech: Baby steps lead to regulatory uncertainty. In M. B. DEBASIS BAGCHI, H. MORIYAMA, and F. SHAHIDI, eds., *Bio-Nanotechnology: A Revolution in Food, Biomedical and Health Sciences.* New York: John Wiley, pp. 720–732.

BENIHOUD, K., YEH, P., and PERRICAUDET. M. 1999. Adenovirus vectors for gene delivery. *Current Opinion in Biotechnology*, **10**, 440–447.

BENYUS, J. M. 1997. *Biomimicry: Innovation Inspired by Nature.* New York: HarperCollins.

BERGTROM, C. 2020. Endocytosis and Exocytosis. *Libretexts Biology.* https://bio.libretexts.org/@go/page/16523.

BETZIG, E., PATTERSON, G. H., SOUGRAT, R., et al. 2006. Imaging intracellular fluorescent proteins at nanometer resolution. *Science*, **313**, 1643–1645.

BHATTACHARJEE, S. 2016. DLS and zeta potential: What they are and what they are not? *Journal of Controlled Release*, **235**, 337–351.

BISCHEL, L. L., BEEBE, D. J., and SUNG, K. E. 2015. Microfluidic model of ductal carcinoma in situ with 3D, organotypic structure. *BMC Cancer*, **15**, 1–10.

BISSELL, M. J., RADISKY, D. C., RIZKI, A., WEAVER, V. M., and PETERSEN, O. W. 2002. The organizing principle: Microenvironmental influences in the normal and malignant breast. *Differentiation*, **70**, 537–546.

BISWAS, D., GANESHALINGAM, J., and WAN, J. C. M. 2020. The future of liquid biopsy. *The Lancet Oncology*, **21**, e550.

BLIND, M., and BLANK, M. 2015. Aptamer selection technology and recent advances. *Molecular Therapy Nucleic Acids*, **4**, e223.

BURDA, C., CHEN, X., NARAYANAN, R., and EL-SAYED, M. A. 2005. Chemistry and properties of nanocrystals of different shapes. *Chemistry Review*, **105**, 1025–1102.

CAMPBELL, N. A., and REECE, J. B. 2002. *Biology*. San Francisco: Benjamin Cummings.

CERVADORO, A., CHO, M., KEY, J., et al. 2014. Synthesis of multifunctional magnetic nanoflakes for magnetic resonance imaging, hyperthermia, and targeting. *ACS Applied Materials & Interfaces*, **6**, 12939–12946.

CHADWICK, A. C., and MUSUNURU, K. 2018. CRISPR-Cas9 genome editing for treatment of atherogenic dyslipidemia. *Arteriosclerosis, Thrombosis, and Vascular Biology*, **38**, 12–18.

CHAN, S. M., OLSON, J. A., and UTZ, P. J. 2006. Single-cell analysis of siRNA-mediated gene silencing using multiparameter flow cytometry. *Cytometry A*, **69**, 59–65.

CHAN, W. H., SHIAO, N. H., and LU, P. Z. 2006. CdSe quantum dots induce apoptosis in human neuroblastoma cells via mitochondrial-dependent pathways and inhibition of survival signals. *Toxicology Letters*, **167**, 191–200.

CHITHRANI, B. D., GHAZANI, A. A., and CHAN, W. C. 2006. Determining the size and shape dependence of gold nanoparticle uptake into mammalian cells. *Nano Letters*, **6**, 662–668.

CHO, Y.-B., LEE, I.-G., JOO, Y.-H., HONG, S.-H., and SEO, Y.-J. 2020. TCR transgenic mice: A valuable tool for studying viral immunopathogenesis mechanisms. *International Journal of Molecular Sciences*, **21**, 1–12.

CONG, L., RAN, F. A., COX, D., et al. 2013. Multiplex genome engineering using CRISPR/Cas systems. *Science*, **339**, 819–823.

COOPER, C. L., and LEARY, J. F. 2014. High-speed flow cytometric analysis of nanoparticle targeting to rare leukemic stem cells in peripheral human blood: Preliminary in-vitro studies. *SPIE Imaging, Manipulation, and Analysis of Biomolecules, Cells, and Tissues XII*. Proceedings of SPIE.

COOPER, C. L., and LEARY, J. F. 2015. Advanced flow cytometric analysis of nanoparticle targeting to rare leukemic stem cells in peripheral human blood in a defined model system. SPIE *Imaging, Manipulation, and Analysis of Biomolecules, Cells, and Tissues XIII*. Proceedings of SPIE.

COOPER, C. L., REECE, L. M., KEY, J., BERGSTROM, D. E., and LEARY, J. F. 2008. Water-soluble iron oxide nanoparticles for nanomedicine. West Lafayette, IN: Purdue e-Pubs.

CORSETTI, J. P., COX, C., LEARY, J. F., et al. 1987. Comparison of quantitative acid-elution technique and flow cytometry for detecting fetomaternal hemorrhage. *Annals of Clinical & Laboratory Science*, **17**, 197–206.

COVEY, S. 1989. *The Seven Habits of Highly Effective People: Restoring the Character Ethic.* New York: Simon & Schuster.

COWLING, T., and FREY, N. 2019. *Macrocyclic and Linear Gadolinium Based Contrast Agents for Adults Undergoing Magnetic Resonance Imaging: A Review of Safety.* Ottawa: Canadian Agency for Drugs and Technology in Health.

CRIBBS, A. P., and PERERA, S. M. W. 2017. Science and bioethics of CRISPR-Cas9 gene editing: An analysis towards separating facts and fiction. *Yale Journal of Biology and Medicine*, **90**, 625–634.

CRISSMAN, H. A., and TOBEY, R. A. 1974. Cell-cycle analysis in 20 minutes. *Science*, **184**, 1297–1298.

CZERNIN, J., and PHELPS, M. E. 2002. Positron emission tomography scanning: Current and future applications. *Annual Review of Medicine*, **53**, 89–112.

DARZYNKIEWICZ, Z., GALKOWSKI, D., and ZHAO, H. 2008. Analysis of apoptosis by cytometry using TUNEL assay. *Methods*, **44**, 250–254.

DARZYNKIEWICZ, Z., JUAN, G., LI, X., et al. 1997. Cytometry in cell necrobiology: Analysis of apoptosis and accidental cell death (necrosis). *Cytometry*, **27**, 1–20.

DEAN, P. N., and Jett., J. H. 1974. Mathematical analysis of DNA distributions derived from flow microfluorometry. *Cell Biology*, **60**, 523–527.

DE JONGE, N., and PECKYS, D. B. 2016. Live cell electron microscopy is probably impossible. *ACS Nano*, **10**, 9061–9063.

DE LA RICA, R., PEJOUX, C., and MATSUI, H. 2011. Assemblies of functional peptides and their applications in building blocks for biosensors. *Advanced Functional Materials*, **21**, 1018–1026.

DE LIMA, R., SEABRA, A. B., and DURAN, N. 2012. Silver nanoparticles: A brief review of cytotoxicity and genotoxicity of chemically and biogenically synthesized nanoparticles. *Journal of Applied Toxicology*, **32**, 867–879.

DE SOUZA, D., NOGUEIRA, C. R., ROSTELATO, B., and ROSTELATO, M. E. C. M. 2019. Review of the methodologies used in the synthesis gold nanoparticles by chemical reduction. *Journal of Alloys and Compounds*, **798**, 714–740.

DI NARDO, F., CAVALERA, S., BAGGIANI, C., GIOVANNOLI, C., and ANFOSSI, L. 2019. Direct vs. mediated coupling of antibodies to gold nanoparticles: The case of salivary cortisol detection by lateral flow immunoassay. *ACS Applied Materials & Interfaces*, **11**, 32758–32768.

DOUDNA, J. A., and STERNBERG, S. H. 2018. *A Crack in Creation: Gene Editing and the Unthinkable Power to Control Evolution.* Boston: Marriner Books.

DREXLER, E. 1991. Molecular machinery and manufacturing with applications to computation. Unpublished PhD thesis, Massachusetts Institute of Technology.

DUAN, H., WANG, D., and LI, Y. 2015. Green chemistry for nanoparticle synthesis. *Chemical Society Reviews*, **44**, 5778–5792.

DUVAL, A., AIDE, A., BELLEMAIN, A., et al. 2007. Anisotropic surface-plasmon resonance imaging biosensor. Proceedings of SPIE.

EATON, P., QUARESMA, P., SOARES, C., et al. 2017. A direct comparison of experimental methods to measure dimensions of synthetic nanoparticles. *Ultramicroscopy*, **182**, 179–190.

EFRONI, I., IP, P. L., NAWY, T., MELLO, A., and BIRNBAUM, K. D. 2015. Quantification of cell identity from single-cell gene expression profiles. *Genome Biology*, **16**, 1–12.

EUSTAQUIO, T., COOPER, C. L., and LEARY, J. F. 2011. Single-cell imaging detection of nanobarcoded nanoparticle biodistributions in tissues for nanomedicine. *SPIE Reporters*,

Markers, Dyes, Nanoparticles, and Molecular Probes for Biomedical Applications III. Proceedings of SPIE, **7910**, 791000-1–79100O-11.

EUSTAQUIO, T., and LEARY, J. F. 2011. Nanobarcoding: A novel method of single nano-particle detection in cells and tissues for nanomedical biodistribution studies. *SPIE Biosensing and Nanomedicine IV.* Proceedings of SPIE, **8099**, 80990V-1–80990V-13.

EUSTAQUIO, T., and LEARY, J. F. 2012a. Nanobarcoding: Detecting nanoparticles in biological samples using in situ polymerase chain reaction. *International Journal of Nanomedicine*, **7**, 5625–5639.

EUSTAQUIO, T., and LEARY, J. F. 2012b. Single-cell nanotoxicity assays of superparamagnetic iron oxide nanoparticles. *Methods in Molecular Biology*, **926**, 69–85.

FEYNMAN, R. 1960. There's plenty of room at the bottom. *Engineering Science*, **23**, 22–36.

FLANNIGAN, D. J., BARWICK, B., and ZEWAIL, A. H. 2010. Biological imaging with 4D ultrafast electron microscopy. *Proceedings of the National Academy of Sciences of the United States of America*, **107**, 9933–9937.

FOTAKIS, G., and TIMBRELL, J. A. 2006. In vitro cytotoxicity assays: Comparison of LDH, neutral red, MTT and protein assay in hepatoma cell lines following exposure to cadmium chloride. *Toxicology Letters*, **160**, 171–177.

FRANGOUL, H., ALTSHULER, D., CAPPELLINI, M. D., et al. 2021. CRISPR-Cas9 gene editing for sickle cell disease and β-thalassemia. *New England Journal of Medicine*, **384**, 252–260.

FREITAS, R. A. 1999. *Nanomedicine. Volume 1: Basic Capabilities.* Georgetown, TX: Landes Bioscience.

FREITAS, R. A., Jr. 2005. What is nanomedicine? *Nanomedicine*, **1**, 2–9.

FRIGERIO, B., BIZZONI, C., JANSEN, G., et al. 2019. Folate receptors and transporters: Biological role and diagnostic/therapeutic targets in cancer and other diseases. *Journal of Experimental & Clinical Cancer Research*, **38**, 1–12.

FU, Z., and XIANG, J. 2020. Aptamer-functionalized nanoparticles in targeted delivery and cancer therapy. *International Journal of Molecular Sciences*, **21**, https://doi.org/10.3390/ijms21239123.

FULKERSON, C. M., DHAWAN, D., RATLIFF, T. L., HAHN, N. M., and KNAPP, D. W. 2017. Naturally occurring canine invasive urinary bladder cancer: A complementary animal model to improve the success rate in human clinical trials of new cancer drugs. *International Journal of Genomics*, **2017**, 1–9.

GAJ, T., GERSBACH, C. A., and BARBAS, C. F. 2013. ZFN, TALEN, and CRISPR/Cas-based methods for genome engineering. *Trends in Biotechnology*, **31**, 397–405.

GARTNER. 2020. Gartner hype cycle methodology. https://www.gartner.com/en/newsroom/press-releases/2020-08-18.

GETTS, D. R., GETTS, M. T., MCCARTHY, D. P., CHASTAIN, E. M. L., and MILLER, S. D. 2014. Have we overestimated the benefit of human(ized) antibodies? *mAbs*, **2**, 682–694.

GIVAN, A. L. 2001. *Flow Cytometry: First Principles.* New York: Wiley-Liss.

GOETZ, C., HAMMERBECK, C., and BONNEVIER, J. 2018. *Flow Cytometry Basics for the Non-Expert.* New York: Springer.

GONG, J., TRAGANOS, F., and DARZYNKIEWICZ, Z. 1995. Growth imbalance and altered expression of cyclins B1, A, E, and D3 in MOLT-4 cells synchronized in the cell cycle by inhibitors of DNA replication. *Cell Growth & Differentiation* **6**, 1485–1493.

GONZÁLEZ, A. L., NOGUEZ, C., BERÁNEK, J., and BARNARD, A. S. 2014. Size, shape, stability, and color of plasmonic silver nanoparticles. *Journal of Physical Chemistry C*, **118**, 9128–9136.

GORCZYCA, W., BRUNO S., DARZYNKIEWICZ, R. J., GONG J., and DARZYNKIEWICZ, Z. 1992. DNA strand breaks occurring during apoptosis: Their early in situ detection by the terminal deoxynucleotidyl transferase and nick translation assays and prevention by serine protease inhibitors. *International Journal of Oncology*, **1**, 639–648.

GRAFTON, M. M., WANG, L., VIDI, P. A., LEARY, J., and LELIEVRE, S. A. 2011. Breast on-a-chip: Mimicry of the channeling system of the breast for development of theranostics. *Integrative Biology*, **3**, 451–459.

GRATZNER, H. G., LEIF, R. C., INGRAHAM, D. J., and CASTRO, A. 1975. The use of antibody specific for bromodeoxyuridine for the immunofluorescent determination of DNA replication in single cells and chromosomes. *Experimental Cell Research*, **95**, 88–94.

GURUNATHAN, S., KIM, E., HAN, J. W., PARK, J. H., and KIM, J. H. 2015. Green chemistry approach for synthesis of effective anticancer palladium nanoparticles. *Molecules*, **20**, 22476–22498.

HAGLUND, E., SEALE, M.-M., and LEARY, J. F. 2009. Design of multifunctional nanomedical systems. *Annals of Biomedical Engineering*, **37**, 2048–2063.

HAGLUND, E. M., SEALE-GOLDSMITH, M.-M., DHAWAN, D., et al. 2008. Peptide targeting of quantum dots to human breast cancer cells. Proceedings of SPIE, **6866**, 68660S-1–68660S-8.

HAMID, R., ROTSHTEYN, Y., RABADI, L., PARIKH, R., and BULLOCK, P. 2004. Comparison of alamar blue and MTT assays for high through-put screening. *Toxicology in Vitro*, **18**, 703–710.

HAYFLICK, L. 1965. The limited in-vitro lifetime of human diploid cell strains. *Experimental Cell Research*, **37**, 614–636.

HILDEBRANDT, C. C., and MARRON, J. M. 2018. Justice in CRISPR-Cas9 research and clinical applications. *AMA Journal of Ethics*, **20**, E826–E833.

HONARY, S., and ZAHIR, F. 2013. Effect of zeta potential on the properties of nano-drug delivery systems: A review (part 2). *Tropical Journal of Pharmaceutical Research*, **12**, 265–273.

HOOD, M. A., MARI, M., and MUNOZ-ESPI, R. 2014. Synthetic strategies in the preparation of polymer/inorganic hybrid nanoparticles. *Materials (Basel)*, **7**, 4057–4087.

HUH, D., MATTHEWS, B. D., MAMMOTO, A., et al. 2010. Reconstituting organ-level lung functions on a chip. *Science*, **328**, 1662–1668.

IANNELLO, A., and AHMAD, A. 2005. Role of antibody-dependent cell-mediated cytotoxicity in the efficacy of therapeutic anti-cancer monoclonal antibodies. *Cancer and Metastasis Reviews* **24**, 487–499.

INTERNATIONAL COUNCIL OF CHEMICAL ASSOCIATIONS. *An Executive Guide: How to Know If and When It's Time to Commission a Life Cycle Assessment*. Amsterdam, Netherlands: ICCA.

JALILI, N., and LAXMINARAYANA, K. 2004. A review of atomic force microscopy imaging systems: Application to molecular metrology and biological sciences. *Mechatronics*, **14**, 907–945.

JETT, J. H., STEVENSON, A. P., WARNER, N. L., and LEARY, J. F. 1980. Quantitation of cell surface antigen density by flow cytometry. In O. D. LAERUM, T. LINDMO, and L. THORUD, eds., *Flow Cytometry and Sorting*. New York: Columbia University Press.

JINEK, M., CHYLINSKI, K., FONFARA, I. E., et al. 2012. A programmable dual-RNA–guided DNA endonuclease in adaptive bacterial immunity. *Science*, **337**, 816–821.

JOHNSON, W. R., WILSON, D. W., and BEARMAN, G. 2006. Spatial-spectral modulating snapshot hyperspectral imager. *Applied Optics*, **45**, 1898–1908.

JULIANO, R. L. 2012. The future of nanomedicine: Promises and limitations. *Science and Public Policy*, **39**, 99–104.

JUNGER, S. 1997. *The Perfect Storm*. New York: Norton.

KANG, H., MINTRI, S., MENON, A. V., et al. 2015. Pharmacokinetics, pharmacodynamics and toxicology of theranostic nanoparticles. *Nanoscale*, **7**, 18848–18862.

KEEFE, A. D., PAI, S., and ELLINGTON, A. 2010. Aptamers as therapeutics. *Nature Reviews Drug Discovery*, **9**, 537–550.

KEY, J., COOPER, C., KIM, A. Y., et al. 2012. In vivo NIRF and MR dual-modality imaging using glycol chitosan nanoparticles. *Journal of Controlled Release*, **163**, 249–255.

KEY, J., DHAWAN, D., COOPER, C. L., et al. 2016. Multicomponent, peptide-targeted glycol chitosan nanoparticles containing ferrimagnetic iron oxide nanocubes for bladder cancer multimodal imaging. *International Journal of Nanomedicine*, **11**, 4141–4155.

KEY, J., DHAWAN, D., KNAPP, D. W., et al. 2012. Design of peptide-conjugated glycol chitosan nanoparticles for near infrared fluorescent (NIRF) in vivo imaging of bladder tumors. *Proceedings of SPIE*, **8233**, 8233R1–8233R10.

KEY, J., KIM, K.., DHAWAN, D., et al. 2011. Dual-modality in vivo imaging for MRI detection of tumors and NIRF-guided surgery using multi-component nanoparticles. *SPIE Nanoscale Imaging, Sensing, and Actuation for Biomedical Applications VIII*. Proceedings of SPIE, **7908**, 790805-1–790805-8.

KEY, J., and LEARY, J. F. 2014. Nanoparticles for multimodal in vivo imaging in nanomedicine. *International Journal of Nanomedicine*, **9**, 711–726.

KIM, J. S., KUK, E., YU, K. N., et al. 2007. Antimicrobial effects of silver nanoparticles. *Nanomedicine*, **3**, 95–101.

KLAUDE, M., ERIKSSON, S., NYGREN, J., and AHNSTRIJM, G. 1996. The comet assay: Mechanisms and technical considerations. *Mutation Research*, **363**, 89–96.

KNAPP, D. W., RAMOS-VARA, J. A., MOORE, G. E., et al. 2014. Urinary bladder cancer in dogs, a naturally occurring model for cancer biology and drug development. *ILAR Journal*, **55**, 100–118.

KOHLER, G., and MILSTEIN, C. 1975. Continuous cultures of fused cells secreting antibody of predefined specificity. *Nature*, **256**, 495–497.

KRPETIC, Z., NATIVO, P., PRIOR, I. A., and BRUST, M. 2011. Acrylate-facilitated cellular uptake of gold nanoparticles. *Small*, **7**, 1982–1986.

KRUTH, H. S. 2011. Receptor-independent fluid-phase pinocytosis mechanisms for induction of foam cell formation with native low-density lipoprotein particles. *Current Opinion in Lipidology*, **22**, 386–393.

KRUTZIK, P. O., IRISH, J. M., NOLAN, G. P., and PEREZ, O. D. 2004. Analysis of protein phosphorylation and cellular signaling events by flow cytometry: Techniques and clinical applications. *Clinical Immunology*, **110**, 206–221.

KRUTZIK, P. O., and NOLAN, G. P. 2003. Intracellular phospho-protein staining techniques for flow cytometry: Monitoring single cell signaling events. *Cytometry A*, **55**, 61–70.

KUMAR, C. S. S. R., ed. 2005. *Nanotechnologies for the Life Sciences*. 1st ed. Weinheim: Wiley-VCH.

LAI, C.-Y., TREWYN, B. G., JEFTINIJA, D. M., et al. 2003. A mesoporous silica nanosphere-based carrier system with chemically removable CdS nanoparticle caps for stimuli-responsive controlled release of neurotransmitters and drug molecules. *Journal of the American Chemical Society*, **125**, 4451–4459.

LEARY, J. F. 1994. Strategies for rare cell detection and isolation. *Methods in Cell Biology*, **42**, 331–358.

LEARY, J. F. 2005. Ultra high-speed sorting. *Cytometry A*, **67**, 76–85.

LEARY, J. F. 2007. *Invited Talk at FDA: "Single-Cell Nanotoxicity Measures of Nanomedical Systems."* March 7.

LEARY, J. F. 2009. Molecular characterization of rare single tumor cells. In D. ANSELMETTI, ed., *Single Cell Analysis: Technologies and Applications*. Weinheim: WILEY-VCH, pp. 197–221.

LEARY, J. F. 2010. Nanotechnology: What is it and why is small so big? *Canadian Journal of Ophthalmology*, **45**, 449–456.

LEARY, J. F. 2013. Nanomedicine: Reality will trump hype! *Journal of Nanomedicine and Biotherapeutic Discovery*, **4**, e125.

LEARY, J. F. 2014. *Quantitative Single-Cell Approaches to Assessing Nanotoxicity in Nanomedical Systems*. Boca Raton, FL: CRC Press.

LEARY, J. F. 2019. Design of sophisticated shaped, multilayered, and multifunctional nano-particles for combined in-vivo imaging and advanced drug delivery. *Nanoscale Imaging, Sensing, and Actuation for Biomedical Applications XVI*. Proceedings of SPIE, **10891**, 108910R-1–108910R-9.

LEARY, J. F., TODD, P. W., WOOD, J. C. S., and JETT, J. H. 1979. Laser flow cytometric light scatter and fluorescence pulse-width and pulse rise-time sizing of mammalian cells. *Journal of Histochemistry and Cytochemistry*, **27**, 315–320.

LEE, J. F., HESSELBERTH, J. R., MEYERS, L. A., and ELLINGTON, A. D. 2004. Aptamer database. *Nucleic Acids Research*, **32**, D95–D100.

LELIEVRE, S. A., VIDI, P.-A., LEARY, J. F., and MALEKI, T. 2018. Disease on a Chip. US Patent 9,969,964.

LEONG, S. S., NG, W. M., LIM, J., and YEAP, S. P. 2018. Dynamic light scattering: effective sizing technique for characterization of magnetic nanoparticles. In S. K. SHARMA, ed., *Handbook of Materials Characterization*. New York: Springer, pp. 77–111.

LI, S.-D., and HUANG, L. 2008. Pharmacokinetics and biodistribution of nanoparticles. *Molecular Pharmaceutics*, **5**, 496–504.

LIANG, X. W., XU, J. Z. P., GRICE, J., et al. 2013. Penetration of nanoparticles into human skin. *Current Pharmaceutical Design*, **19**, 6353–6366.

LIAO, D. L., WU, G. S., and LIAO, B. Q. 2009. Zeta potential of shape-controlled TiO_2 nanoparticles with surfactants. *Colloids and Surfaces A: Physicochemical and Engineering Aspects*, **348**, 270–275.

LOAIZA, O. A., JUBETE, E., OCHOTECO, E., et al. 2011. Gold coated ferric oxide nano-particles based disposable magnetic genosensors for the detection of DNA hybridization processes. *Biosensors and Bioelectronics*, **26**, 2194–2200.

LOO, C., LIN, A., HIRSCH, L., et al. 2004. Nanoshell-enabled photonics-based imaging and therapy of cancer. *Technology in Cancer Research & Treatment*, **3**, 33–40.

LOW, P. S., SINGHAL, S., and SRINIVASARAO, M. 2018. Fluorescence-guided surgery of cancer: Applications, tools and perspectives. *Current Opinion in Chemical Biology*, **45**, 64–72.

LU, G., and FEI, B. 2014. Medical hyperspectral imaging: A review. *Journal of Biomedical Optics*, **19**, 10901.

LU, Y., and LOW, P. S. 2002. Folate-mediated delivery of macromolecular anticancer thera-peutic agents. *Advanced Drug Delivery Reviews*, **54**, 675–693.

LUISONI, S., and GREBER, U. F. 2016. Biology of adenovirus cell entry: Receptors, pathways, mechanisms. In D. T. CURIEL, ed., *Adenoviral Vectors for Gene Therapy*. 2nd ed. New York: Academic Press, 27–58.

LVOV, Y., ARIGA, K., ICHINOSE, I., and KUNITAKE, T. 1995. Assembly of multicomponent protein films by means of electrostatic layer-by-layer adsorption. *Journal of the American Chemical Society*, **117**, 6117–6123.

LVOV, Y., DECHER, G., and MOEHWALD, H. 1993. Assembly, structural characterization, and thermal behavior of layer-by-layer deposited ultrathin films of poly(vinyl sulfate) and poly(allylamine). *Langmuir*, **9**, 481–486.

MALIK, P., SHANKAR, R., MALIK, V., SHARMA, N., and MUKHERJEE, T. K. 2014. Green chemistry based benign routes for nanoparticle synthesis. *Journal of Nanoparticles*, **2014**, 1–14.

MALIKOVA, H., and HOLESTA, M. 2017. Gadolinium contrast agents: Are they really safe? *Journal of Vascular Access*, **18**, 1–7.

MCCOY, C. P., BRADY, C., , COWLEY, J. F., et al. 2010. Triggered drug delivery from biomaterials. *Expert Opinion on Drug Delivery*, **7**, 605–616.

MCCOY, C. P., ROONEY, C., EDWARDS, C. R., JONES, D. S., and GORMAN, S. P. 2007. Light-triggered molecule-scale drug dosing devices. *JACS Communications*, **129**, 9572–9573.

MCKELVEY-MARTIN, V. J., GREEN, M. H. L., SCHMEZER, P., et al. 1993. The single cell gel electrophoresis assay (comet assay): A European review. *Mutation Research*, **288**, 47–63.

MENG, F., WANG, J., PING, Q., and YEO, Y. 2018. Quantitative assessment of nanoparticle biodistribution by fluorescence imaging, revisited. *ACS Nano*, **12**, 6458–6468.

MONTANTE, S., and BRINKMAN, R. R. 2019. Flow cytometry data analysis: Recent tools and algorithms. *International Journal of Laboratory Hematology*, **41** Suppl 1, 56–62.

MONTEIRO-RIVIERE, N. A., and TRAN, C. L. 2014. *Nanotoxicology: Progress toward Nanomedicine*. Boca Raton, FL: CRC Press.

MORALES, C. S., VALENCIA, P. M., THAKKAR, A. B., SWANSON, E., and LANGER, R. 2012. Recent developments in multifunctional hybrid nanoparticles: Opportunities and challenges in cancer therapy. *Frontiers in Bioscience*, **E4**, 529–545.

MOURDIKOUDIS, S., PALLARES, R. M., and THANH, N. T. K. 2018. Characterization techniques for nanoparticles: Comparison and complementarity upon studying nanoparticle properties. *Nanoscale*, **10**, 12871–12934.

MOUSAVI, S. M., ZAREI, M., HASHEMI, S. A., et al. 2020. Gold nanostars-diagnosis, bioimaging and biomedical applications. *Drug Metabolism Reviews*, **52**, 299–318.

MURCIA, M. J., and NAUMANN, C. A. 2005. Biofunctionalization of fluorescent nanoparticles. In C. S. S. R. KUMAR, ed., *Biofunctionalization of Nanomaterials*. Weinheim: Wiley-VCH, pp. 1–40.

MURPHY, R. F. 2005. Location proteomics: A systems approach to subcellular location. *Biochemical Society Transactions*, **33**, 535–538.

NARAYAN, R., NAYAK, U. Y., RAICHUR, A. M., and GARG, S. 2018. Mesoporous silica nanoparticles: A comprehensive review on synthesis and recent advances. *Pharmaceutics*, **10**, 1–49.

NATIONAL RESEARCH COUNCIL. 1983. *Risk Assessment in the Federal Government: Managing the Process*. Washington, DC: National Academies Press. https://doi.org/10.17226/366.

NIE, S., and EMORY, S. R. 1997. Probing single molecules and single nanoparticles by surface-enhanced Raman scattering. *Science*, **275**, 1102–1106.

NOLAN, J. P., and SEBBA, D. S. 2011. Surface-enhanced Raman scattering (SERS) cytometry. *Methods in Cell Biology*, **102**, 515–532.

OBERDORSTER, G., OBERDORSTER, E., and OBERDORSTER, J. 2005. Nanotoxicology: An emerging discipline evolving from studies of ultrafine particles. *Environmental Health Perspectives*, **113**, 823–839.

OKSEL, C., SUBRAMANIAN, V., SEMENZIN, E., et al. 2016. Evaluation of existing control measures in reducing health and safety risks of engineered nanomaterials. *Environmental Science: Nano*, **3**, 869–882.

OSTLING, D., and JOHANSON, K. J. 1984. Micro-electrophoretic study of radiation-induced DNA damages in individual mammalian cells. *Biochemical and Biophysical Research Communications*, **123**, 291–298.

PARDI, N., HOGAN, M. J., PORTER, F. W., and WEISSMAN, D. 2018. mRNA vaccines: A new era in vaccinology. *Nature Reviews Drug Discovery*, **17**, 261–279.

PARK, S.-J., SEUNGSOO, K. S., LEE, S., et al. 2000. Synthesis and magnetic studies of uniform iron nanorods and nanospheres. *Journal of the American Chemical Society*, **122**, 8581–8582.

PEDNEKAR, P. P., GODIYAL, S. C., JADHAV, K. R., and KADAM, V. J. 2017. Mesoporous silica nanoparticles: A promising multifunctional drug delivery system. In A. FICAI and A. M. GRUMEZESCU, eds., *Nanostructures for Cancer Therapy*. New York: Elsevier, pp. 593–621

PENNISI, E. 2013. The CRISPR craze. *Science*, **341**, 833–836.

PEREZJUSTE, J., PASTORIZASANTOS, I., LIZMARZAN, L., and MULVANEY, P. 2005. Gold nanorods: Synthesis, characterization and applications. *Coordination Chemistry Reviews*, **249**, 1870–1901.

PERFETTO, S. P., CHATTOPADHYAY, P. K., and ROEDERER, M. 2004. Seventeen-colour flow cytometry: Unravelling the immune system. *Nature Reviews Immunology*, **4**, 648–655.

PIERPONT, T. M., LIMPER, C. B., and RICHARDS, K. L. 2018. Past, present, and future of rituximab: The world's first oncology monoclonal antibody therapy. *Frontiers in Oncology*, **8**, 1–23.

PONS, T., PIC, E., LEQUEUX, N., et al. 2010. Cadmium-free $CuInS_2/ZnS$ quantum dots for sentinel lymph node imaging with reduced toxicity. *ACS Nano*, **4**, 2531–2538.

PREVO, B. G., ESAKOFF, S. A., MIKHAILOVSKY, A., and ZASADZINSKI, J. A. 2008. Scalable routes to gold nanoshells with tunable sizes and response to near-infrared pulsed-laser irradiation. *Small*, **4**, 1183–1195.

PROW, T., GREBE, R., MERGES, C., et al. 2006. Nanoparticle tethered antioxidant response element as a biosensor for oxygen induced toxicity in retinal endothelial cells. *Molecular Vision*, **12**, 616–625.

PROW, T., SMITH, J. N., GREBE, R., et al. 2006. Construction, gene delivery, and expression of DNA tethered nanoparticles. *Molecular Vision*, **12**, 606–615.

PROW, T. W. 2004. *Nanomedicine: Targeted nanoparticles for the delivery of biosensors and therapeutic genes*. PhD dissertation, University of Texas Medical Branch.

PROW, T. W., ROSE, W. A., WANG, N., et al. 2005. Biosensor-controlled gene therapy/drug delivery with nanoparticles for nanomedicine. *SPIE Advanced Biomedical and Clinical Diagnostic Systems III*. Proceedings of SPIE, **5692**, 199–208.

PROW, T. W., SALAZAR, J. H., ROSE, W. A., et al. 2004. Nanomedicine: Nanoparticles, molecular biosensors, and targeted gene/drug delivery for combined single-cell diagnostics and therapeutics. *SPIE Advanced Biomedical and Clinical Diagnostic Systems II*. Proceedings of SPIE.

PUTNAM, W. P., and YANIK, M. F. 2009. Noninvasive electron microscopy with interaction-free quantum measurements. *Physical Review A*, **80**, 1–4.

RASTINEHAD, A. R., ANASTOS, H., WAJSWOL, E., et al. 2019. Gold nanoshell-localized photothermal ablation of prostate tumors in a clinical pilot device study. *Proceedings of the National Academy of Sciences of the Unites States of America*, **116**, 18590–18596.

ROSENBLATT, J. I., HOKANSON, J. A., MCLAUGHLIN, S. R., and LEARY, J. F. 1997. Theoretical basis for sampling statistics useful for detecting and isolating rare cells using flow cytometry and cell sorting. *Cytometry*, **27**, 233–238.

ROSENBLUM, D., GUTKIN, A., KEDMI, R., et al. 2020. CRISPR-Cas9 genome editing using targeted lipid nanoparticles for cancer therapy. *Science Advances*, **6**, 1–12.

ROSI, N. L., and MIRKIN, C. A. 2005. Nanostructures in biodiagnostics. *Chemical Reviews*, **105**, 1547–1562.

RUDOLPH, N. S., OHLSSON-WILHELM, B. M., LEARY, J. F., and ROWLEY, P. T. 1985. Single-cell analysis of the relationship among transferrin receptors, proliferation, and cell cycle phase in K562 cells. *Cytometry*, **6**, 151–158.

SACHS, K., SARVER, A. L., NOBLE-ORCUTT, K. E., et al. 2020. Single-cell gene expression analyses reveal distinct self-renewing and proliferating subsets in the leukemia stem cell compartment in acute myeloid leukemia. *Cancer Research*, **80**, 458–470.

SARGENT, J. F. 2010. *The National Nanotechnology Initiative: Overview, reauthorization, and appropriations issues.* CRS Report for Congress. Washington, DC: Congressional Research Service.

SCHLAKE, T., THESS, A., FOTIN-MLECZEK, M., and KALLEN, K. J. 2012. Developing mRNA-vaccine technologies. *RNA Biology*, **9**, 1319–1330.

SCHWARTZBERG, A. M., OLSON, T. Y., TALLEY, C. E., and ZHANG, J. Z. 2006. Synthesis, characterization, and tunable optical properties of hollow gold nanospheres. *Journal of Physical Chemistry B*, **110**, 19935–19944.

SEALE-GOLDSMITH, M.-M., and LEARY, J. F. 2009. Nanobiosystems. *WIREs Nanomedicine and Nanobiotechnology*, **1**, 553–567.

SEALE, M.-M., HAGLUND, E., COOPER, C. L., REECE, L. M., and LEARY, J. F. 2007. Design of programmable multilayered nanoparticles with in situ manufacture of therapeutic genes for nanomedicine. *SPIE: Advanced Biomedical and Clinical Diagnostic Systems V.* Proceedings of SPIE, **6430**, 643003-1–643003-7.

SEALE, M.-M., and LEARY, J. F. 2009. Nanobiosystems. In J. R. BAKER, ed., *Wiley Interdisciplinary Reviews: Nanomedicine and Nanobiotechnology.* New York: Wiley, pp. 553–567.

SEALE, M.-M., ZEMLYANOV, D., COOPER, C. L., et al. 2007. Multifunctional nanoparticles for drug/gene delivery in nanomedicine. *SPIE: Nanoscale Imaging, Spectroscopy, Sensing, and Actuation for Biomedical Applications IV.* Proceedings of SPIE, **6447**, 64470E-1–64470E-9.

SHADIDI, M., and SIOUD, M. 2003. Identification of novel carrier peptides for the specific delivery of therapeutics into cancer cells. *FASEB Journal*, **17**, 256–258.

SHAPIRO, H. M. 2001. *Practical Flow Cytometry.* Hoboken, NJ: John Wiley & Sons.

SIKULU, M., DOWELL, K. M., HUGO, L. E., et al. 2011. Evaluating RNAlater® as a preservative for using near-infrared spectroscopy to predict *Anopheles gambiae* age and species. *Malaria Journal*, **10**, 1–7.

SLOWING, I. I., TREWYN, B. G., GIRI, S., and LIN, V. S. Y. 2007. Mesoporous silica nanoparticles for drug delivery and biosensing applications. *Advanced Functional Materials*, **17**, 1225–1236.

SRIDHAR, V., GAUD, R., BAJAJ, A., and WAIRKAR, S. 2018. Pharmacokinetics and pharmacodynamics of intranasally administered selegiline nanoparticles with improved brain delivery in Parkinson's disease. *Nanomedicine*, **14**, 2609–2618.

STETEFELD, J., MCKENNA, S. A., and PATEL, T. R. 2016. Dynamic light scattering: A practical guide and applications in biomedical sciences. *Biophysical Reviews*, **8**, 409–427.

SUN, X., ROSSIN, R., TURNER, J. L., et al. 2005. An assessment of the effects of shell cross-linked nanoparticle size, core composition, and surface PEGylation on in vivo biodistribution. *Biomacromolecules*, **6**, 2541–2554.

SUSI, T., PICHLER, T., and AYALA, P. 2015. X-ray photoelectron spectroscopy of graphitic carbon nanomaterials doped with heteroatoms. *Beilstein Journal of Nanotechnology*, **6**, 177–192.

SZANISZLO, P. 2007. *Gene Expression Microarray Analysis of Small, Purified Cell Subsets.* PhD dissertation, University of Texas Medical Branch.

SZANISZLO, P., WANG, N., SINHA, M., et al. 2004. Getting the right cells to the array: Gene expression microarray analysis of cell mixtures and sorted cells. *Cytometry A*, **59**, 191–202.

SZOLLOSI, J., DAMJANOVICH, S., and LASZLO, M. 1998. Application of fluorescence resonance energy transfer in the clinical laboratory: Routine and research. *Cytometry (Communications in Clinical Cytometry)*, **34**, 159–179.

TEKADE, R. K. 2019. *Basic Fundamentals of Drug Delivery.* New York: Academic Press.

THOMAS, C. R., FERRIS, D. P., LEE, J.-H., et al. 2010. Noninvasive remote-controlled release of drug molecules in vitro using magnetic actuation of mechanized nanoparticles. *Journal of the American Chemical Society*, **132**, 10623–10625.

TOY, R., and KARATHANASIS, E. 2016. Paramagnetic nanoparticles. In Z.-R. LU and S. SAKUMA, eds., *Nanomaterials in Pharmacology*. New York: Springer, pp. 113–136.

TRON, L., SZOLLOSI, J., and DAMJANOVICH, S. 1984. Flow cytometric measurement of fluorescence resonance energy transfer. *Biophysical Journal*, **45**, 939–946.

TSIEN, R. Y. 1998. The green fluorescent protein. *Annual Review of Biochemistry*, **67**, 509–544.

TUCHIN, V. V., TARNOK, A., and ZHAROV, V. P. 2009. Towards in vivo flow cytometry. *Journal of Biophotonics*, **2**, 457–460.

VALENCIA, L. C., GARCÍA, A., RAMÍREZ-PINILLA, M. P., and FUENTES, J. L. 2011. Estimates of DNA damage by the comet assay in the direct-developing frog *Eleutherodactylus johnstonei* (Anura, Eleutherodactylidae). *Genetics and Molecular Biology*, **34**, 681–688.

VAN DAM, G. M., THEMELIS, G., CRANE, L. M., et al. 2011. Intraoperative tumor-specific fluorescence imaging in ovarian cancer by folate receptor-alpha targeting: First in-human results. *Nature Medicine*, **17**, 1315–1319.

VAN DER MEER, A. D., and VAN DEN BERG, A. 2012. Organs-on-chips: Breaking the in vitro impasse. *Integrative Biology*, **4**, 461–470.

VAN ENGELAND, M., NIELAND, L. J. W., RAMAEKERS, F. C. S., SCHUTTE, B., and REUTELINGSPERGER, C. P. M. 1998. Annexin V-affinity assay: A review on an apoptosis detection system based on phosphatidylserine exposure. *Cytometry*, **31**, 1–9.

VIDI, P. A., MALEKI, T., OCHOA, M., et al. 2014. Disease-on-a-chip: Mimicry of tumor growth in mammary ducts. *Lab on a Chip*, **14**, 172–177.

WAJIMA, T., YANO, Y., FUKUMURA, K., and OGUMA, T. 2004. Prediction of human pharmacokinetic profile in animal scale up based on normalizing time course profiles. *Journal of Pharmaceutical Sciences*, **93**, 1890–1900.

WANG, L., WANG, K., SANTRA, S., et al. 2006. Watching silica nanoparticles glow in the biological world. *Analytical Chemistry*, 646–654. https://pubs.acs.org/doi/pdf/10.1021/ac0693619.

WANI, M. Y., HASHIM, M. A., NABI, F., and MALIK, M. A. 2011. Nanotoxicity: Dimensional and morphological concerns. *Advances in Physical Chemistry*, **2011**, 1–15.

WEI, Q., SONG, H. M., LEONOV, A. P., et al. 2009. Gyromagnetic imaging: Dynamic optical contrast using gold nanostars with magnetic cores. *Journal of the American Chemical Society*, **131**, 9728–9734.

WHEELESS, L., REEDER, J. E., and O'CONNELL, M. J. 1990. Slit-scan flow analysis of cytologic specimens from the female genital tract. *Methods in Cell Biology*, **33**, 501–507.

WHITE-SCHENK, D., SHI, R., and LEARY, J. F. 2015. Nanomedicine strategies for treatment of secondary spinal cord injury. *International Journal of Nanomedicine*, **10**, 923–938.

WORLD HEALTH ORGANIZATION. 2019. International nonproprietary names (INN) for biological and biotechnological substances: A review. In WHO, ed., *WHO/EMP/RHT/TSN/2019.1*. Geneva, Switzerland: WHO.

XIAO, Y., FORRY, S. P., GAO, X., et al. 2010. Dynamics and mechanisms of quantum dot nanoparticle cellular uptake. *Journal of Nanobiotechnology*, **8**, 1–9.

YADAV, K. S., RAJPUROHIT, R., and SHARMA, S. 2019. Glaucoma: Current treatment and impact of advanced drug delivery systems. *Life Sciences*, **221**, 362–376.

YANG, X., BASSETT, S. E., LI, X., et al. 2002. Construction and selection of bead-bound combinatorial oligonucleoside phosphorothioate and phosphorodithioate aptamer libraries designed for rapid PCR-based sequencing. *Nucleic Acids Research*, **30**, 1–8.

YANG, X., LI, N., and GORENSTEIN, D. G. 2011. Strategies for the discovery of therapeutic aptamers. *Expert Opinion on Drug Discovery*, **6**, 75–87.

YANG, X., LI, X., PROW, T. W., et al. 2003. Immunofluorescence assay and flow-cytometry selection of bead-bound aptamers. *Nucleic Acids Research*, **31**, e54.

YU, Q., YAO, Y., ZHU, X., et al. 2020. In vivo flow cytometric evaluation of circulating metastatic pancreatic tumor cells after high-intensity focused ultrasound therapy. *Cytometry A*, **97**, 900–908.

YU, W. W., CHANG, E., DREZEK, R., and COLVIN, V. L. 2006. Water-soluble quantum dots for biomedical applications. *Biochemical and Biophysical Research Communications*, **348**, 781–786.

ZARBIN, M., MONTEMAGNO, C., LEARY, J., and RITCH, R. 2011. Artificial vision. *Panminerva Medica*, **53**, 167–177.

ZARBIN, M. A., MONTEMAGNO, C., LEARY, J. F., and RITCH, R. 2010a. Nanotechnology in ophthalmology. *Canadian Journal of Ophthalmology*, **45**, 457–476.

ZARBIN, M. A., MONTEMAGNO, C., LEARY, J. F., and RITCH, R. 2010b. Nanomedicine in ophthalmology: The new frontier. *American Journal of Ophthalmology*, **150**, 144–162.

ZARBIN, M. A., MONTEMAGNO, C., LEARY, J. F., and RITCH, R. 2012. Regenerative nanomedicine and the treatment of degenerative retinal diseases. *WIREs Nanomedicine and Nanobiotechnology*, **4**, 113–137.

ZEISS. 2019. *Zeiss Super Resolution Microscopy*. Essential Knowledge Briefings. 3rd ed. Jena, Germany: Zeiss.

ZHANG, L. W., and MONTEIRO-RIVIERE, N. A. 2009. Mechanisms of quantum dot nanoparticle cellular uptake. *Toxicological Sciences*, **110**, 138–155.

ZHAO, Z., ZHOU, Z., BAO, J., et al. 2013. Octapod iron oxide nanoparticles as high-performance T(2) contrast agents for magnetic resonance imaging. *Nature Communications*, **4**, 2266.

Zhou, J., and Rossi, J.J. 2014. Cell-type-specific, aptamer-functionalized agents for targeted disease therapy. *Molecular Therapy–Nucleic Acids*, **3**, e169.

ZHOU, Y., PENG, Z., SEVEN, E. S., and LEBLANC, R. M. 2018. Crossing the blood-brain barrier with nanoparticles. *Journal of Controlled Release*, **270**, 290–303.

ZWEIFEL, D. A., and WEI, A. 2005. Sulfide-arrested growth of gold nanorods. *Chemistry of Materials*, **17**, 4256–4261.

Index